STUDIES IN

THE GEOGRAPHY OF ANCIENT AND MEDIEVAL INDIA

D. C. SIRCAR

Carmichael Professor and Head of the Department of Ancient Indian History and Culture, University of Calcutta

MOTILAL BANARSIDASS

DELHI :: VARANASI :: PATNA

Second Edition, 1971 Revised and Enlarged

ISBN: 978-81-2084-116-1

Also available at:

MOTILAL BANARSIDASS
41 U.A. Bungalow Road, Jawahar Nagar, Delhi 110 007
8 Mahalaxmi Chamber, Warden Road, Mumbai 400 026
120 Royapettah High Road, Mylapore, Chennai 600 004
Sanas Plaza, Subhash Nagar, Pune 411 002
16 St. Mark's Road, Bangalore 560 001
8 Camac Street, Calcutta 700 017
Ashok Rajpath, Patna 800 004
Chowk, Varanasi 221 0011

PRINTED IN INDIA
BY JAINENDRA PRAKASH JAIN AT SHRI JAINENDRA PRESS,
A-45 NARAINA, PHASE I, NEW DELHI 110 028
AND PUBLISHED BY NARENDRA PRAKASH JAIN FOR
MOTILAL BANARSIDASS PUBLISHERS PRIVATE LIMITED,
BUNGALOW ROAD, DELHI 110 007

STUDIES IN THE GEOGRAPHY OF ANCIENT AND MEDIEVAL INDIA

Dedicated to
the Memory of my Teacher
the Late Prof. H. C. Raychaudhuri
to whom I owe my Interest in the
Study of Ancient Indian Geography

PREFACE

It is a pleasure to me that my *Studies in the Geography of Ancient and Medieval India* (1960), in spite of its defects, was favourably received by the students of the subject, so that its first edition was exhausted much earlier than was expected.

A decade has now elapsed after the appearance of the first edition of the work, in which about 30 articles were presented in 22 Chapters and 2 Appendices. The number of my articles on geographical topics published during this period is about a dozen and a half. Most of these have been either incorporated in the old Chapters or presented in new Chapters. The old arrangement has also been altered in a few cases. In the present edition, about 45 articles have been presented in 29 Chapters. As in the earlier edition, sometimes more papers than one have been clubbed together for the facility of presentation, and a few articles have been omitted. Among the omissions, mention may be made of 'Capital of the Later Aulikaras' (*Indian Historical Quarterly*, Vol. XXXVI, June-September, 1960, pp. 192 ff.) and 'Mahī-sāgara-saṅgama' (i.e. the *tīrtha* at the confluence of the Mahī and the Gulf of Cambay; *Purāṇa*, Vol. VI, No. 1, January 1964, pp. 215 ff.). I am grateful to the authorities of the publications, in which the various articles of mine originally appeared.

The sources of the papers incorporated in the different Chapters of the present edition are indicated below.

I—*Sarūpa Bhāratī* (Lakshman Sarup Commemoration Volume), Hoshiarpur, 1954, pp. 315 ff.; cf. *Journal of the* [*Royal*] *Asiatic Society* [*of Bengal*], Calcutta, Letters, Vol. V, 1939, pp. 407 ff.

II—*Journal of Indian History*, Trivandrum, Vol. XLVI, Part I, April, 1968, pp. 19 ff.; *Indian Culture*, Calcutta, Vol. VII, 1940, p. 111.

III—*Indian Historical Quarterly*, Calcutta, Vol. XXI, 1945, pp. 297 ff.

IV—*Ibid.*, Vol. XXVII, 1951, pp. 215 ff.

V—*Indian Culture*, Vol. VIII, 1941, pp. 32 ff.

VI—*Indian Historical Quarterly*, Vol. XXVIII, 1952, pp. 123 ff.; cf. *Hindusthan Review*, Patna, May-June, 1041, pp. 617 ff.

VII—*Bhāratiya Vidyā*, Bombay, Vol. V, 1944, pp. 34 ff.; *Indian Historical Quarterly*, Vol. XXIII, 1947, pp. 62 ff. ; *Indian Studies Past and Present*, Calcutta, Vol. III, No. 3, April-June, 1962, pp. 355 ff.

VIII—*Indian Museum Bulletin*, Calcutta, Vol. II, No. 1, January, 1967, pp. 7 ff.; *Journal of the Asiatic Society*, Letters, Vol. XIX, 1953, pp. 33 ff.

IX—*Journal of Indian History*, Vol. XLI (Silver Jubilee Number), 1963, pp. 263 ff.

X—*Ibid.*, Vol. XXXIV, 1956, pp. 263 ff. (cf. *Orissa Historical Research Journal*, Bhubaneswar, Vol. IV, Nos. 3-4, 1955-56, pp. 51 ff.); *Journal of the Oriental Institute*, Baroda, Vol. XIII, No. 4, June, 1964, pp. 329 ff.

XI—*Indian Historical Quarterly*, Vol. XXI, 1945, pp. 61 f.; Vol. XXII, 1946, pp. 233 ff.; Vol. XXIII, 1947, pp. 65 ff.; Vol. XXIV, 1948, pp. 78 f.; *Actes du Congrès International Oriental*, Tome XXI, Paris, 1948, pp. 199 ff.

XII—*Purāṇa*, Varanasi, Vol. V, No. 2, July, 1963, pp. 251 ff.; cf. Vol. VI, No. 1, January, 1964, pp. 215 ff.; *Itihāsa* (Bengali), New Series, Vol. I, No. 3, Kārttika-Pauṣa, 1373 B.S., pp. 235 ff. (cf. *Journal of Ancient Indian History*, Calcutta, Vol. I, 1967-68, pp. 196-97); *Bhāratiya Vidyā*, Vol. XXV, pp. 1 ff.

XIII—*Proceedings of the Indian History Congress*, Bombay, 1947, pp. 91 ff.

XIV—*Journal of Indian History*, Vol. XLIII, Part 3, December, 1965, pp. 693 ff.; also Part 2, August, 1965, pp. 343 ff.

XV—*P. V. Kane Presentation Volume*, Poona Oriental Series, Poona, 1941, pp. 469 ff.; *Journal of the Numismatic Society of India*, Varanasi, Vol. VIII, 1946, pp. 135 ff.

XVI—*Indian Historical Quarterly*, Vol. XXII, 1946, p. 315; Vol. XXIV, 1948, p. 78; *Journal of the Numismatic Society of India*, Vol. XII, 1950, pp. 50ff.

XVII—*Indian Historical Quarterly*, Vol. XXVI, 1950, pp. 291 ff.; *Bhāratīya Vidyā*, Vols. XX-XXI (Munshi Indological Fellcitation Volume), 1963, pp. 274ff.

XVIII—*Journal of the Bihar Research Society*, Patna, Vol. XL, Part I, 1954, pp. 8 ff.

XIX—*Journal of the Andhra Historical Research Society*, Rajahmundry, Vol. IX, Part 3, 1935, pp. 1 ff.

XX—*Indian Historical Quarterly*, Vol. XXI, 1945, pp. 56ff.

XXI—*Journal of Indian History*, Vol. XXXII, 1954, pp. 283ff.

XXII—*Siddha Bhāratī* (Siddheshwar Varma Presentation Volume), Hoshiarpur, 1950, pp. 291 ff.

XXIII—*Indian Historical Quarterly*, Vol. XXXVI, 1960, pp. 194ff.; *Bhāratīya Vidyā*, Vol. XI, 1950, pp. 183 ff.

XXIV—*Bhāratīya Vidyā*, Vol. VI, 1945, pp. 123 ff.; also p. 240.

XXV—*Ācārya Bhikṣu Smṛti Grantha*, I.J. Terapanthi Mahasabha, Calcutta, 1961, pp. 27 f.; *Journal of Indian History*, Vol. XXXVII, 1959, pp. 407ff.

XXVI—*Calcutta Review*, New Series, Vol. I, No. I, July-September, 1969, pp. 69 f. ; see also *Journal of Ancient Indian, History*, Vol. I, 1967-68, p. 203.

XXVII—*India's Contribution to World Thought and Culture* (Vivekananda Commemoration Volume), edited by Lokesh Chandra and others, Madras, 1970, pp. 41ff.

XXVIII—*Indian Archives*, New Delhi, Vol. V, 1951, pp. 60 ff.

XXIX—*The Bhakti Cult and Ancient Indian Geography*, edited by Sircar, Calcutta University, 1970, pp. 160 ff.

645, New Alipore, Calcutta-53, D. C. SIRCAR
May 1, 1971.

EXTRACT FROM THE *FOREWORD* OF THE FIRST EDITION

Out of the papers I have published in various periodicals, a fairly large number deal with problems relating to the geography of ancient and medieval India. Since some of these were prescribed by a few universities for the post-graduate students in the Department of Ancient Indian History and Culture, my friends were pressing me for a long time to publish the important among my geographical studies in the form of a monograph so that they might be easily available to the students of the subject. The present volume is the outcome of that request. My thanks are due to the celebrated publishers, Messrs. Motilal Banarsidass, who kindly offered to undertake the publication of the book.

A few of my geographical studies have been excluded from the volume for particular reasons. A notable omission is the *Śākta Pīṭhas* which is a religio-geographical study originally published in the *Journal of the* [*Royal*] *Asiatic Society* [of *Bengal*], Letters, Vol. XIV, 1948, pp. 1-108. It has been excluded because it is a rather lengthy dissertation and also because copies are available in the market.

Many of the Chapters of the present volume were published as independent discussions of particular topics. But, in some of them, more than one article have been brought together. This is particularly the case when different aspects of a problem were discussed on different occasions, especially in the course of controversies. In a few cases, discussions on more problems than one have also been clubbed together for the sake of facility of presentation. Thus about 30 articles, published during the period of about quarter of a century since 1935, have been presented in the book in the twenty-two Chapters and two Appendices.

CONTENTS

CHAPTER I

CAKRAVARTI-KṢETRA

1. *A Hyperbolic and Conventional Element in the Description of Indian Rulers.*

It is well known that the poets at the courts of medieval Indian rulers often grossly exaggerated the achievements and status of their patrons. Sometimes even a small landlord enjoying the right of collecting taxes from a few villages only was endowed with epithets befitting an emperor. Thus several manuscripts copied about the beginning of the 18th century for Rājanārāyaṇa, the insignificant zamindar of Kāśijoḍā in the Midnapur District, West Bengal, represent the landlord as दोर्दण्ड-प्रबल-प्रताप-परम, प्रबलप्रतापान्वितराजाधिराज and महाराजाधिराज.[1] To illustrate the tendency to exaggerate in the court poets of the early medieval kings of India, one may refer to the description of Candella Dhaṅga (c. 950-1002 A. D.) in an inscription from Khajuraho.[2]

It is implied that the Candella king crushed the rulers of Kāñcī, Andhra, Rāḍhā and Aṅga and had the queens of the defeated monarchs imprisoned in his capital. The claim is, however, absurd on the face of it. In the first place, it is extremely doubtful whether Dhaṅga at all came into conflict with the four kings mentioned, even if we take the rulers of Aṅga and Rāḍhā to have been merely viceroys of the mighty Pāla emperor of Eastern India. Secondly, if Dhaṅga actually came into con-

1. Cf. *Vaṅgiya Sāhitya Pariṣat Patrikā*, Vol. LVIII, pp. 17-18. Similarly Rājā Kṛṣṇacandra of Nadia is described as *bhūp-augh-ārcita-caraṇa*, *kṣmāpati-vṛnda-sevita*, etc. (K. C. Ray, *Kṣitiśavaṁśāvalīcarita*, pp. 227 ff.).

2. *Ep. Ind.*, Vol. I, p. 145, verse 46 :

का त्वं कांचीनृपतिवनिता का त्वमंध्राधिपस्त्री
का त्वं राढापरिवृढवधूः का त्वमंगेन्द्रपत्नी ।
इत्यालापाः समरजयिनो यस्य वैरिप्रियाणां
कारागारे सजलनयनेंदीवराणां बभूवुः ॥

flict with the above-mentioned kings, it is more doubtful that he came off victorious in all the four cases. Thirdly, even if he may have succeeded in defeating the four kings, it is really impossible to think that he succeeded also in carrying off their wives. Fourthly, supposing that he actually captured the queens of Kāñcī, Andhra, Rāḍhā and Aṅga, he would have naturally placed them in his own harem instead of his prison.

It may, however, be admitted that such hopeless exaggeration is less noticeable in the description of the Indian rulers of the earlier period of history. For this reason, the earlier the king is, the greater is our reliance in his claims, in spite of the obvious fact that there is always a considerable amount of exaggeration in the royal *praśastis* composed by the court poets of Indian monarchs. But definite statements such as the mention of the personal names of adversaries are generally more trustworthy than vague claims. Whatever be the nature of exaggeration, there is certainly an amount of truth in the claims put forward in such records as the Allahabad pillar inscription of Samudragupta and the Tirumalai inscription of Rājendracoḷa I.[1]

Although exaggeration with reference to the achievements of Indian kings is a remarkable feature of royal *praśastis* of the medieval period, it is by no means absent even in the earliest records, both literary and epigraphic. There are the following two *gāthās* in the *Śatapatha Brāhmaṇa*, XIII. 5. 4. 11 and 13 : (1) अष्टासप्तर्ति भरतो दौःषन्तिर्यमुनामनु गङ्गायां वृत्रघ्नेऽबध्नात् पञ्चपञ्चाशतं हयान् इति । (2) परःसहस्रानिन्द्रायाश्वान् मेध्यान् य आहरद् विजित्य पृथिवीं सर्वामिति । According to the first *gāthā*, Bharata, son of Duṣyanta,[2] performed seventyeight Aśvamedhas on the Yamunā and fiftyfive on the Gaṅgā, i.e. one hundred and thirtyfive horse-sacrifices in all. But according to the second *gāthā*, Bharata celebrated more than one thousand Aśvamedhas after having conquered the whole earth. Critics can hardly fail to note the difference between the definite nature of the statements in the first *gāthā* and the vagueness of the number 'more than one thousand' and of the expression 'after having conquerred the whole earth' in the second *gāthā*.

1. *Select Inscriptions*, 1942, pp. 254 ff.; *Ep. Ind.*, Vol. IX, pp. 229 ff.
2. *Duḥṣanta* is an older form of *Duṣyanta*.

In connection with the tradition regarding Bharata's conquest of the whole earth, attention may be drawn to the Purāṇic legends according to which the dominions of Bharata and his descendants, the Bhāratas, extended over that tract in the south of Jambu-dvīpa, which later came to be called Bhārata-varṣa after that monarch.[1] Again, the great Maurya emperor Aśoka (c. 272-232 B.C.) claims in the Dhauli version of his Fifth Rock Edict to have employed the Superintendents of the Religious Department, called Dharma-mahāmātra, throughout the earth (*sava-puṭhaviyaṁ*), although all the other versions of the above Rock Edict read *sarvatra vijite* (i.e. everywhere in the dominions of Aśoka) in that context.[2] It is interesting to note in this connection that Aśoka's Minor Rock Edict I seems to refer to his empire as Jambu-dvīpa while he is described in Buddhist tradition as a *Dvīpa-cakravartin*, i.e. the ruler of the entire Jambu-dvīpa.[3] As is well known, Aśoka's empire included India with the exception of Assam and certain small kingdoms to the south of Mysore but together with parts of Afghanistan, and was no doubt only a small portion of the earth. In the fourth and fifth centuries A.D., the Gupta emperors, who held sway over the major part of Northern India, but succeeded in extending their political influence over parts of Southern India as well claimed to have either conquered or ruled over 'the whole earth'. Thus the prosperity of Samudragupta (c. 335-76 A.D.) is

1. Cf. *Mahābhārata*, I. 73. 131; *Vāyu*, 45.76; *Viṣṇu*, II. 3. 1. There are diverse traditions regarding the origin of the name Bhārata-varṣa. According to many of the Purāṇas, it was derived from Bharata, son of Ṛṣabha and grandson of Nābhi who was a descendant of Vaivasvata Manu (*Bhāgavata*, XI. 2. 15 ff.; *Brahmāṇḍa*, 34. 44; etc.). In certain Purāṇic passages (*Matsya*, 114. 5; *Brahmāṇḍa*, 49. 10), however, the name Bhārata-varṣa is said to have been derived from Bharata which was just a second name of Manu himself. But the *Bhārati-prajā*, *Bhārati-santati*, etc., of the Purāṇic passages appear to refer to the Bhārata people (probably the Bharatas of the Vedic literature; cf. Raychaudhuri, *Studies in Indian Antiquities*, p. 79) represented clearly as the descendants of Bharata Dauṣyanti in the *Mahābhārata*. Traditions, moreover, represent Bharata Dauṣyanti as a *cakravartin* or *sārvabhauma*, i.e. a universal monarch or emperor (*Mahābhārata*, I. 73. 129; cf. *Bhāgavata*, IX. 20. 33).

2. *Select Inscriptions*, 1942, p. 24.

3. *Ibid.*, p. 50; Malalasekera, *Dictionary of Pali Proper Names*, s.v. *Asoka*.

mentioned as *sarva-pṛthivī-vijaya-janita*;[1] the minister of Candragupta II (c. 376-414 A.D.), who accompanied his master in the latter's campaign against the Śakas of Malwa, describes the monarch as having the aim of *kṛtsna-pṛthvī-jaya*;[2] Kumāragupta I (c. 414-55 A.D.) is said to have ruled over the *pṛthivī* that was encircled by the four seas;[3] Skandagupta (c. 455-67 A.D.) is described as the conqueror of the whole earth (*pṛthivīṁ samagrām*);[4] and Budhagupta (c. 576-95 A.D.) is represented as the ruler of the earth.[5]

Although often the word 'earth' was used to indicate the dominions even of a petty ruler, the expression 'whole earth' was used to signify the kingdom of an imperial, or at least an independent, monarch.

2. *Imperial Rulers of India and their Sphere of Influence.*

The word *cakravartin* means an imperial ruler, often also called *samrāṭ* (literally 'paramount ruler'), *eka-cchatra* (literally, 'one who alone enjoys the umbrella or the insignia of sovereignty'), or *sārvabhauma* which literally means 'the ruler of all land (i.e. the whole earth)', but actually indicates a paramount ruler of a territory without owing allegiance to any overlord. The literal meaning of *cakravartin* is likewise 'one moving everywhere without obstruction' or 'a ruler, the wheels of whose chariot roll everywhere without obstruction'. Two types of this class of rulers are indicated by the words *diśām-pati*, i.e. 'the ruler of the quarters', and *dig-vijayin*, i.e. 'the conqueror of the quarters', the word *diś* being used to signify the territories lying in all the four sides of the capital or original kingdom of the monarch in question. The distinction between the *dig-vijayin*, i.e. the ruler who acquired an empire by means of conquest, and the *diśām-pati*, i.e. one who inherited an empire from his forefathers is, however, not clearly maintained in literature.[6]

1. *Select Inscriptions*, Vol. I, 1942, p. 259.
2. *Ibid.*, p. 272.
3. *Ibid.*, p. 293.
4. *Ibid.*, p. 301.
5. *Ibid.*, p. 323.
6. Three classes of the *cakravartin* are mentioned in Pali literature; viz. (1) *cakravāla-cakravartin*, (2) *dvīpa-cakravartin* and (3) *pradeśa-cakravartin*. The first is the ruler of all the four *dvīpas* or continents constituting the earth,

According to the Kauṭilīya *Arthaśāstra*, the land which extends north to south from the Himālaya to the sea and measures east to west a thousand *yojanas* is the *kṣetra* (i.e. the sphere of influence) of a *cakravartin* (i.e. an imperial ruler of India)."[1] Rājaśekhara's *Kāvyamīmāṁsā* says that a *samrāṭ* (practically the same as a *cakravartin*) is one who conquers the whole land from the southern sea (the Indian Ocean) to the Himavat (Himālaya).[2] In the same breath, Rājaśekhara also says that the sphere of influence of a *cakravartin* is the land measuring one thousand *yojanas* and lying between Kumārīpura (Cape Comorin) and Bindusaras (in the Himālayas according to the Purāṇas) and that the conqueror of the above *kṣetra* is a *cakravartin*.[3] Thus the sphere of influence of an Indian imperial ruler was the Purāṇic Bhārata-varṣa which is described as lying between the Himālaya and the sea (*Himālayād=ā samudram*) or as bounded in the south, west and east by the sea and in the north by the Himavat 'resembling the string of a bow'.[4] This land

according to the Buddhists; the second of only one of the *dvīpas* and the third merely of a portion of a *dvīpa*. The classification is, however, apparently theoretical.

1. IX. 1 : देशः पृथिवी। तस्यां हिमवत्समुद्रान्तरमुदीचीनं योजनसहस्र-परिमाणमतिर्यक् चक्रवर्तिक्षेत्रम्।

2. G. O. S. ed., p. 92: तत्रेदं भारतं वर्षम्। अस्य च नव भेदाः। इन्द्रद्वीपः कसेरुमान् ताम्रपर्णी गभस्तिमान् नागद्वीपः सौम्यो गन्धर्वो वरुणः कुमारीद्वीपश्चायं नवमः। पञ्चशतानि जलं पञ्च स्थलमिति विभागेन प्रत्येकं योजनसहस्रावधयो दक्षिणात्समुद्रादद्रिराजं हिमवन्तं यावत्परस्परमगम्यास्ते। तान्येतानि यो जयति स सम्राडित्युच्यते। cf. below, p. 29.

3. *Loc. cit.*: कुमारीपुरात्प्रभृति बिन्दुसरोऽवधि योजनानां दशशती चक्रवर्त्तिक्षेत्रम्। तां विजयमानश्चक्रवर्ती भवति।

4. Raychaudhuri, *Studies in Indian Antiquities*, pp. 77-79. See the *Mārkaṇḍeya Purāṇa*, LVII. 59 :

दक्षिणापरतो ह्यस्य पूर्वेण च महोदधिः।
हिमवानुत्तरेणास्य कार्मुकस्य यथा गुणः॥

Cf. *Vāyu*, XLV. 75-76 :

उत्तरं यत्समुद्रस्य हिमवद्दक्षिणं च यत्।
वर्षं तद्भारतं नाम यत्रेयं भारती प्रजा॥

is sometimes referred to as the country 'extending from the Himālaya to Rāma's bridge'[1] or 'from the Himālaya to the sea-shore.'[2] In some Purāṇic passages, e.g., in *Kūrma*, I. 35. 41, as well as records like some versions of Minor Rock Edict I of Aśoka, the same land seems to be called Jambu-dvīpa,[3] although according to other popular conceptions Bhārata-varṣa was merely a part of Jambu-dvīpa.

That the sphere of influence of an Indian imperial ruler did not extend beyond the limits of Bhārata-varṣa, seems to be supported by a statement of Arrian that 'a sense of justice, they say, prevented a king from attempting conquest beyond the limits of India.'[4]

The conception of a mighty king or his generals performing *dig-vijaya*, that is to say, conquering 'the whole earth', permeates the entire body of the epic and Purāṇic literature. Epic heroes like Karṇa and the brothers of Yudhiṣṭhira are described as conquering the quarters or the whole earth. Karṇa completed his *dig-vijaya* by conquering all the countries in the north, east, south and west and is described as returned to Nāga-sāhvaya (Hastināpura, capital of his patron Duryodhana) 'after having subdued the whole earth'.[5] It is, however, interesting to note that the lists of countries lying in the four different directions and conquered by the epic *dig-vijayins* are practically the same as the lists of countries and peoples of Bhārata-varṣa as found in the geographical sections of the epics, Purāṇas and other works. Stories of the epic *dig-vijayas* were no doubt very popular in the age of the Gupta emperors. This is proved not only by the references to epic heroes in records dating from the second century A.D. but also by epigraphic and literary records of the Gupta period. Epigraphic references to the Gupta monarchs either as rulers or as conquerors of 'the whole earth' have already been indicated above. Raghu's conquest of all the quarters in the *Raghuvaṁśa* (Canto IV) of Kālidāsa who lived

1. Ray, *DHNI*, Vol. II, p. 1171.
2. *Corp. Ins. Ind.*, Vol. III, p. 205.
3. *Sel. Ins.*, 1942, p. 50.
4. Mc'Crindle, *Ancient India as described by Megasthenes and Arrian*, p. 209.
5. *Mahābhārata*, III. 253. 22; cf. II. 26. 32.

in the fourth-fifth century A.D. is obviously modelled on the epic description of the *dig-vijayas* of Karṇa, the Pāṇḍava brothers and others. Mallinātha in his commentary on the *Raghuvaṁśa*, IV. 85, rightly explains the conquest of quarters as indicating *eka-cchatratva* or universal sovereignty. Raghu began his conquest in the eastern countries on the eastern sea or the Bay of Bengal. Thence he marched towards the south and conquered the Pāṇḍya king. He moved then towards the west and subdued the Pārasīkas or Persians. Then he reached the north where he conquered the Hūṇas on the Vaṅkṣu or Oxus and the lord of Prāgjyotiṣa or Kāmarūpa (modern Assam, in the Himālaya). The description of Raghu's *digvijaya* in the *Raghuvaṁśa* was imitated by a *Praśastikāra* at the Paramāra court, who describes the Paramāra king Lakṣmadeva as having defeated Aṅga and Kaliṅga in the east, Cola and also Pāṇḍya in the Tāmraparṇī valley in the south, and the Turuṣkas on the banks of the Vaṁkṣu in the north.[1] The substitution of the Hūṇas by the Turuṣkas is interesting to note. The 'whole earth' in such cases, it may be pointed out again, does not mean the entire land between the North Pole and the South Pole. It simply refers to Bhārata-varṣa regarded as the sphere of influence of an imperial monarch. But even that is also in a conventional sense.

Although the *cakravarti-kṣetra* comprised the whole of Bhārata-varṣa, the claim of the conquest or rule over it on behalf of a historical monarch must naturally be regarded as conventional. Numerous Indian rulers are actually known to have made the claim ; but, in none of the cases, the king in question can be taken as the lord of the whole of Bhārata-varṣa extending from the Himalayas to the Indian Ocean. These kings were rulers directly of only a part of India, although they conventionally claimed suzerainty over the whole of the country. Sometimes contemporary monarchs are found to make similar claims. There is, therefore, no doubt that the kings claiming conquest of or rule over the whole earth or all the quarters were merely imperial rulers as opposed to those of the subordinate rank. It is also to be remembered that a ruler was often regarded as a *dig-vijayin* type of *cakravartin* even if he succeeded in conquering a petty rival. The frequent representation of a prince inheriting

1. See Bhandarkar's List, No. 170.

his father's kingdom as one who conquered the kingdom by the prowess of his own arms is a convention of the same type. By way of illustration, it may be pointed out that, in the Omgodu grant[1] of Skandavarman II, his grandfather Skandavarman I, son and successor of Kumāraviṣṇu described as the performer of an Aśvamedha sacrifice, is called *sva-viry-ādhigata-rājya* probably owing to the fact that Skandavarman I succeeded in overcoming an obstacle that stood in his way to the paternal throne.

3. *Boundaries of the Cakravarti-kṣetra.*

We have seen above that the *kṣetra* (sphere of influence) of a *cakravartin* (an imperial ruler of India), often mentioned as 'the whole earth', was regarded as bounded by the Himālaya or Bindusaras (in the Himalayas) in the north and by the Indian Ocean, Cape Comorin or Rāma's bridge in the south. It has also been shown that sometimes it was described as bounded by four seas on all the four sides. The conception of the *catuḥ-samudra* is no doubt conventional, as Bhārata-varṣa is bounded only on the east, south and west by seas (viz. the Bay of Bengal, the Indian Ocean and the Arabian Sea respectively). There is no sea in the north of this land. H.C. Raychaudhuri suggests that any lake or lakes to the north of India in Central Asia may have been wrongly conceived as a sea and this fact may have been responsible for the conception of India having four seas in four directions; the northern sea, however, may have actually been the deserts of Central Asia often represented in literature as 'the sea of sand.'[2] The conception of the *catuḥ-samudra* is also found in the Vedic literature where the expression is usually explained by the word *antarikṣa*, i.e. sky in the four directions. According to the Purāṇas,[3] the world consists of seven concentric *dvīpas* or islands each of which is encircled by a sea, the central island called Jambu-dvīpa being surrounded

1. *Ep. Ind.*, Vol. XV, pp. 249 ff.

2. The suggestion is supported by the *Rājataraṅgiṇī*, IV. 172, 279, 294; VIII. 2763. See also Stein's trans., Vol. I, p. 75.

3. Cf. *Mārkaṇḍeya*, 54. 6-7.

by the salt sea.[1] Bhārata-varṣa was the southernmost part of this Jambu-dvīpa, although, as we have seen, sometimes the two names are used almost synonymously. The mythical conception of the seven seas may have owed its origin to the Vedic *sapta-sindhu*, i.e. the seven rivers of North-Western Bhārata-varṣa. Whatever that may be, the Purāṇic conception of the earth being encircled by the sea coupled with the Vedic idea of the *catuḥ-samudra* may have been responsible for the later conception of the *cakravarti-kṣetra* (called 'the whole earth') bounded by the seas on all the four directions. In the literary and epigraphic records, often definite localities (which are mythical in some cases) are mentioned in the north, east, south and west of 'the whole earth' or *cakravarti-kṣetra* and, as expected, they practically correspond to the boundaries of ancient Bhārata-varṣa.

Bāṇa's *Kādambarī*[2] gives the following boundaries of the kingdom of an imaginary emperor named Candrāpīḍa: Gandhamādana (on which Badarikāśrama in the Himalayas is situated) in the north, Setubandha in the south, the mythical Udaya or Sun-rise mountain in the east and the mythical Mandara (apparently located here in the western or Arabian sea just as the mythical Sun-set mountain in a similar context in the *Harṣacarita*) in the west. In the *Harṣacarita*[3] of the same author, the boundaries given in connection with king Harṣavardhana's attempt to conquer the quarters are : Gandhamādana in the north, Suvela (the hill on which the city of Laṅkā was believed to have been situated) in the south, the mythical

1. There is another Purāṇic conception according to which the earth consists of four great *dvīpas* resembling four petals of a lotus. The pericarp of this earth-lotus is the Meru or Sumeru mountain. Similar is the conception of the Buddhists as found in the Pali literature : 'In each *cakravāla*, between the *cakravāla-parvata* and the outermost of the rocky circles which environ Meru, lies a vast ocean. In this ocean are situated, equidistant from each other, four *mahā-dvīpas*, 'great islands' or continents. On the north is Uttara-Kuru, on the south Jambu-dvīpa, on the east Pūrva-Videha and on the west Apara-Goyāna. Of these, Jambu-dvīpa is larger than the rest, being a hundred thousand *yojanas* in diameter, and includes India.' Cf. *Vāyu Purāṇa*, Ch. 34; Childers, *Pali-Eng. Dictionary*, s. v. *mahādīpo*; also below pp. 17 ff.

2. Ed. Siddhāntavāgīśa, Calcutta, pp. 194-95.

3. N. S. Press ed., p. 217.

Sun-rise mountain in the east and the mythical Sun-set mountain in the west.

Similarly, in the colophon of Vijñāneśvara's *Mitākṣarā*, commentary on the *Yājñavalkyasmṛti*, the dominions of the author's patron, the Later Cālukya monarch Vikramāditya VI (c. 1076-1127 A.D.) of Kalyāṇa, are described as bounded by Rāma's bridge in the south, Himālaya in the north, western sea (Arabian Sea) in the west and the eastern sea (Bay of Bengal) in the east.[1]

We have also a large number of inscriptions supporting the same traditions. The Meharauli inscription[2] of Candra, who is apparently no other than the Imperial Gupta monarch Candragupta II (c. 376-414 A.D.), describes the king as having conquered the Vaṅgas [of Southern Bengal] in the east, crossed the seven mouths of the Sindhu (or Indus, all of which fall into the Arabian Sea) in the west, defeated the Bālhikas (inhabiting the Balkh region on the Oxus) in the north and subdued certain people on the shores of the southern sea (Indian Ocean). A Mandasor inscription[3] describes king Yaśodharman (532 A.D.) of Malwa as having subdued the whole land bounded by the Lauhitya (Brahmaputra) in the east, the Mahendra (i.e. Mahendragiri in the Tirunelveli District) in the south, the Himālaya in the north and the western sea (Arabian Sea) in the west.

1. आ सेतोः कीर्त्तिराशे रघुकुलतिलकस्या च शैलाधिराजाद्
आ च प्रत्यक्पयोधेश्चटुलतिमिकुलोत्तुङ्गरिङ्गत्तरङ्गात् ।
आ च प्राचः समुद्रान्नतनृपतिशिरोरत्नभाभासुरांघ्रिः
पायादाचन्द्रतारं जगदिदमखिलं विक्रमादित्यदेवः ॥

2. *Sel. Ins.*, 1942, p. 276 :
यस्योद्वर्त्तयतः प्रतीपमुरसा शत्रून्समेत्यागतान्
वङ्गेष्वाहववर्त्तिनोऽभिलिखिता खड्गेन कीर्तिर्भुजे ।
तीर्त्वा सप्त मुखानि येन समरे सिन्धोर्ज्जिता वाह्लिका
यस्याद्याप्यधिवास्यते जलनिधिर्वीर्यानिलैर्दक्षिणः ॥

3. *Ibid.*, p. 394:
आ लौहित्योपकंठात्तलवनगहनोपत्यकादा महेन्द्राद्
आ गङ्गाश्लिष्टसानोस्तुहिनशिखरिणः पश्चिमादा पयोधेः ।

Both the above records belong to the age of the Imperial Guptas when the convention of the *cakravartin* and his *kṣetra* appears to have developed its popularity. There are numerous similar references in later records.

King Devapāla (c. 810-50 A.D.) is described in his records as the only lord of the whole land bounded by the source of the Gaṅgā (Himālaya) in the north, Rāma's bridge in the south, Varuṇa's abode (western sea) in the west and Lakṣmī's abode (eastern sea) in the east. A similar list of the boundaries of the *cakravarti-kṣetra* seems to be quoted in connection with the victories of the army of Devapāla's father Dharmapāla (c. 770-810 A.D.) where mention is made of Kedāra (in the Himalayas) in the north, the Gaṅgā-sāgara confluence in the east and Gokarṇa (in the North Kanara District of Mysore) and other holy places in the south and west.[2] The Karhad plates[3] of the Rāṣṭrakūṭa king Kṛṣṇa III (c. 939-67 A.D.) of Mānyakheṭa (Malkhed in the Gulbarga Dist., Mysore) say that he subdued the rulers of the whole land bounded by the eastern sea in the east, the western sea in the west, the Himaśaila (Himālaya) in the north and Siṁhala-dvīpa (i.e. Ceylon) in the south. In the records[4] of the Paramāra kings of Malwa, king Bhoja (c. 1000-55 A.D.) is said to have enjoyed the earth extending from the Kailāsa (in the Himālaya) in the north, the Malaya mountain (Travancore hills) in the south, the mythical Sun-set mountain in the west and the mythical Sun-rise mountain in the east. In the inscriptions of the Vijayanagara rulers, Vīranarasiṁha (c. 1506-09 A.D.),

1. *Gauḍalekhamālā*, p. 38 :

 आ गङ्गागममहितात्सपत्नशून्याम्
 आ सेतोः प्रथितदशास्यकेतुकीर्त्तेः ।
 उर्व्वीमा वरुणनिकेतनाच्च सिन्धोर्
 आ लक्ष्मीकुलभवनाच्च यो बुभोज ॥

2. *Ibid.*, p. 36, verse 7.
3. *Ep. Ind.*, Vol. IV, pp. 284-85 :

 अनमन्ना पूर्वापरजलनिधिहिमशैलसिंहलद्वीपात् ।
 यं जनकाज्ञावशमपि मंडलिनश्चंडदंडभयात् ॥

4. *Ibid.*, Vol. I, p. 235 :

 आ कैलासान्मलयगिरितोऽस्तोदयाद्रिद्वयादा
 भुक्ता पृथ्वी पृथुनरपतेस्तुल्यरूपेण येन ॥

elder brother of the celebrated Kṛṣṇadevarāya (1509-29 A.D.), is described as having ruled over a kingdom bounded by the Setu (i.e. Rāma's bridge) in the south, the mythical Sumeru mountain (to the north of Bhārata-varṣa) in the north, the mythical Sun-rise mountain in the east and the western mountain (the mythical Sun-set mountain) in the west.[1] The boundaries of the *cakravarti-kṣetra* given in connection with the achievements of Kṛṣṇadevarāya himself, often in the same inscription, are : the eastern mountain (i.e. the mythical Sun-rise mountain) in the east, the mythical Sun-set mountain in the west, the Himācala in the north and the Setu in the south.[2]

It should be remembered that, whenever a king is described as the ruler or conqueror of a territory indicated by four boundaries that may be placed about the borders of Bhārata-varṣa, the claim has to be regarded as merely conventional. There is usually no truth in the claim beyond that the king in question regarded himself as an imperial, or at least an independent, ruler. The conventional nature of such claims is further demonstrated clearly by a verse describing the victories of the Pāla army in the whole area bounded by the eastern land in the east, the valley of the Malaya (Travancore hills) in the south, the Maru country (Marwar or the Rajaputana Desert) in the west and the Prāleyādri (Himālaya) in the north, as found in the records of different monarchs of the Pāla family of Bengal and Bihar.[3]

1. *Mys. Arch. Surv.*, An. Rep., 1941, pp. 185-86 :

   आ सेतोरा सुमेरोरवनिसुरनुतः स्वैरमा चोदयाद्रेर्
   आ पाश्चात्याचलादखिलहृदयमावर्ज्य राज्यं शशास ।

2. *Ibid.*, p. 187 :

   आ पूर्वाद्रेरथास्तक्षितिधरकटकादा च हिमाचलान्ताद्
   आसेतोरर्थिसार्थश्रियमिह सफलीकृत्य कीर्त्या समिद्धैः ॥

3. *Gauḍalekhamālā*, p. 95 :

   देशे प्राचि प्रचुरपयसि स्वच्छमापीय तोयं
   स्वैरं भ्रान्त्वा तदनु मलयोपत्यकाचन्दनेषु ।
   कृत्वा सान्द्रैर्मरुषु जडतां शीकरैरभ्रतुल्याः
   प्रालेयाद्रेः कटकमभजन्यस्य सेनागजेन्द्राः ॥

   Cf. *Ep. Ind.*, Vol. XXIX, p. 4. Cf. the following passage in an inscription of the Somavaṁśīs of Kosala and Utkala (*IHQ*, Vol. XXII, p. 301):

   सेतुप्रान्तवनान्तरे हिमवतः पर्यन्तभूसीमनि
   प्रागम्भोधितटावनिषु कटके पूर्वेतरक्ष्माभृतः ।

Now exactly the same verse is found to have been employed by the Pāla court poets in describing the achievements of no less than four monarchs, viz. Gopāla II (middle of the tenth century), his son Vigrahapāla II (end of the tenth century) and the latter's son Mahīpāla I (biginning of the eleventh century) and great-grandson Vigrahapāla III (middle of the eleventh century).

It will be seen that the following boundaries of the *çakra-varti-kṣetra* are indicated by the epigraphic and literary records discussed above : (1) north—Himālaya, Bindusaras, Gandha-mādana, Kailāsa, Kedāra, the mythical Sumeru mountain, Prāgjyotiṣa or Kāmarūpa, the Vaṅkṣu or Oxus and Bālhīka; (2) south—Indian Ocean, Cape Comorin, Mahendragiri, Setuban-had Rāmeśvara, Suvela, Siṁhala-dvīpa and Malaya; (3) east—Vaṅga, Lauhitya, the mythical eastern or Sun-rise mountain, Bengal, the eastern country and Gaṅgā-sāgara-saṅgama ; (4) west—the mouths of the Sindhu, Arabian Sea, the mythical Mandara, western or Sun-set mountain, the Rajputana Desert and Pārasīka (Persia). As already pointed out above, these were the traditional boundaries of Bhārata-varṣa, conceived as the sphere of influence of an imperial ruler of India, and they had really nothing to do with the actual territories of the king who is described as the lord or conqueror of the land within the above boundaries.

4. *Conception of two Cakravarti-kṣetras.*

Besides the above *cakravarti-kṣetra* comprising the whole of Bhārata-varṣa, sometimes North India and South India were regarded as separate spheres of influence of the imperial monarchs respectively of Northern and Southern India. In view of the conventional nature of the claim of the status of a *cakra-vartin*, it is interesting to note that sometimes the same king is represented as the lord of the whole of the *cakravarti-kṣetra* in one case, but of the partial *cakravarti-kṣetra* comprising either North or South India in another. We have referred above to the description of the Pāla king Devapāla of Bengal and Bihar as the only lord of the whole land bounded by the Himālaya, Rāma's bridge, the Bay of Bengal and the Arabian

Sea. Elsewhere[1], however, the same monarch is said to have subdued the whole land bounded by the father of the Revā (Narmadā), i.e. the Vindhya, in the south; the father of Gaurī, i.e. the Himālaya, in the north; and the two seas (eastern and western, i.e. the Bay of Bengal and the Arabian Sea) in the east and west. These boundaries of North India are no doubt based on Manu's celebrated definition of Āryāvarta.[2] Similar claims are known to have been made on behalf of many other imperial rulers of Northern India. Thus Vīsala or Vigraharāja IV (c. 1153-64 A.D.) of the Cāhamāna dynasty of Ajmer and Śākambharī is described as having subdued Āryāvarta lying between the Vindhya and the Himādri.[3]

The Śātavāhana kings of the Deccan claimed to have been the lords of Dakṣiṇāpatha, a name often applied to the whole of South India lying beyond the Vindhyas and the Narmadā. The great Śātavāhana monarch Gautamīputra Śātakarṇi (c. 106-30 A.D.) actually ruled over the western part of South India from the Kṛṣṇā in the south to Ākara and Avanti in Malwa in the north. But he claims at the same time to have been the lord of the Vindhya, Ṛkṣavat, Pāriyātra, Mahendra and Malaya mountains which covered practically the whole of the Vindhyan range as well as the Eastern and Western Ghats. That Gautamīputra Śātakarṇi claimed a sort of vague suzerainty over the whole of South India is further indicated by the claim that his chargers drank the waters of the three seas, no doubt meaning the eastern, southern and western seas, i.e. the Bay of Bengal, the Indian Ocean and the Arabian sea. A Śātavāhana king is mentioned in Bāṇa's *Harṣacarita* as the 'lord of the three seas.'[4]

South India conceived as a separate *cakravarti-kṣetra* is mentioned as bounded by the three seas also in the inscriptions

1. *Gauḍalekhamālā*, p. 72 :

   आ रेवाजनकान्मतंगजमदस्तिम्यच्छिलासंहतेर्
   आ गौरीपितुरीश्वरेन्दुकिरणैः पुष्यत्सितिम्नो गिरेः ।
   मार्त्तण्डास्तमयोदयारुणजलादा वारिराशिद्वयान्
   नीत्या यस्य भुवं चकार करदां श्रीदेवपालो नृपः ॥

2. *Manusmṛti*, II. 22.
3. *Ind. Ant.*, Vol. XIX, pp. 215 ff.
4. See *Sel. Ins.*, 1942, pp. 196 ff.; *Harṣacarita*, *op. cit.*, p. 251.

of a later date. The Cālukyas of Badami claimed to be lords of the land lying within the three seas : त्रि-समुद्र-मध्यवर्ति-भुवन-मण्डलाधीश्वर.[1] The Mulgund inscription of Pañcāla similarly represents him as the lord of the whole country bounded by the eastern, western and southern seas.[2] The inscriptions of the Eastern Cālukyas, however, refer to the southern *cakravarti-kṣetra* as the land between the Narmadā and the Setu.[3]

Among other epigraphic references to South India as the special sphere of influence of South Indian monarchs, mention may be made of a Kadamba inscription,[4] in which the Kadamba king Ravivarman (c. 490-538 A.D.) is represented as the lord of the earth as far as the Narmadā in the north.

5. *The two Cakravarti-kṣetras mentioned together.*

We have seen above that the Paramāra king Bhoja (c. 1000-55 A.D.) is sometimes represented as the lord of the land bounded by the Kailāsa, the Malaya and the mythical Sun-set and Sun-rise mountains. There is another tradition which says that the same Paramāra king ruled for a little over fiftyfive years over 'Dakṣiṇāpatha together with Gauḍa.'[5] There is no doubt that Dakṣiṇāpatha has been mentioned here in the sense of the whole of South India, often conceived as a separate *cakravarti-kṣetra*. That the claim is merely conventional is shown by the fact that Bhoja's dominions did not include any considerable part of South India. But scholars have so long failed to grasp the correct significance of the word *Gauḍa* mentioned in relation to Dakṣiṇāpatha in the above context.

I have little doubt that Gauḍa has been mentioned here to indicate the northern *cakravarti-kṣetra* or Āryāvarta, so that Bhoja is represented as the lord of the entire Cakravarti-kṣetra comprising both Dakshiṇāpatha and North India. Thus both the partial *cakravarti-kṣetras* are mentioned here side by side to signify the whole of Bhārata-varṣa conceived as the main *cakravarti-kṣetra*. It is interesting to note in this connection that there is some

1. Cf. *Ep. Ind.*, Vol. XIX, pp. 64, etc.
2. *Bombay Gazetteer*, Vol. I, Part ii, pp. 307, 432.
3. Venkataramanayya, *The Eastern Cālukyas of Veṅgi* p. 7.
4. *Ep. Ind.*, Vol. XXXIII, p. 88.
5. Ray, *op. cit.*, p. 858.

evidence in support of the name Gauḍa being sometimes applied to indicate the whole of Nothern India. The Brāhmaṇical society of South India is usually divided into five classes called the *Pañca-Drāviḍa.* These classes are : (1) Drāviḍa (Tamil), (2) Karṇāṭa (Kannaḍa), (3) Gurjara (Gujarati), (4) Mahārāṣṭra and (5) Tailaṅga (Telugu). This classification is based on a linguistic division of the South Indian Brāhmaṇas. Although the population of North India can hardly be divided into five linguistic groups with propriety, an attempt was made in imitation of the South Indian classification to classify the Brāhmaṇa society of Northern India into the same number of subdivisions. Strangely, however, the common name applied to the five classes of North Indian Brāhmaṇas was Gauḍa as Drāviḍa is the general name of southern Brāhmaṇas. The divisions of the Northern Brāhmaṇas are : (1) Sārasvata (associated with the valley of the Sarasvatī in the Punjab), (2) Kānyakubja (in Uttar Pradesh), (3) Gauḍa (in Bengal), (4) Maithila (in North Bihar) and Utkala (in Orissa). In these cases, therefore, the name Drāviḍa indicates South India, while Gauḍa signifies North India. We know that the name Gauḍa, originally applied to a part of Bengal, was often used to indicate all the countries of Eastern India. Thus the East Indian style of Sanskrit composition as well as the medieval East Indian alphabet came to be called after Gauḍa. A further expansion of the connotation of the name to indicate the whole of North India is suggested by the classification of Brāhmaṇas referred to above.[1]

1. Some of the points discussed here were raised by me first in *JRASB*, Letters, Vol. V, 1939, pp. 407ff. As regards the last paragraph, cf. Chapter VI below. The conception of Pañca-Gauḍa or the Five Gauḍas is noticed in an inscription of 926 A.D. (*Ep. Ind.*, Vol. XXXII, p. 48) as well as in the *Rājataraṅgiṇī* (IV. 468) composed about 1150 A.D.

Chapter II

CATUR-DVĪPĀ AND SAPTA-DVĪPĀ VASUMATĪ

I

The dominant cosmographical conception of the Purāṇas is that of the *Sapta-dvīpā Vasumatī*, i.e. the earth consisting of seven concentric island-continents. Although there is some difference in the reading of the names and their order in the list, the earlier and authoritative Purāṇas offer the following names of the islands :

1. Jambu having Mount Meru or Sumeru at the centre and surrounded by the ocean of Lavaṇa (salt);
2. Plakṣa surrounding the Lavaṇa Ocean and surrounded by the ocean of Ikṣu (sugarcane juice);
3. Śālmali surrounding the Ikṣu Ocean and surrounded by the ocean of Surā (wine);
4. Kuśa surrounding the Surā Ocean and surrounded by the ocean of Sarpis (clarified butter);
5. Krauñca surrounding the Sarpis Ocean and surrounded by the ocean of Dadhi (curds);
6. Śāka surrounding the Dadhi Ocean and surrounded by the ocean of Dugdha (milk); and
7. Puṣkara surrounding the Dugdha ocean and surrounded by the Ocean of Jala (water).[1]

The earliest reference to the *Sapta-dvīpā Vasumatī* conception seems to be found in the *Mahābhāṣya* of Patañjali who flourished at the court of Puṣyamitra Śuṅga (c. 187-151 B.C.),

1. जम्बुप्लक्षाह्वयौ द्वीपौ शाल्मलिश्चापरो महान् ।
कुशः क्रौञ्चस्तथा शाकः पुष्करश्चैव सप्तमः ॥
एते द्वीपाः समुद्रैस्तु सप्त सप्तभिरावृताः ।
लवणेक्षुसुरासर्पिर्दधिदुग्धजलैः समम् ॥
(*Agni Purāṇa*, 108, 1-3).

though the work as we have it today seems to contain some later interpolations.[1]

Some of the earlier Purāṇas confuse the seven-island conception of the earth with another conception of the world consisting of four island-continents. Thus, in the *Vāyu Purāṇa*, we have the following passages :

1. XXXIII. 4—पृथिवी सर्वा सप्तद्वीपसमन्विता;
2. XXXIII. 24—सप्तद्वीपनिवासिभिः;
3. XXXIII. 31—सप्तद्वीपे तु वक्ष्यामि;
4. XXXIV. 7—सप्तद्वीपं तु वक्ष्यामि;
5. XXXIV. 46—महाद्वीपास्तु विख्याताश्चत्वारः;
6. XXXIV. 55-56—चत्वारो यस्य वै देशाः;
7. XLI. 83—चतुर्महाद्वीपवती सेयमुर्वी;
8. XLI. 85—चत्वारो नैकवर्णाढ्या महाद्वीपाः;
9. XLI. 86—चतुर्महाद्वीपा नानाद्वीपसमाकुला पृथिवी.

According to this *Catur-dvīpā Vasumatī* conception, the earth was shaped like a lotus having Mt. Meru or Sumeru as its *karṇikā* (pericarp)[2] and the four island-continents as its four petals. These four *dvīpa*-petals of the earth-lotus on the four sides of Mt. Meru are the following :

1. Kuru or Uttara-Kuru in the north;
2. Jambu or Bhārata in the south;
3. Bhadrāśva in the east, and
4. Ketumāla in the west.[3]

1. Kielhorn's ed., Vol. I, p. 9; cf. *IHQ*, Vol. XV, pp. 633 ff; below, p. 24, note 1.

2. In connection with Mt. Meru, mention is made of four *Viṣkambha-parvatas* or supporting ranges, viz. Mandara in the east, Gandhamādana in the south, Vipula in the west and Supārśva in the north, which have respectively the following lakes on them—Aruṇoda, Mānasa, Sitoda or Śītoda and Bhadra. See *Matsya Purāṇa*, Chapter 113; *Vāyu Purāṇa*, Chapters 35-36; etc. Another eight are sometimes mentioned as *Maryādā-parvata* or boundary ranges, e. g., Jaṭhara and Devakūṭa in the east and Niṣadha and Pāripātra in the west both groups extending from Mt. Nīla to Mt. Niṣadha, as also Kailāsa and Himavat in the south and Śṛṅgavat and Jārudhi in the north, both groups extending from sea to sea. See *Mārkaṇḍeya Purāṇa*, Chapter 54. Sometimes the Mālyavat is mentioned as the range extending from Mt. Nīla to Mt. Niṣadha. See *Matsya Purāṇa*, 113. 34-35. For the *varṣa-parvatas*, see below, p. 20.

3. See, e.g., *Mahābhārata*, VI. 6. 12-13; *Vāyu Purāṇa*, XXXIV. 37 ff.; *Viṣṇu Purāṇa*, II. 2. 38.

Of these four-continent and seven-continent theories, the first may be regarded as earlier on the following grounds. In the first place, the number four associating the continents with the four directions (viz. north, east, south and west) is quite natural, while the number seven is regarded by scholars as conventional even in the *Ṛgveda*.[1] Secondly, as we shall see below, the Buddhist writers conceived the great mountain at the centre of the earth as having seven concentric circles of rock around it, and these appear to have later developed into the Puranic theory of the seven concentric islands forming the earth. Thirdly, the early Pali works of the Buddhists alluding to the four-continent earth appear to be earlier than the Epico-Puranic sections on geography and cosmography which were mostly compiled about the early centuries of the Christian era, many of them as late as the 4th century A.D. or later.[2]

According to the Pali Buddhist works, there are enumerable *cakravālas* forming the world and each of them has at the centre a mountain called Meru. Between the Cakravāla-parvata and the outermost of the seven rocky circles surrounding Mt. Meru lies a vast ocean and the four *Mahādvīpas* are situated in the said ocean, equidistant from each other. These are the following :

1. Kuru or Uttara-Kuru in the north;
2. Jambu in the south;
3. Pūrva-Videha in the east; and
4. Apara-Goyāna (or Godāna) in the west.[3]

It will be seen that, while the names of the first and second are the same as in the Epico-Puranic list, those of Nos. 3 and 4 are different in the Brāhmanical and Buddhist accounts. The name Pūrva-Videha would mean 'Eastern Videha' or more probably 'Videha in the east'. It may be that the Buddhists preferred this name because the Buddha's birth-place, which is one of the four greatest Buddhist *tīrthas*, was situated in the ancient Videha country in Eastern India. While the Buddha was born at Lumbinī-grāma near Padaria in the Nepalese

1. Cf. *Vedic Index*, Vol. II, p. 424.
2. Cf. below, pp. 28-29. For the seven seas, cf. above, p. 9.
3. See Childers, *Pali-Eng. Dict.*, s. v. *mahādīpo*, etc.; Malalasekera, *Dict. Pali Prop. N.*, s. v. *Jambu-dīpa*, etc., cf. above, p. 9, note 1.

Tarai, the city of Mithilā, the ancient capital of Videha, stood on the site of modern Janakpur in the same Tarai region, though the Videha country comprised primarily the Tirhut area of Northern Bihar.[1]

As regards the names Bhadrāśva and Ketumāla in the Epico-Puranic list, they link the four-continent theory with the seven-continent one. Jambu-dvīpa, the central island according to the *Sapta-dvīpā Vasumatī* conception, was divided into nine divisions or *varṣas*, of which three lay to the south of Mt. Meru, three to its north, one around Mt. Meru at the centre, and one each to its east and west as follows :

I. (in the centre)—
   1. Meru or Ilāvṛta lying around Mt. Meru.

II. (to the south of Meru- or Ilāvṛta-varṣa around Mt. Meru)—
   2. Bhārata in the extreme south bounded in the north by Mt. Himavat and by the Salt Ocean on the three other sides.
   3. Kimpuruṣa to the north of Mt. Himavat and to the south of Mt. Hemakūṭa, the southern boundary of Hari-varṣa.
   4. Hari to the north of Mt. Hemakūṭa and to the south of Mt. Niṣadha which is the southern boundary of Ilāvṛta-varṣa around Mt. Meru.

III. (to the north of Mt. Meru and of Meru- or Ilāvṛta-varṣa)—
   5. Ramyaka to the north of Mt. Nīla the northern boundary of Meru- or Ilāvṛta-varṣa, and to the south of Mt. Śveta.
   6. Hiraṇmaya to the north of Mt. Śveta, which is the northern boundary of Ramyaka-varṣa, and to the south of Mt. Śṛṅgin or Śṛṅgavat.
   7. Kuru or Uttara-Kuru to the north of Mt. Śṛṅgin or Śṛṅgavat, the northern boundary of Hiraṇmaya-varṣa and bounded on the other three sides by the Lavaṇa ocean.

IV. 8. Bhadrāśva to the east of Meru- or Ilāvṛta-varṣa around Mt. Meru.

1. The words *pūrva* and *apara* were prefixed to the names of the eastern and western continents on the analogy of *uttara* in Uttara-Kuru. The *Vaijayantī* mentions Pūrvagandhika and Aparagandhika for Pūrva-Videha and Apara-Goyāna. Cf. Sircar, *Cosm. Geog. E. Ind. Lit.*, p. 105, note.

V. 9. Ketumāla to the west of Meru- or Ilāvṛta-varṣa around Mt. Meru.[1]

It seems that Jambu-dvīpa was originally divided into seven *varṣas* (Nos. 1-7) to which Bhadrāśva and Ketumāla (Nos. 8-9) were later added. One has to note in this connection that the Jain writers adopted the seven-fold division of Jambu-dvīpa and gave the names of Nos. 1-7 from south to north as follows :

1. Bharata (not *Bhārata*) ;
2. Haimavata (for Kimpuruṣa) ;
3. Hari;
4. Videha or Mahāvideha (for Meru or Ilāvṛta; cf. Pūrva-Videha of the Buddhists);
5. Ramyaka;
6. Hairaṇyavata (same as Hiraṇmaya); and
7. Airāvata (for Uttara-Kuru).[2]

The Jains also adopted the Puranic conception of the earth consisting of a number of concentric islands each surrounded by an ocean, and it appears that, originally, they thought of seven concentric islands as in the Purāṇas,[3] though the number gradually increased so that different works gave the numbers and names of the island-continents differently.

At the beginning of the Jain lists of islands and oceans, we have the following seven pairs of names :

1. Jambu-dvīpa surrounded by the Lavaṇa-water ocean;
2. Dhātakī-khaṇḍa surrounding the Lavaṇa ocean and surrounded by the Kāla-water Ocean ;

1. See, e.g., *Vāyu Purāṇa*, Chapters 33-34; *Matsya Purāṇa*, Chapter 113; *Mārkaṇḍeya Purāṇa*, Chapter 54. Priyavrata, king of Jambu-dvīpa, divided the kingdom among his sons—Nābhi (Hima- or Bhārata-varṣa), Kimpuruṣa (Hemakūṭa or Kimpuruṣa-varṣa), Harivarṣa (Hari- or Naiṣadha-varṣa), Ilāvṛta (Meru- or Ilāvṛta-varṣa), Ramya (Nīla- or Ramyaka-varṣa), Kuru (Śṛṅgavad- or Uttara-Kuru-varṣa), Bhadrāśva (Bhadrāśva- or Mālyavad-varṣa) and Ketumāla (Ketumāla or Gandhamādana-varṣa).

2. See Kirfel, *Die Kosmographie der Inder*, p. 215.

3. Note that the *Tiloyapaṇṇatti* once (V. 30) specially mentions the following seven separately out of the 32 Seas—(1) Kāla, (2) Vāruṇī, (3) Lavaṇa, (4) Ghṛta, (5) Kṣīra, (6) Puṣkara and (7) Svayambhū-ramaṇa.

3. Puṣkara-dvīpa surrounding the Kāla Ocean and surrounded by the Puṣkara-water Ocean ;
4. Vāruṇīvara-dvīpa surrounding the Paṣkara Ocean and having beyond it the Vāruṇī-water Ocean (cf. Surā-samudra of the Purāṇas, the words *surā* and *vāruṇī* being synonymous) ;
5. Kṣīravara-dvīpa lying beyond the Vāruṇī Ocean and having beyond it the Kṣīra-water ocean (cf. Dugdha-samudra of the Purāṇas, the words *kṣīra* and *dugdha* being synonymous) ;
6. Ghṛtavara-dvīpa lying beyond the Kṣīra Ocean and having beyond it the Ghṛta-water Ocean (cf. Sarpiḥ-samudra of the Purāṇas, the words *ghṛta* and *sarpis* being synonymous) ; and
7. Kṣaudravara-dvīpa lying beyond the Ghṛta Ocean and having beyond it the Kṣaudra-water Ocean (cf. Ikṣu-samudra of the Purāṇas, the word *kṣaudra* meaning 'honey').

Of the seven names of islands, Nos. 4-7 are coined after the names of the seas around each of them, which remind us of the Oceans of similar names in the Puranic lists, although No. 4 is also reminiscent of the Vāruṇa-dvīpa which was one of the nine divisions of Bhārata-varṣa according to the Purāṇas.[1] The names Jambu and Puṣkara are well-known from the Puranic lists while Dhātakī-khaṇḍa is known to have been the name of a division of Puṣkara-dvīpa according to the Purāṇas.[2]

According to the *Tiloyapaṇṇatti* (V. 11-26) which is a post-Gupta work, the earth consists of 16 inner and 16 outer *dvīpas*, each of them having an ocean beyond it. The 16 inner

1. इन्द्रद्वीपः कसेरुमांस्ताम्रपर्णो गभस्तिमान् ।
नागद्वीपस्तथा सौम्यो गान्धर्वो वारुणस्तथा ॥
अयं तु नवमस्तेषां द्वीपः सागरसंवृतः ॥

(*Mārkaṇḍeya Purāṇa*, 57. 6-7). Some late Purāṇas mention Kaṭāha and Siṁhala in place of Saumya and Gāndharva and apply the name Kumāra, Kumārī or Kumārikā to Sāgarasaṁvṛta (*Vāmana Purāṇa*, 13. 10-11; *Garuḍa Purāṇa*, 55. 5; etc.). The ninth *dvīpa* is regarded as the Indo-Pakistan subcontinent.

2. Savana, who had received Puṣkara-dvīpa from his father Priyavrata, divided his kingdom between his two sons Mahāvīta and Dhātaki whose portions became known as Mahāvīta-varṣa and Dhātakī-khaṇḍa (*Vāyu Purāṇa*, 33. 14-15).

dvīpas are the seven already enumerated above together with nine others whose names are also applied to the oceans beyond each of them. They are the following—(1) Nandīśvara, (2) Aruṇavara, (3) Aruṇābhāsa, (4) Kuṇḍalavara, (5) Śaṅkhavara, (6) Rucakavara, (7) Bhujagavara, (8) Kuśavara and (9) Krauñcavara. It will be seen that the last two names were borrowed from the Puranic lists. The names of the 16 outer *dvīpas*, applied also to the oceans beyond each of them, are as follows —(1) Manaḥśilā, (2) Haritāla, (3) Sindūra, (4) Śyāma, (5) Añjanavara, (6) Hiṅgula, (7) Rūpyavara, (8) Kāñcanaka, (9) Vajravara, (10) Vaiḍūrya, (11) Nāgavara, (12) Bhūtavara, (13) Yakṣavara, (14) Devavara, (15) Ahīndravara, and (16) Svayambhūramaṇa. Of these names, Manaḥśilā reminds us of Manaḥśilātala located by the Buddhists in the Himalayan region,[1] while Rūpya and Kāñcana are no doubt the same as the Suvarṇa-Rūpyaka-dvīpa of the *Rāmāyaṇa* and the Chryse and Argyre of the Classical writers,[2] which were beyond the Bay of Bengal.

How the number of the islands and oceans were increased arbitrarily can be seen from their lists in later works which mention the following names of *dvīpas* (also applied to the allied oceans) beyond the Nandīśvara-dvīpa surrounded by the Nandīśvara or Nandīśvaroda Ocean—(1) Aruṇa (cf. the name of the Aruṇoda Lake on the *Viṣkambha-parvata* called Mandara in the Purāṇas), (2) Aruṇavara, (3) Aruṇavarāvabhāsa, (4) Kuṇḍala, (5) Kuṇḍalavara, (6) Kuṇḍalavarāvabhāsa, (7) Śaṅkha, (8) Śaṅkhavara (9) Śaṅkhavarāvabhāsa, (10) Rucaka, (11) Rucakavara, (12) Rucakavarāvabhāsa, (13) Hāra, (14) Hāravara, (15) Hāravarāvabhāsa, (16) Ardhahāra, (17) Ardhahāravara, (18) Ardhahāravarāvabhāsa, (19) Kanakāvali, (20) Kanakāvalivara, (21) Kanakāvalivarāvabhāsa, (22) Ratnāvali, (23) Ratnāvalivara, (24) Ratnāvalivarāvabhāsa, (25) Muktāvali, (26) Muktāvalivara, (27) Muktāvalivarāvabhāsa, (28) Ājina, (29) Ājinavara, (30) Ājinavarāvabhāsa, (31) Sūrya, (32) Sūryavara, (33) Sūryavarāvabhāsa, (34) Deva, (35) Nāga, (36) Yakṣa, (37) Bhūta, and (38) Svayambhūramaṇa.[3] It will be seen that,

1. Malalasekera, *op. cit.*, s.v.
2. See R.C. Majumdar, *Suvarṇadvīpa*, Part I, pp. 54-55. The names Suvarṇadvīpa and Suvarṇabhūmi are famous in early Indian literature.
3. Kirfel, *op. cit.*, pp. 256-61.

while the *Tiloyapaṇṇatti* list offers 32 islands, the later list has no less than 46. We have also to note the attempt to create 22 new names out of 11 names with the expressions *vara* and *varāvabhāsa* suffixed thereto, even though a few names are found not to have been modified in the same way. There are also numerous fantastic details about all the islands and oceans in the Jain works, which have really nothing to do with geography. But we may thank the Jain authors for their power of imagination and passion for useless description in which they appear to have excelled the Puranic writers.

II

One of the *dvīpas* (literally, 'an island' or 'a land lying between two rivers') constituting the Puranic *Sapta-dvīpā Vasumatī* (i.e. the earth consisting of seven *dvīpas*) is called by the name Śaka-dvīpa or Śāka-dvīpa.[1] This land is generally identified with the country inhabited by the Śaka people, which was called Scythia by the Greeks, the Greek name for the people being 'Scythian.' Three different settlements of the Śakas are mentioned in the old Persian inscriptions, one of which lay in the plains of the rivers Jaxartes or Syr Darya and Oxus or Amu Darya in Central Asia and these Śakas are believed by some scholars to have settled later in the valley of the Helmund in the eastern part of Iran.[2] Both these old and new homelands of the Śaka people were called Scythia, while East Iran is known to have been called Śakastān (modern Seistān, medieval Sijistān) by the Persians and Śaka-dvīpa or Śāka-dvīpa by the Indians though Śaka-dvīpa or Śāka-dvīpa in the original Puranic conception of *Sapta-dvīpā Vasumatī* may have also included the Śaka

1. See above, p. 17; cf. *Agni Purāṇa*, Chapter 108, verse 1-2:

जंबुप्लक्षाह्वयो द्वीपौ शाल्मलिश्चापरो महान् ।
कुशः क्रौंचस्तथा शाकः पुष्करश्चेति सप्तमः ॥

The conception, as we have seen, occurs in Patañjali's *Mahābhāṣya* (ed. Kielhorn), Vol. I, p. 9. The *Harṣacarita* (ed. Parab, p. 185) mentions *aṣṭādaśa-dvīpā medinī*. For the traditions about the earth of 9 and 13 *dvīpas*, see Sircar, *Cosm. Geog. E. Ind. Lit.*, p. 38, note 17. Cf. above, pp. 17 ff.

2. See *The Age of Imperial Unity*, p. 120; Dey, *Geographical Dictionary*, s.v.

settlement in the Oxus and Jaxartes valleys in Central Asia. There is another name in the list of the seven *dvīpas*, viz. Kuśa-dvīpa, which likewise seems to be mentioned in the old Persian inscriptions.

The name of a country calle d Kush and a people called Kushiya (i.e. the people of Kush) are found in several old Persian inscriptions. The Hamadan inscription[1] of Darius (*c.* 522-486 B.C.), for instance, gives the following boundaries of the empire of that great monarch : 'from the Scythians that are beyond Sogdiana (i.e. the Bukhara region between the Oxus and the Jaxartes)—from there—as far as Kush, from Hindu (i.e. Sindhu or India meaning the land on the river Indus) —therefrom—as far as Sparda (i.e. Sardis in Asia Minor).' Some scholars identify Kush with Ethiopia while others place it in Central Egypt.[2] As Mudrāya (i.e. Egypt) and Kush are both mentioned in the list of satrapies of the ancient Persian emperors, the identification of Kush with Ethiopia seems to be more reasonable. There is no doubt that the country called Kush was situated somewhere in North-Eastern Africa beyond Mudrāya or Egypt. It is possible to connect this country with Kuśa-dvīpa which is, like Śaka-dvīpa or Śāka-dvīpa, one of the seven *dvīpas* constituting the Puranic *Sapta-dvīpā Vasumatī*.

1. *Select Inscriptions*, 1942, pp. 8-9; *hacā Sakaibish tyaiy para Sugdam amata yatā ā Kushā hacā Hindauv amata yātā ā Sparda*=Sanskrit *सचा *शकेभिः (=शकेभ्यः) त्ये (=ते=ये) परं सुग्दात्, अमुतः (=अमुष्मात् स्थानात्) यावत् आकुशात् (=यावत् कुशं), *सचा *सिंधौ (—सिंधोः—from India, *i.e.* the land on the Sindhu or Indus), अमुतः यावत् आस्पर्दात्. The members of the priestly class of ancient Persia (Iran) called *Magi* (Greek *Magos*), who settled in India, became known, after their absorption in the Indian social system, as the Maga Brāhmaṇas of Śaka-dvīpa or the Śākadvīpīya Brāhmaṇas. Cf. R. G. Bhandarkar, *Vaiṣṇavism, Śaivism and Minor Religious Systems*, Poona ed., pp. 218 ff.

2. Hall, *Ancient History of the Near East*, 7th ed., p. 225.

CHAPTER III[1]

PURANIC LIST OF PEOPLES

F.E. Pargiter made some interesting remarks in regard to the historical sections common to several of the Purāṇas. His views[2] may be summarised as follows :

1. Special abbreviations for Chapters III and IV—
Ac—*Abhidhānacintāmaṇi*.
Agn—*Agni Purāṇa*.
Bgvt—*Bhāgavata Purāṇa*, Baṅgabāsī Office, Calcutta.
Bhv—*Bhaviṣya Purāṇa*.
Bmd—*Bramāṇḍa Purāṇa*, Baṅgabāsī Office, Calcutta.
Bmh—*Brahma Purāṇa*, Baṅgabāsī Office, Calcutta.
Bs—*Bṛhatsaṁhitā*—*a* Baṅgabāsī Office, Calcutta; *b* Restored text of the Ms. consulted by Al-Bīrūnī (c. 1030 A.D.).
Grd—*Garuḍa Purāṇa*, Baṅgabāsī Office, Calcutta.
Hv—*Harivaṁśa*.
Km—Rājaśekhara's *Kāvyamīmāṁsā*, G.O.S. ed., Baroda.
Krm—*Kūrma Purāṇa*—*a* Baṅgabāsī Office, Calcutta ; *b* Bibliotheca Indica, ASB, Calcutta ; *c* Veṅkaṭeśvara Press, Bombay.
Mbh—*Mahābhārata*, Baṅgabāsī Office, Calcutta.
Mk—*Mārkaṇḍeya Purāṇa*—*a* ed. M. C. Pal, Calcutta, 1890; *b* Bibliotheca Indica, ASB, Calcutta, 1862; *c* Baṅgabāsī Office, Calcutta; *d* Veṅkaṭeśvara Press, Bombay.
Mts—*Matsya Purāṇa*—*a* Baṅgabāsī Office, Calcutta; *b* Veṅkaṭeśvara Press, Bombay; *c* Ānandāśrama Press, Poona; *d* Restored text of the Ms consulted by Al-Bīrūnī (c. 1030 A.D.).
Pdm—*Padma Purāṇa*.
Rām—*Rāmāyaṇa*, Nirṇayasāgara Press, Bombay.
Sv—*Śiva Purāṇa*.
Tks—Puruṣottama's *Trikāṇḍaśeṣa*.
Vj—Yādavaprakāśa's *Vaijayantī*.
Vmn—*Vāmana Purāṇa*—*a* ed. M. C. Pal, Calcutta, 1893; *b* Veṅkaṭeśvara Press, Bombay.
Vsn—*Viṣṇu Purāṇa*.
Vmn Dhm—*Viṣṇudharmottara*, Veṅkaṭeśvara Press, Bombay.
Vy—*Vāyu Purāṇa*—*a* Baṅgabāsī Office, Calcutta; *b* Restored text of the Ms consulted by Al-Bīrūnī (c. 1030 A.D.); *c* Bibliotheca Indica, ASB, Calcutta; *d* Veṅkaṭeśvara Press, Bombay; *e* Ānandāśrama Press, Poona.

2. Cf. *The Purāṇa Text of the Dynasties of the Kali Age*, Oxford University Press, 1913, pp. xxvi f.

The *Bhv* was the first Purāṇa to give an account of the dynasties of the Kali age and the *Mts*, *Vy* and *Bmd* got their account from it, though they as well as some other Purāṇas existed before the *Bhv*. Metrical accounts of the dynasties of North India grew up gradually, composed in Ślokas in a literary Prakrit and recited by bards and minstrels. After the introduction of writing about the 7th century B.C., the accounts were composed and written down in or near Magadha in Māgadhī or Pāli. The *Bhv* appropriated the Prakrit metrical accounts and converted the Prakrit Ślokas into Sanskrit in the form of a prophecy uttered by Vyāsa. This reshaping was carried through generally but not yet completely, and revisions of the text with new inclusions were taking place from time to time. The account of the Andhras (Śātavāhanas) originally composed in North India and written in Kharoṣṭhī was incorporated into the text about the middle of the 3rd century A.D. The *Mts* borrowed the accounts from the *Bhv* during the last quarter of the 3rd century. These *Bhv* accounts, which were still in Kharoṣṭhī and cannot now be traced in the extant text of the Purāṇa, were revised during the first quarter of the 4th century and the revised text was borrowed by the *Vy* now represented by a Ms (*e* Vāyu of Pargiter) in the India Office Library. The text was revised again about the beginning of the second quarter of the 4th century and this version was adopted by the *Vy* and soon after by the *Bmd* (which may have copied from the *Vy*) and now constitutes their general versions. The *Vsn* next utilised the accounts about the end of the 4th century and condensed it in Sanskrit prose. About the 8th or 9th century, the *Bgvt* (which in the main condensed the matter in new Ślokas) drew its materials from the *Bmd* and more particularly from the *Vsn*. The *Grd*, whose date is uncertain, utilised the same materials but has a bald list of the kings in new Sanskrit Ślokas. Since those times a quiet process of small emendations in details has been at work in these Purāṇas.

Some of the theories of Pargiter have been criticised. If we believe that the Brāhmī alphabet evolved out of the prehistoric Indus valley script, we should only speak of the growth of the popularity of writing and not of the introduction of writing about the 7th century B. C. The Andhras (Śātavāhanas) of the Deccan had nothing to do with the north-western region of India where the Kharoṣṭhī script was prevalent; it is, therefore, difficult to believe that the historical accounts of this royal family were originally composed in northern (north-western ?) India and written in Kharoṣṭhī. The definite epochs suggested by Pargiter for the composition, revision or transmission of the particular texts are again no better than tentative. But from a study of the sections on cosmography and geography which are common to several of the Purāṇas, it appears that Pargiter is probably right as regards the chronological sequence of at least some of the Purāṇas with which he deals. As to the suggestion that the original of the historical section of the Purāṇas was composed in Prakrit, we know that the theory has been contested by Keith,[1] though Winternitz[2] seems to be inclined to agree with Pargiter. It is interesting to note that the geographical sections exhibit Prakrit influence in such forms of names as भारुकच्छ वेदभ, ऋष्यवन्त, शुक्तिमन्ती, etc. It may be pointed out that the geographical sections, especially the list of *janapadas*, found in the *Bmd*, *Mk*, *Mts*, *Vmn* and *Vy* practically follow the same draft. The original draft seems to have been compiled for *Mts*. The *Vy* text appears to be the earliest copy from that of the *Mts*. The *Mk* probably copied the text from the *Vy* at a slightly later date; but a little independence of the copyist in the arrangement of names in a few cases is noticed in this text. The *Bmd* furnishes us with a still later copy of the *Vy* text. The date of this copy is possibly not far removed from the time of Al-Bīrūnī (c. 1030 A.D.) who is known to have consulted a Ms of the *Vy* in giving a list of peoples after the Purāṇas. The *Krm*, which like some other Purāṇas, copies the smaller draft of the *janapada* list from the *Vsn*, has a fuller list of rivers probably after the *Vy*. The smaller drafts of the lists of both the peoples and rivers appear to have been compiled from the bigger drafts in the *Mts* and *Vy* for the first time for the *Vsn*. Like that of the *Bmd*, the *Krm* text does not seem to be far removed in point of time from the date of Al-Bīrūnī. The *Vmn* appears, however, to be the latest copy of the modified draft as found in the other texts. But in several cases the compiler of the *Vmn* shows a tendency to arrange the names rather independently, although in many cases he does not go much

1. *JRAS*, 1914, pp. 1021 ff.; 1915, pp. 328 ff.
2. *Hist. Ind. Lit.*, Vol. I, p. 524, note.

beyond the originals before him. The new reading introduced by later copyists of the original draft or its modifications apear to have been, in some cases, offered by way of improvement or correction, and it is sometimes difficult to determine the reading of the original draft. The fact, however, that even the late *Vmn* often leaves the mistakes of the other copies as they are probably points to the absence of any determined and persistent effort on the part of the copyists to deviate from the texts before them. It should be remembered that numerous mistakes had already crept into the *Vy* text as early as c. 1030 A.D. when Al-Bīrūnī consulted a *Vy* Ms. The geographical section of the *Grd* exhibits considerable freedom, though it seems to have utilised the *Mk* section known as the *Kūrma-vibhāga*. The date of the *Grd* section seems to be close to that of the *Vmn*, as both add to the old description of Kumārī-dvīpa as पूर्वे किराता यस्यान्ते पश्चिमे यवनास्तथा, a new line placing the Turuṣkas in the north and the Andhras in the south. As these Turuṣkas and Andhras appear to be no other than the Turkish Musalmans and the Kākatīyas respectively, the *Vmn* and *Grd* sections on geography could not possibly have been compiled much earlier than the 13th century. The *Sv* also contains the same reference.

We have seen above that the Indian sub-continent together with some parts of Central Asia (sometimes called Bhārata-varṣa and sometimes Sāgarasaṁvṛta or Kumārī or Kumāra *dvīpa* which was regarded as one-ninth of Bhārata-varṣa) was divided into two divisions, viz. Northern (Āryāvarta) and Southern (Dākṣiṇātya or Dakṣiṇāpatha). The dividing line between the two was often mentioned as the river Narmadā. The Northern division was subdivided into four territories, viz. Madhyadeśa or the Central region; Prācya or Pūrva-deśa, the Eastern region; Aparānta or Paścād-deśa, the Western region (which included some areas lying to the south of the Narmadā); and Udīcya or Uttarāpatha, the Northern region comprising the northern and north-western areas. Sometimes the Himalayan and Vindhyan territories were separately mentioned.[1]

1. For some of these points, cf. also below, Chapter XIII, notes.

TEXT

Mk, Ch. 57 (Ch. 54 in some versions), vv. 34-59; *Vy*, Ch. 45, vv. 109-36; cf. Sachau, *Alb. Ind.*, I, pp. 299-300; *Bmd*, Ch. 49, vv. 44-71 ; *Mts*, Ch. 114, vv. 34-56 ; *Vmn*, Ch. 13, vv. 36-58. Cf. Kirfel, *Bhāratavarṣa*, pp. 43-49; C.A. Lewis in *Purāṇa*, Vol. IV, Nos. 1 and 2; Sircar, *Cosm. Geog. E. Ind. Lit.*, pp. 71ff.

I

Madhya-deśa

[1]तास्विमे कुरुपाञ्चाला:[2] शाल्वाश्चैव स-जाङ्गला:[3] ।
शूरसेना भद्रकारा[4] बोधा: सह-पटच्चरा:[5] ।।

1. The following two lines are omitted in Mk. Mbh (VI.9. 39-40), however, has— तत्रेमे कुरुपाञ्चाला: शाल्वा माद्रेय-जाङ्गला: । शूरसेना: पुलिन्दाश्च बोधा मालास्तथैव च ।। मत्स्या: कुशल्या: सौशल्या: कुन्तय: कान्ति-कोशला: । चेदि-मत्स्य-करूषाश्च भोजा: सिन्धु-पुलिन्दका: ।। As to the *janapadas* of the Madhyadeśa, Vmn has rather freely—यत्स्था: (*a*—यस्या:) कुशूद्रा: किल कुण्डलाश्च (*a*—कण्ड०) पञ्चालकाश्चैव (*a*—०काश्च) सह कौशिकैश्च । वृका: शका (*a*—शाका) वर्वर-कौरवाश्च कलिङ्ग-वङ्गाङ्गजनास्तथैते ।। मर्मका मध्यदेशीया (*a*—०देशा वा) आभीरा: (*a*—अभीरा:) शाठ्यधानका: ।।

2. Mts—तास्विमे कुरुपाञ्चाला:; Vy, Bmd—तास्त्विमे कुरुपाञ्चाला:. For the Kuru country proper with its capitals at Hastināpura in the present Meerut District and Indraprastha in the Delhi region, see Raychaudhuri, *PHAI*, 1938, pp. 19-20, 113. For the Pāñcālas with their capitals at modern Rāmnagar (ancient Ahicchatra) in the Bareilly District and Kampil (ancient Kāmpilya) in the Farrukhābād District, see *ibid.*, pp. 114 f.

3. Vy, Bmd, Mts—शाल्वाश्चैव स-जाङ्गला:; Mbh— शाल्वा माद्रेय-जाङ्गला:. For the six branches of the Śālva tribe, cf. *Vj*, 1.3.1. 38-39—साल्वास्तु कारकुत्सी (Ac—क्षी) यास्तेषां त्ववयवा: परे । उदुम्बरास्तिलखला महा-(द्र) कारा युगन्धरा: । हु (भु) लिगा: शरदण्डाश्च षट् साल्वावयवा इमे ।।
For the traditional verse referring to the constituting elements of the Śālva or Kārakukṣīya tribe that lived in the Punjab and the adjoining regions of U.P. and Rajasthan, quoted by the grammarians, cf. *Pre-Aryan and Pre-Dravidian in India*, p. 80. Jāṅgala is apparently the same as Kuru-jāṅgala near the Sarasvatī. See Raychaudhuri, *op. cit.*, pp. 19 f.

4. Vy, Bmd, Mts—शूरसेना भद्रकारा:; Mbh—शूरसेना: पुलिन्दाश्च. The Bhadrakāras are apparently the same as the Madrakāras who were a branch of the Śālva tribe (supra, note 3) and inhabited the north-eastern part of the Punjab or the adjoining regions. The Śūrasenas were a branch of the Yādavas and lived about Mathurā. See Raychaudhuri, *op. cit.*, pp. 117 f.

5. Vy *acde*, Bmd—बोधा: शतपथेश्वरै:; Vy *b*—बोधा: सह पथेश्वरै:;

मत्स्याः किराताः कुल्याश्च[1] कुन्तयः काशि-कोशलाः[2] ।
आवन्ताश्च भुलिङ्गाश्च[3] मगधाश्चान्धकैः सह[4] ॥
मध्यदेश्या जनपदाः[5] प्रायशोऽमी प्रकीर्त्तिताः[6] ॥

Mts—बाह्याः सहपटच्चराः; Mbh—बोधा मालास्तथैव च. The Paṭaccaras are known from Grd, I. 55. 11—पञ्चालाः कुरवो मत्स्या यौधेयाः सपटच्चराः । कुन्तयः शूरसेनाश्च मध्यदेशजनाः स्मृताः । For the Yaudheya tribe of Grd, see Raychaudhuri, *op. cit.*, pp. 457-58. The territories of the Yaudheyas included the Bayana region of Bharatpur and Johiyabar on the Satlaj. The Paṭaccaras are believed to have occupied parts of the Allahabad and Banda Districts, or a tract near Gwalior. The Bodhas probably lived on the eastern border of the Punjab.

1. Vy *a*—वत्साः किसल्वाः कुल्याश्च, *b*—वत्साः किसथाः कुल्याश्च, *c d e*—वत्साः किसष्णाः कुल्याश्च; *c* (v.l.) *e* (v.l.)—वत्साः किसष्टाः कुल्याश्च; *e* (v.l.) वत्साः किसज्याः कुल्याश्च; Bmd—वत्साः कुसट्टाः कुल्याश्च; Mk—मत्स्याश्वकूटाः कुल्याश्च; Mts—मत्स्याः किराताः कुल्याश्च. The first name is either Matsya (Jaipur-Alwar region) or Vatsa (Allahabad region). Cf. Raychaudhuri, *op. cit.*, pp. 111 f. Kirāta=Himalayan mountaineers; cf. the name of the Kirāntis who once reigned over a large portion of Nepal. The Kulyas were probably the people of the present Kulait region on the upper Rāvī.

2. Vy, Mk, Bmd, Mts—कुन्तलाः काशिकोशलाः; Mts *c* (v.l.)—कुन्तयः काशिकोशलाः; Mbh—कुन्तयः कान्तिकोशलाः. The reading कुन्तयः is supported by Grd, I. 55.11, quoted at p. 30, note 5. The Kuntis were a branch of the Yādava tribe living apparently not far from the Mathurā region. The Kāśis lived about Vārāṇasī (Banaras). This Kośala is Uttara-Kośala with its capital at Ayodhyā (in the Fyzabad Dist.). See Raychaudhuri, *op. cit.*, pp. 8 ff., 84 ff.

3. Mts—आवन्ताश्च कलिङ्गाश्च; Mk *a c* (v.l.)—अथर्वाश्च कलिङ्गाश्च, *b*—अथर्वाश्चार्कलिङ्गाश्च, *c*—अथवाश्चाकलिङ्गाश्च, *d*—अर्वुदाश्चार्कलिङ्गाश्च; Bmd.—प्रथमाश्च कलिङ्गाश्च; Vy *ae*—अर्थपाश्च तिलङ्गाश्च, *cde* (v.l.)—अथ पार्श्वे तिलङ्गाश्च, *b*—अर्थयाश्वपुःलिङ्गाश्च (sic —अर्थपाश्च भुलिङ्गाश्च). The Bhuliṅgas were a branch of the Śālva tribe; cf. p. 30, note 3. The Āvantas are apparently the same as the Avantis who lived in West Malwa and had their chief city at Ujjayinī. Cf. Raychaudhuri, *op. cit.*, pp. 122-23.

4. Mk—मलकाश्च वृकैः सह; Bmd, Vy *acde*—मगधाश्च वृकैः सह, *b*—मघकाश्च वृकैः सह; Mts—मूकाश्चैवान्धकैः सह. Magadha=Patna and Gaya Districts of Bihar; cf. Raychaudhuri, *op. cit.*, pp. 94 ff. The Andhakas were a branch of the Yādava tribe and possibly lived originally not far from Mathurā. Later they appear to have migrated towards Kathiawar.

5. Mk—मध्यदेश्या जनपदाः; Vy, Bmd, Mts—मध्यदेशा जनपदाः. Cf. p. 46, note 5.

6. Mk, Vy, Bmd—प्रायशोऽमी प्रकीर्त्तिताः; Mts—प्रायशः परिकीर्तिताः.

[1]सह्यस्य उत्तरार्द्धे तु[2] यत्र गोदावरी नदी[3]।
पृथिव्यामिह कृत्स्नायां[4] स प्रदेशो मनोरमः[5] ॥

II *Udīcya* (*Uttarāpatha*)

वाह्लीका वाटधानाश्च[6] आभीराः कालतोयकाः[7]।
अपरान्ताश्च शूद्राश्च[8] पह्लवाश्चर्मखण्डिकाः[9] ॥

1. The following lines including those referring to Govardhana are omitted in some of the *Vy* Mss. The verse is quoted in *Km*.

2. Vy *a*—सह्यस्य उत्तरार्द्धे तु, *cde*—सह्यस्य चोत्तरार्द्धे तु, *e* (v.l.)—सह्यस्य चोत्तरार्द्धे तु; Mk *a*—सह्यस्य चोत्तरे यस्तु, *bcd*—सहस्य चोत्तरे यास्तु; Bmd—सह्यस्य उत्तरान्ते तु; Mts *a*—सह्यस्यानन्तरे चैते. Sahya=the Western Ghats excluding the Travancore hills.

3. Mk, Vy, Bmd—यत्र गोदावरी नदी; Mts— तत्र गोदावरी नदी. The Purāṇic passage further refers to the holy place called Govardhana near Nasik on the Godāvarī river. Cf. note on the Godāvarī below, Ch. IV, Sec. V. Vy *a* reads after the above—

तत्र गोवर्धनो नाम सुरराजेन निर्मितः ।
रामप्रियार्थं स्वर्गोऽयं वृक्षा ओषधयस्तथा ॥
भरद्वाजेन मुनिना तत्प्रियार्थेऽवतारिताः ।
अन्तःपुरजनोद्देशस्तेन जज्ञे मनोरमः ॥

Bmd has practically the same reading and Mts only slight variation. This tradition regarding the creation of a pleasure garden by Bharadvāja for Rāma's wife seems to refer to the stay of Dāśarathi Rāma and his wife Sītā in the Nasik region and to support the location of Janasthāna and Pañcavaṭī in the same area. The modified text of the Mk*a* has half a verse in place of the said two stanzas, which speaks of Govardhana as the city of Bhārgava (i.e. Paraśurāma) apparently through confusion.

4. Vy, Bmd—पृथिव्यामिह कृत्स्नायां; Mk, Mts— पृथिव्यामपि कृत्स्नायां.

5. Mk, Vy, Bmd, Mts— स प्रदेशो मनोरमः.

6. Mk, Vy *ac*, Bmd, Mts, Vmn—वाह्लीका वाटधानाश्च, Vy *b*—वाह्लीका वाढवानाश्च, *de* (v.l.)—वाह्लीका वाढधानाश्च. Vālhīka=modern Balkh area in the northern part of Afghanistan. The Vāṭadhānas lived in the Punjab-Rajasthan region and had a settlement at Madhyamikā near Chitor (Mbh, II. 29. 7). See Moti Chandra, *Geog. Ec. Stud.*, pp. 27-28, 31-32.

7. Mk, Vy., Bmd, Mts—आभीराः कालतोयकाः; Vmn *a*—अभीराः कालतोयदाः, *b*—अभिराः कालतोयदाः. For the Ābhīras who lived to the north of the Rajputana desert, see Sircar, *Suc. Sāt.*, pp. 242 f. We may also think of Abiravān between Herat and Kandahar which may have been the original home of the Ābhīras. In the 3rd century A.D., there was an Ābhīra kingdom in the north-western Deccan. Kālatoyaka=Kalāt in Baluchistan.

8. Mk—अपरान्ताश्च शूद्राश्च; Vy *acde*, Bmd—अपरीताश्च शूद्राश्च; Vy *b* omits शूद्र; Mts *a*—पुरन्ध्राश्चैव शुद्राश्च, *bc*—पुरन्ध्राश्चैव शूद्राश्च, *c* (v.l.)—परान्धाश्चैव शूद्राश्च; Vmn—अपरान्तास्तथा शूद्राः. The Śūdras lived near the Ābhīras; cf. Sircar, *op. cit.*, p. 242. Aparānta=Aparīta=Purandhra may be Greek Paraitakene in the lower Helmund valley. As the Śūdras are mentioned below separately (p. 34, note 2), we may possibly suggest Kṣudra or Kṣudraka (Greek Oxydrakai) living about the Montgomary Dist. See Raychaudhuri, *op. cit.*, p. 205. Otherwise there would be two Śūdra territories (cf. Sodrai in Northern Sind).

9. Mk *ad*, Vy *be* (v.l.)—पह्लवाश्चर्मखण्डिकाः; Vy *acde*—पह्नवाश्चर्म-

गान्धारा यवनाश्चैव[1] सिन्धु-सौवीर-मद्रकाः[2] ।
शतद्रुजाः कुणिन्दाश्च[3] पारदा हारहूणकाः[4] ॥
रमठा रुद्धकटकाः[5] कैकेया दशमानकाः[6] ।

खण्डिकाः; Mk *bc*, Bmd—पल्लवाश्चर्मखण्डिकाः; Mts—पल्लवाश्चात्तखण्डिकाः; Vmn—पल्लवाश्च सखेटकाः. Pahlava=the Pahlavis or Sassanians of Persia. Carmakhaṇḍika=people of Samarkand.

1. Mk *acd*, Vy, Bmd, Mts, Vmn—गान्धारा यवनाश्चैव; Mk *b*—गान्धारा गबलाश्चैव.Gandhāra=the Rawalpindi and Peshawar Districts. Cf. Raychaudhuri, *op. cit.*, pp. 124 f. The ancient capitals of the country were Takṣaśilā in the Rawalpindi Dist. and Puṣkalāvatī (Charsadda near Peshawar), and Udabhāṇḍapura (Und near Attock) in the early medieval period. Yavana= Indo-Greek settlements in West Pakistan and adjoining lands, one of them in the Kandahar region according to the evidence of the Graeco-Aramaic edicts of Aśoka. Cf. Sircar, *op. cit.*, p. 31; below, Ch. XIV.

2. Mk, Bmd, Mts *abe*—सिन्धुसौवीरमद्रकाः; Vy *acde*, Mts *c* (v.l.), Vmn—सिन्धुसौवीरभद्रकाः; Vy *b*—सिन्धुसौवीरमघ्रकाः. According to *Alb. Ind.*, I, pp. 259-60, 300, Sauvīra included Mūltān and Jahrāvār which latter lay about fifty miles below the junction of the Jhelam and the Chenab. Sindhu lay to the west and Sauvīra to the east of the lower Indus. Madra= district round Sialkot (ancient Śākala).

3. Mk—शतद्रुजाः कलिङ्गाश्च; Vy *acde*—शकह्रदाः कुलिन्दाश्च; *e* (v.l.)—शकाः ह्रदाः कुणिन्दाश्च; Bmd—शका हूणाः कुलिन्दाश्च; Mts—शका द्रुह्याः पुलिन्दाश्च, *c* (v.l.)—शतद्रव(वः) पुलिन्दाश्च; Vy *b*—शकाद्रिह्रा ललिताश्च. The ancient Kuṇindas are supposed to be represented by the modern Kunets of Kūlū and the Simla region; but they formerly appear to have extended upto the Saharanpur and Ambala Districts where their coins have been discovered in large numbers. See Smith, *Cat.*, p. 161. Satadruja= people of the Satlaj valley.

4. Bmd—पारवा हारहूणकाः; Mk—पारदा हारभूषिकाः; Vy *acde*—परिता हारपूरिकाः; Mts—पारदा हारमूर्त्तिकाः, *c* (v.l.)—परेता हारमूषिकाः. Omitted in Vy *b*. The reading हारहूण is supported by Mbh, II. 32. 12, while Bs, XIV. 33, suggests हारहूर which is also supported by Km. The Pāradas were the Parthians of Khorasan, who settled in West Pakistan. The Hārahūṇas or Hārahūras were probably a branch of the Huns who subjugated the whole land from Central Asia to Central India about the end of the 5th century A.D. Moti Chandra locates the Hārahūras in the Herat region (*op. cit.*, p. 66).

5. Mk—माठरा बहुभद्राश्च, Vy *a*—रमठा रुद्धकटकाः; *cde*—रमटा रुद्धकटकाः; Bmd.—रमणा रुद्धकटकाः; Mts—रामठाः कण्टकाराश्च; Vmn—माठरोदकधाराश्च. Omitted in Vy *b*. Lévi locates the Ramaṭhas between Ghazni and Wakhan (*J.A.*, 1918, p. 126). The Ruddhakaṭakas may have lived in the gold-producing Rudok area of Tibet.

6. Mk—कैकेया दशमालिकाः; Bmd—केकषा दशमालिकाः. Vy *ac*—केकया दशमाणिकाः; *de*—केकशा दशमानिकाः; Mts *a*—कैकया दशमानकाः;

क्षत्रियोपनिवेशाश्च[1] तथा शूद्रकुलानि च[2] ॥
[3]काम्बोजा दरदाश्चैव[4] वर्वरा अङ्गलौकिकाः[5] ।
चीनाश्चैव तुषाराश्च[6] बहुला बाह्यतो नराः[7] ॥

b—कैकेया दशनामकाः, c—कैकेय्या दशनामकाः, c (v.l.)—कैकेय्या दशनामिकाः; Vmn a—कैकेया दशमास्तथा, b—कैकेया दंशनास्तथा; Vy e (v.l.)—केशपा दशमाणिकाः. Omitted in Vy b. For the Kekayas, cf. Raychaudhuri, *op. cit.*, pp. 52 ff. They lived in the Jhelam District of West Pakistan and had their capital at Girivraja (Girjāk or Jalālpur on the Jhelam). The Daśamānakas may have lived in the Dasht valley in Kalāt.

1. Mk, Vy *acde*, Bmd—क्षत्रियोपनिवेशाश्च; Mts—क्षत्रियोपनिवेश्याश्च; Vmn—क्षत्रियाः प्रतिवैश्याश्च. Omitted in Vy *b*. For the Kṣatṛs (Greek Xathroi) living near the confluence of the Chenab and the Indus, see Raychaudhuri, *op. cit.*, p. 207.

2. Mk, Vy *acde*, Bmd—वैश्यशूद्रकुलानि च; Mts—वैश्याः शूद्रकुलानि च; Vmn—तथा शूद्रकुलानि च. Omitted in Vy *b*. See p. 32, note 8 above.

3. The following two lines are omitted in Mts and partly in Vy *b*.

4. Mk, Vy *acde*, Bmd, Vmn—काम्बोजा दरदाश्चैव. The Kāmbojas had settlements in the land extending from Kashmir to Kandahar. See Ch. XII below. The Daradas lived round Daratpurī in the upper valley of the Kishengaṅgā in Northern Kashmir. Some scholars locate the Kāmbojas near Badakhshan beyond the Hindu Kush. See Chapter XII below.

5. Mk *abc*—वर्वरा हर्षवर्द्धनाः; Vy *acde*—वर्वराः प्रियलौकिकाः; Mk *d*, Bmd—वर्वरा अंगलौकिकाः; Vmn *a*—वर्वराश्चाङ्गलौकिकाः, *b*—वर्वराश्चाङ्गलोकिकाः. Aṅgaloka is also mentioned elsewhere in the Purāṇas; cf. Ch. IV, Sec. X-B. Varvara reminds one of Barbaricum, a port on the principal mouth of the Indus in the early centuries of the Christian era. But apparently the same people is also located elsewhere; cf. *loc. cit.* Aṅgaloka may be the land of the Agalassoi of the Greeks. But they lived in the land between the Jhelam and the Chenab.

6. Mk *ad*, Bmd, Vy *e* (v.l.)—चीनाश्चैव तुषाराश्च; Vy *acde*—पीनाश्चैव तुषाराश्च; Mk *bc*—चीनाश्चैव तुखाराश्च; Vmn—वेणाश्चैव तुषाराश्च. चीन—the Tibeto-Chinese peoples. तुखार, तुषार—cf. Tukhāristān of the Arabs which included the old Bactria and indicated the mountainous country on both sides of the middle Oxus as far as Badakhshan, though the country to the south of the river was usually understood by the name (Wellhausen, *The Arab Kingdom and its Fall*, p. 430). For the Cīnas, see Jarret, *'Ain-i-Akbarī*, II, pp. 118-19; also below, Chapters IV (Sec. X-B and C) and V (Country No. 33).

7. Mk *abc*—बहुला बाह्यतो नराः; Vy *acde*, Mk *d*—पह्लवा बाह्यतोदराः; Bmd—पह्लवाश्च क्षतोदराः; Vmn—बहुधा बाह्यतो दराः; Vy *b*—मल्लकोदरा (?), *c* (v.l.)—रत्नधाराः क्षतोदराः .

क्षत्रयोऽथ भरद्वाजा:[1] प्रस्थलाश्च दसेरका:[2] ।
लम्पाकास्तल(श्चाव?)गानाश्च[3] चूलिका जागुडै: सह[4] ॥
[5]औपधा(औरसा?)श्चानिभद्राश्च[6] किरातानाञ्च जातय:[7] ।
तोमरा हंसमार्गाश्च[8] काश्मीरास्तङ्गणास्तथा[9] ॥

1. Mk, Vy *acde*, Bmd—आत्रेयाश्च भरद्वाजा:; Mts *a*—क्षत्रयोथ भरद्वाजा:; ∞—अत्रयोथ भरद्वाजा:; Vmn—आत्रेया: सभरद्वाजा:; Vy *b*—आत्रेयाश्च भरद्वका:.

2. Mk *acd*—पुष्कलाश्च कशेरुका:, *b*—पुष्कलाश्च कुशेरुका:; Vy *acde*, Bmd—प्रस्थलाश्च कसेरुका:; Mts—प्रस्थला: सदसेरका:, *c* (v.l.)—प्रस्थला: सह-सौभका: ; Vmn, Mk *c* (v.l.)—प्रस्थलाश्च दशेरका:. Both Prasthala and Puṣkalāvataka are mentioned in the list of Bs, 16. 26. Rām, IV. 43. 11 locates Prasthala in the north. Puṣkala=people of Puṣkalāvatī, modern Prang-Charsabda-Mir Ziyarat region near Peshawar. Daseraka=Maru, i.e. the Marwar region in Rajasthan.

3. Mk—लम्पाका: शूलकाराश्च; Vy *acde*, Bmd—लम्पाकास्तनपाश्चैव; Mts *a*—लम्पाकास्तलनागाश्च, *bc*—लम्पाकास्तलगानाश्च, *c* (v.l.)—लम्पाकास्त-ङ्गणाश्चैव; Vy *e* (v.l.)—लम्पाकास्तानकाश्चैव, लम्पाकास्तनकाश्चैव, लम्पाका-स्तनसाश्चैव, *b*—लम्पाकास्तालकूनाश्च. Lampāka=modern Laghman in Afghanistan. The Āvagānas or Afghans are mentioned in Bs.

4. Mk—चुलिका जागुडै: सह; Vy *acde*, Bmd—पीडिका जुहुडै: सह; Vy *b*—सूलिका जगरै: सह, *e* (v.l.)—चूडिका जुहुडै: सह; Mts—सैनिका: सह जाङ्गलै:; Vmn—चूडिकास्तङ्गणै. सह. Jāguḍa=Southern Afghanistan with its capital at Ghazni. The Cūlikas were the Sogdians living to the north of the Oxus. See Bagchi in *JDL*, Vol. XXI.

5. The following three lines are omitted in Mts and Vy *b*.

6. Mk *a*—औपधाश्चालिभद्राश्च, *bd*—औपधाश्चानिमद्राश्च, —*c* औषधा-श्चानिभद्राश्च; Vy *a*—अपसाश्चालिमद्राश्च; Bmd Vy *cde*—अपगाश्चालि-मद्राश्च; Vmn—अलसाश्चालिभद्राश्च. Aurasa=people of Urasā, modern Hazara District. See below, Ch. IV, Sec. X-D.

7. Mk, Vy *acde*, Bmd, Vmn—किरातानाञ्च जातय:. Cf. p. 31, note 1.

8. Mk—तामसा हंसमार्गाश्च; Vy *acde*, Bmd—तोमरा हंसमार्गाश्च; Vmn—तामसा कर्ममार्गाश्च. See below, p. 45, note 5. Haṁsamārga=Humza in North-Western Kashmir according to some.

9. Mk *ac*, Vy *acde*, Bmd—काश्मीरास्तङ्गणास्तथा; Mk *b*—काश्मीरा-स्तुङ्गणास्तथा, *d*—काश्मीरास्तुमणास्तथा; Vmn—सुपार्श्वा गणकास्तथा. Kāśmira=people of the upper Vitastā valley. A settlement of the Taṅgaṇa or Ṭaṅgana people probably had its headquarters at Ṭaṅgaṇāpura near Joshimath in the Garhwal District, U.P. (*Ep. Ind.*, Vol. XXXI, p. 286).

कुलूता वाहु(बाही?)काश्चैव[1] ऊर्णा(हूणा?) दार्वास्तथैव च[2]।
एते देशा उदीच्यास्तु[3] प्राच्यान् देशान् निबोधत[4] ॥

III

Prācya (*Pūrva-deśa*)

अङ्गा वङ्गा मुद्गरकाः[5] अन्तर्गिरिबहिर्गिरी[6]।

1. Mk—शूलिकाः कुहकाश्चैव; Vy *a*, Bmd—चुलिकाश्चाहुकाश्चैव; Vy *cde*—चूलिकाश्चाहुकाश्चैव; Vmn *a*—कुलूतो कुहिकाश्चूर्णाः, *b*—कुलूताः कुहिकाश्चूर्णा; Vy *e* (v.l.)—चूलिका बाहुकाश्चैव. For the Kulūtas of Kūlū in the Kangra Dist., see Allan, *Cat. C.A.I.*, p. *c*; for the Bāhīkas of the Punjab, see Sircar, *Sel. Ins.*, I, 1942, p. 276 note; also below, Ch. XV.

2. Mk *ac*—जर्णा दर्वास्तथैव च, *d*—ऊर्णा दार्वास्तथैव च; Bmd, Mk *b*—ऊर्णा दर्वास्तथैव च; Vmn—तूर्णपादाः सकुक्कुटाः; Vy *acde*—पूर्णदर्वास्तथैव च. The reading suggested by another passage is हूणा दार्वा० (p. 45, note 8; cf. Kirfel, *op. cit.*, p. 45). The *Harṣacarita* places the Hūṇas in Uttarāpatha apparently somewhere about the Western Panjab. Dārva=people of Dārvābhisāra which roughly comprised the Punch and Naushera regions between the Jhelam and the Chenab.

3. Mk *abc*—एते देशा ह्य् दीच्यास्तु, *d*—एते दशा ह्यु दीच्यास्तुः; Vy *acde*, Bmd—एते देशाह्यु दीच्याश्च; Mts—एते देशा उदीच्यास्तु. Vmn has rather freely—माण्डव्याः पानवीयाश्च उत्तरापथवासिनः. Cf. below, p. 46, note 5.

4. Vy *acde*, Bmd, Mts—प्राच्यान् देशान् निबोधत; Mk—प्राच्यान् देशान् निबोध मे. Cf. below, p. 46, note 6.

5. Mk *abc*—अघ्रारका मुदकराः, अभ्रा वका मुद्गरकाः, *c* (v.l.)—अन्ध्रवाश्च मुद्गरकाः; Vy *acde*, Bmd—अन्ध्रवाकाः सुजरकाः; Vy *b*—अन्ध्रवाका मुद्रकरकाः; Mts *ab*—अङ्गा वङ्गा मद्गुरकाः, *c*—अङ्गा वङ्गा भद्गुरकाः. v.l.—अङ्गा वङ्गा भद्रवकाः, अङ्गा वङ्गा मद्गुरुकाः; Vmn—अङ्गा वङ्गा भद्गुरवाः, Aṅga and Vaṅga may have been separated from Āṅgeya and Vāṅgeya through confusion (cf. below). Mudgaraka (modern Monghyr) is the same as Mudgara of Km and Mudgagiri of inscriptions. Aṅga=East Bihar, and Vaṅga=South and South-East Bengal. For the Aṅga capital Campā near Bhagalpur, see Sircar, *Cosm. Geog. E. Ind. Lit.*, pp. 152, 155. For the Vaṅga capital near about the mouth of the Bhāgīrathī, see below, Chapter XIII.

6. Mk *abc*—अन्तर्गिर्या बहिर्गिराः, *c* (v.l.)—अन्तर्गिरबहिर्गिराः; Vy *acde*, Bmd, Mk *d*—अन्तर्गिरिबहिर्गिराः; Mts—अन्तर्गिरिबहिर्गिरी; Vmn—स्वन्तर्गिरिबहिर्गिराः; Vy *b*—प्रात्रगिरबहिर्गिराः. The Antargiri and Bahirgiri may be located towards the north of Prāgjyotiṣa (Assam) on the strength of Mbh, II. 27. 1-3.

[1]तथा प्रवङ्ग-वाङ्गेया[2] मलदा मल्लवर्त्तकाः[3]॥
[4]ब्रह्मोत्तरा प्रविजया[5] भार्गवाङ्गेय-मल्लकाः[6]।
प्राग्ज्योतिषाश्च पौण्ड्राश्च[7] विदेहास्ताम्रलिप्तकाः[8]॥

1. The line is omitted in Mts *ab*.

2. Mk *ac*—यथा प्रवङ्गा रङ्गेयाः, *b*—तथा प्रवङ्गा रङ्गेयाः; *c* (v.l.) तथा सवङ्गा वङ्गेयाः, *d*—तथा प्लवङ्गा रंगेयाः; Vy *acde*—तथा प्रवंगवंगेयाः, *b*—तथा प्रथंगवंगेयाः; Bmd—तथा प्रवंगा वङ्गाश्च; Vmn—तथा प्रवंगवाङ्गेयाः; Mts *c*—ततः प्लवंगमातङ्गाः. The correct reading of the names and their identification are doubtful.

3. Mk *abc*—मानदा मानवर्त्तिकाः, *c* (v.l.)—मालदा मालवर्त्तिकाः, *d*—मालदा मलवर्त्तिकाः; Vy *acde*—मालदा मालवर्त्तिनः, *e* (v.l.)—मलदा मलवर्त्तिनः, मालवा मलवर्त्तिनः, *b*—मालवा मालवर्त्तिकाः, Mts *c*—यमका मल्लवर्णकाः; Bmd—मालदा मालवणिकाः; Vmn—मांसादा वलदन्तिकाः. The Maladas (people of Malda in West Bengal?) and Mallavartakas are also known from *Km*. Cf. p. 42, note 8 below.

4. This line is omitted in Vy *b*.

5. Mk *acd*—ब्राह्मोत्तराः प्रविजयाः; Vy *acde*, Bmd, Vmn *a*. Mk *b*—ब्रह्मोत्तराः प्रविजयाः; Vmn *b*—ब्रह्मोत्तराः प्राविजयाः; Mts *abc*—सुह्मोत्तराः प्रविजयाः, *c* (v.l.)—सह्मोत्तराः प्रविजयाः. The reading सुह्मोत्तर is not supported by Km which reads सुह्म (the ancient name of Rāḍha in South-West Bengal) and ब्रह्म or ब्रह्मोत्तर side by side. Can Brahmottara be associated with the name of Burma which is derived from the Mrama, one of the three tribes that migrated into Burma and ultimately lent their name as a national designation to all Burmese? But cf. below, Ch. IV, Sec. E, and for a Barmhattar in Sarkar Satgaon, see Jarret, *op. cit.*, p. 141. Kirfel (*op.cit.*, p. 46) has also प्रतिजया and प्रावृषेया. It is difficult to be sure about the name and its location.

6. Mk *abc*—भार्गवा ज्ञेयमल्लकाः, *d*—भार्गवा गेयमल्लकाः; Vy *acde*—Bmd—भार्गवा गेयमर्थकाः; Mts *abc*—मार्गवा गेयमालवाः; *c* (v.l.)—भार्गवांगेयमालवाः; भार्गवाश्चार्यगालवाः, मार्गवा ये च मालवाः; Vmn *a*—भार्गवांगेयमर्यकाः, *b*—भार्गवांगेयमर्षकाः. The correct reading of the names and their location are doubtful.

7. Mk—प्राग्ज्योतिषाश्च मद्राश्च; Vy—प्राग्ज्योतिषाश्च मुण्डाश्च; Bmd—प्रागज्योतिषाश्च पौण्ड्राश्च; Mts—प्रागज्योतिषाश्च पुण्ड्राश्च; Vmn—प्राग्ज्योतिषाः पृषघ्राश्च. Prāgjyotiṣa=modern Gauhati area of Assam and the adjoining region originally known as an Udīcya territory. Pauṇḍra=people of North Bengal with their city called पुण्ड्रनगर (later पुण्ड्रवर्द्धन, modern Mahāsthān in the Bogra District).

8. Mk, Vy *acde*, Bmd, Mts, Vmn—विदेहास्ताम्रलिप्तकाः; Vy *b*—आबिकास्ताम्रलिप्तिकाः. Videha=a people living in North Bihar and the adjoining region with their capital at Mithilā, modern Janakpur in the

मल्ला मगध-गोनर्दा:[1] प्राच्या जनपदाः स्मृताः[2] ॥

IV

Dakṣiṇāpatha (Dākṣiṇātya)

[3]अथापरे जनपदा[4] दक्षिणापथवासिनः[5] ।
पाण्ड्याश्च केरलाश्चैव[6] चोलाः कुल्यास्तथैव च[7] ॥
सेतुका मूषिकाश्चैव[8] कुमारा वनवासका[9] ।

Nepalese Tarai. Tāmraliptaka=people of the district round modern Tamluk in South-West Bengal

1. Mk *abc*—मल्ला मगधगोमन्ताः; *d*—मल्ला मगधगोमेदाः; Vy—माला मगधगोविन्दाः; Bmd—मालामगधगोनन्दाः; Mts *a*—शाल्वमगधगोनर्द्दाः,*bc*—शाल्वमागधगोनर्द्दाः; Vmn—माला मगधमानन्दाः. For the Mallas of the district round Kuśīnagara (Kasia in the Deoria District), see Raychaudhuri, *op. cit.*, pp. 106 f. Note that Magadha was sometimes located in Madhyadeśa (p. 31, note 4) and sometimes in Prācya. *Vj* which locates the Andhras and Sālvas (cf. p. 36, note 5,) in East India apparently relied on such erroneous texts of the Purāṇas. For Gonarda which belonged to East India only theoretically, see *JAHRS*, Vol. IX, Part iii, pp. 1 ff., and below, Ch. XIX.

2. Mk, Vy *a*, Bmd, Mts—प्राच्या जनपदाः स्मृताः; Vmn—प्राच्या जनपदास्त्विमे; Vy *cde*—प्राच्यां जनपदाः स्मृताः.

3. The following line is omitted in Vmn.

4. Mk, Vy, Bmd—अथापरे जनपदाः; Mts—तेषां परे जनपदाः.

5. Mk, Vy, Bmd, Mts—दक्षिणापथवासिनः. Cf. p. 46, note9.

6. Mk *ac*, Vmn—पुण्ड्राश्च केरलाश्चैव; Mk *b*—पुण्ड्राश्च केवलाश्चैव; Vy, Bmd, Mts, Mk *d*—पाण्ड्याश्च केरलाश्चैव. The Pāṇḍyas and Keralas lived about the Madurai-Ramanathapuram-Tirunelveli region and the Malayalam-speaking area respectively.

7. Mk *ac*—गोलाङ्गूलास्तथैव च,*b*—गोलाङ्गूलास्तथैव च,*d*—चोलाः कुल्यास्तथैव च; Vy, Bmd—चौल्याः कुल्यास्तथैव च; Mts—चोलाः कुल्यास्तथैव च; Vmn—चोडाकुल्याश्च राक्षस.For the reading गोलाङ्गूलwhich seems to be a later emendation, see *Bs*, 16. 3. The Colas lived about the Tanjavur and Tiruchirappalli Districts of Madras State. Kulya=Kollam (Quilon) ?

8. Mk—शैलषा मूषिकाश्चैव; Vy, Bmd—सेतुका मूषिकाश्चैव; Mts—सेतुकाः सूतिकाश्चैव; Vmn—जानुका मूषिकादाश्च. Setuka=people of Setubandha-Rāmeśvara. Mūṣika=probably the people living on the banks of the Musi now running by the town of Hyderabad (Raychaudhuri, *op. cit.*, p. 80). Another river of this name runs through the Nellore and Guntur Districts.

9. Mk *abc*—कुसुमा नाम वासकाः, *d*—कुमारा वनवासकाः; Vy *acde*—कुमना वनवासिकाः; Bmd—कुनासा वानवासकाः; Mts—कुपथा वाजिवासिकाः,

महाराष्ट्रा माहिषकाः[1] कलिङ्गाश्चैव सर्व्वशः[2] ॥
कावेराः सह चैषीका[3] आटव्याः शवराश्च ये ॥[4]
पुलिन्दा विन्ध्यमूलीका (या)[5] विदर्भा दण्डकैः सह[6] ॥
[7]पौरिका मौलिका (मौलका?) श्चैव[8] अश्मका भोगवर्द्धनाः[9] ।

. (v.l.)—कुपथा वाजिकाः शकाः, कुपथा वाजिकाशिकाः; Vmn—कुमारादा महाशकाः; Vy *b*—रुमना वनवासिकाः. The Vanavāsakas were the inhabitants of modern Banavasi in the North Kanara District of Mysore. Kumāra=people of Cape Comorin called Kumāra-dvīpa.

1. Mk, Vy, Bmd, Vmn *a*—महाराष्ट्रा माहिषकाः; Mts *a*—नवराष्ट्रा माहिषकाः; *bc*—नवराष्ट्रा माहिषिकाः; Mk *d*, Vmn *b*—महाराष्ट्रा माहिषिकाः. Mahārāṣṭra=modern Marāṭhā country. Māhiṣaka=people of the ancient Mahiṣa-viṣaya in Mysore or of Māhiṣmatī on the Narmadā. See Ch. XVI below.

2. Mk, Vy, Bmd, Mts, Vmn—कलिङ्गाश्चैव सर्व्वशः; Mts *c* (v.l.)—कपिलाश्चैव सर्व्वशः. The Kaliṅga country originally comprised the present Puri and Ganjam Districts of Orissa and the adjoining regions.

3. Mk—आभीराः सहवैशिक्याः; Vy *abe* (v.l.), Bmd—आभीराः सहचैषीकाः; Vy *cde*—अभीराः सहचैषीकाः; Mts—कारूषाश्च सहैषीकाः; *c* (v.l.)—कावेराश्च सहोषीकैः; Vmn—आभीराः सहवैसक्याः. Kāvera=people living on the Kāveri.

4. Mk *abc*—आढक्याः शवराश्च ये; Mk *d*, Vy *b*—आटव्याः शवराश्च ये; Vy *acde*, Bmd—आटव्याश्च वराश्च ये; Mts—आटव्याः शबरास्तथा; Vmn—आरण्याः शबराश्च ये. Śavara foresters=the Saoras of Ganjam and Visakhapatnam.

5. Mk *abc*—पुलिन्दा विन्ध्यमौलेयाः, *d*—पुलिन्दा विन्ध्यमालेयाः; Vy *ace*, Bmd—पुलिन्दा विन्ध्यमूलीकाः; Vy *ad*—पुलिन्द्रा विन्ध्यमूलीकाः; Mts—पुलिन्दा विन्ध्यपुषिकाः; Vmn—पुलिन्दा विन्ध्यशौलेयाः. The reference is to the Pulindas of the Vindhyan region.

6. Mk, Vy *bcde*, Mts *bc*, Vmn *b*—वैदर्भा दण्डकैः सह; Vy *a*, Bmd, Mts *a*—विदर्भा दण्डकैः सह; Vmn *a*—वेदभो दण्डकैः सह. The Prakritism in the form वेदभो is interesting to note. Vidarbha=modern Berar and the adjoining regions. Daṇḍaka is the old name of parts of the Marāṭhā country including especially the Nasik District.

7. The following lines together with the first line of the next Section are omitted in Mts.

8. Mk—पौरिका मौलिकाश्चैव; Vy *a*—पौनिका मौलिकाश्चैव, *cde*—पानिका मौनिकाश्चैव, *b* recognises मूलिक only; Bmd—शौलिका मौलिकाश्चैव. Paurika=people of Purī in the Konkan. Maulika=people of Mūlaka, the district round Paithan on the Godavari in the Aurangabad District, Maharashtra.

9. Mk, Vy *ab*, Bmd, Vmn *b*—अश्मका भोगवर्द्धनाः; Vy *cde*—अस्मका

नै(ऋ?)षिकाः कुन्तला अन्ध्रा[1] उद्भिदा नलकालिकाः[2] ।
दाक्षिणात्याश्च वै देशा[3] अपरान्तान् निबोधत[4] ॥

V

Aparānta (*Paścād-deśa*)

[5]शूर्पारकाः कोलवना[6] दुर्गास्तालीकटैः सह[7] ।

भोगवर्द्धनाः; Vmn *a*—अलका भोगवर्द्धनाः. Aśmaka was contiguous to Mūlaka and had its capital at Paudanya, identified by Raychaudhuri with Bodhan in the Nizamabad District of Andhra Pradesh. Bhogavardhana reminds us of the Bhokardan Taluk of the Aurangabad District, Maharashtra.

1. Mk *ac*—नैषिकाः कुन्तला अन्ध्राः, *b*—नैषिकाः कुन्तला अन्धाः, *d*—नैषिकाः कुन्तला आन्ध्राः; Vy *ace*—नैणिकाः कुन्तला अन्ध्राः, *b*—नैतिकाः कुन्तला अन्ध्राः; *d*—नैणिकाः कुन्तला आन्ध्राः; Bmd—मैन्दिकाः कुन्तला अन्ध्राः; Vmn—नैमिकाः कुन्दला आन्ध्राः. Kuntala=people of the region around the North Kanara District. The Andhra people lived in the lower valleys of the Krishna and the Godavari rivers. Originally they appear to have spread over large parts of the Central and Northern Deccan. For the Ṛṣikas on the Krishna, see Sircar, *Sel. Ins.*, 1942, p. 198, note.

2. Mk—उद्भिदा वनदारकाः; Vy *acde*, Bmd—उद्भिदा नलकालिकाः, *b*—उद्भिरा नलकालिकाः; Vmn *a*—उच्छिदा नलकारकाः, *b*—उलिदा नलकारकाः. नलकालकाः=नलकालुपाः ? =the Nalas and Ālupas ? उद्भिद=आभीर ? The Ābhīras ruled in Maharashtra, but conquered the Nagarjunikonda valley for sometime about 278 A.D. (*Ep. Ind.*, Vol. 34, pp. 197 ff.).

3. Mk—दाक्षिणात्यास्त्वमी देशाः; Vy, Bmd—दाक्षिणात्याश्च वै देशाः. Vmn has rather freely—दाक्षिणात्या जनपदास्त्विमे शालकटांकट (*b*—कंटकट). Al-Bīrūnī wrongly recognises दाक्षिणात्य and वैदिश (sic—वै देशाः) as two *janapadas* of the South and refers to the following countries of the West as those belonging to the South.

4. Mk *abc*—अपरान्तान् निबोध मे; *d*—अपरांस्तान् निबोध मे; Vy, Bmd—अपरांस्तान् निबोधत । Cf. below, p. 46, note 10.

5. The following line is omitted in Mts.

6. Mk—सूर्यारकाः कालिबलाः; Vy *acd*—शूर्पाकाराः कोलवनाः, *e*—सूर्पाकाराः कालवनाः; Bmd, Vy *b*—शूर्पारकाः कोलवनाः; Vmn—शूर्पारका वारिधानाः. Śūrpāraka=Sopara to the north of Bombay. Kolavana=Kolāpura (Kolhapur) or Koḍagu (Coorg).

7. Mk *abc*—दुर्गाश्चानीकटैः सह, *d*—दुर्गाश्चामीकटैः सह; Vy *acde*—दुर्गाः कालीतकैः सह, *b*—दुर्गास्तिल्लीतकैः सह; Bmd—दुर्गाः कालीकटैः सहः; Vmn—दुर्गास्तालीकटैः सह. The reading तालीकट is supported by Bs, XIV. 11 and Mbh, II. 31. 65 (where तालाकट is a mistake for तालीकट). Tālikaṭa=Talakāḍ,

पुलेयाश्च सुरा(मुर?)लाश्च[1] रूपसास्तामसैः सह[2] ॥
[3]तथा तुरमिनश्चैव (तुम्बवनाश्चैव?)[4] सर्वे चैव कारस्कराः[5] ।
नासिक्याद्याश्च ये चान्ये[6] ये चैवोत्तरनर्मदाः[7] ॥

the Western Gaṅga capital which was really in the south. Durga=people of the valley of the Durgā, a tributary of the Sābarmatī in Gujarat.

1. Mk—पुलिन्दाश्च सुमीनाश्च; Vy *acde*, Bmd—पुलेयाश्च सुरालाश्च; Vy *b*—पुलेयाश्च करालाश्च (?); Mts—कुलीयाश्च सिरालाश्च; Vmn—पुलीयाश्चासिनीलाश्च. For a river called मुरला in which the अपरान्त elephants sported, see *Sel. Ins.*, 1942, p. 453n. The Murala people is known to have come into contact with Pratihāra Mahīpāla I according to Rājaśekhara's *Bālabhārata* or *Pracaṇḍapāṇḍava* (N.S.P. ed., I. 7) and with Kalacuri Karṇa according to the Bheraghat inscription (*Ep. Ind.*, II, p. 11). Paramāra Sindhurāja (c. 995-1000 A.D.) is represented in the *Navasāhasāṅkacarita* (X. 14-20) as having defeated the Muralas. The reading of the other names and their location are uncertain.

2. Mk—रूपपाः स्वापदैः सह; Vy *acde*, Bmd—रूपसास्तापसैः सह; Vmn—तापसास्तामसास्तथा; Vy *b*—रूपकास्तामसैः सह. The reading of the names and their location are uncertain.

3. Vmn has for the following two lines—कारस्करास्तु भमिनो नासिकान्ताः सुनर्मदाः.

4. Mk—तथा कुरुमिनश्चैव; Vy *acde*, Bmd—तथा तु रसिताश्चैव; Mts—तथा तैत्तिरिकाश्चैव; Vy *b*—तथा तरूपणाश्चैव, *e* (v.l.)—तथा तु रमिनश्चैव. Tumbavana=Tumain in the Guna District of Madhya Pradesh.

5. Mk—सर्वे चैव कठाक्षराः; Vy b, Mts, Mk *c* (v.l.)—सर्वे कारस्करास्तथा; Vy *acde*—सर्वे चैव परक्षराः, *e* (v.l.)—सर्वे चैव परस्कराः; Bmd—सर्वे चैवापराक्षराः; cf. Vmn—कारस्करास्तु भमिनः. Mk *d* has, after this, the following additional lines—कारस्करा लोहजंघा वाजेया राजभद्रकाः। तोसलाः कोसलाश्चैव त्रैपुरा वैदिशस्तथा ॥ तुषारास्तम्बुराश्चैव सर्वे चैव करस्कराः ॥ The second of the three lines is taken out of the following Section. The Kāraskaras are known from *Baudh. Dh. S.*, I. 2. 14—आरट्टान् कारस्करान् पुण्ड्रान् सौवीरान् वंगकलिङ्गान् प्राणूनानिति च गत्वा पुनस्तोमेन यजेत सर्वपृष्ठया वा॥ Dey's *Geog. Dict.* identifies Kāraskara with Karkal near Mangalore. But it is doubtful.

6. Mk—नासिक्यावाश्च ये चान्ये; Vy, Bmd—नासिक्याद्याश्च ये चान्ये; Vy *e* (v.l.)—नासिकाद्याश्च ये चान्ये, नासिकाद्याश्च ये चान्ये; Mts *a*—वासिकाश्चैव ये चान्ये, *bc*—वासिक्याश्चैव ये चान्ये. Nāsikya=people of Nasik in Maharashtra.

7. Mk, Vy *b*—ये चैवोत्तरनर्मदाः; Bmd, Mts, Vy *cde* (v.l.)—ये चैवान्तरनर्मदाः; Vy *ae*—ये वै चान्तरनमदाः. Uttara-Narmada=people living on the northern bank probably of the lower Narmadā.

भारु(भृगु)कच्छाः स-माहेयाः[1] सह सारस्वतैरपि[2] ।
कच्छीयाश्च सुराष्ट्राश्च[3] आनर्त्ताश्चार्बुदैः सह[4] ॥
इत्येते अपरान्ताश्च[5] शृणुध्वं विन्ध्यवासिनः[6] ॥

VI

Vindhyapṛṣṭha

[7]मलदाश्च कारूषाश्च[8] मेकलाश्चोत्कलैः सह[9] ।

1. Mk—भीरुकच्छाः समाहेयाः; Vy *acd*—भानुकच्छाः समाहेयाः; Bmd—Mts,Vy *a* (v.l.)—भारुकच्छाः समाहेयाः; Vmn—दारुकच्छाः सुमाहेयाः; Vy *be*—भानुकच्छाः समाहेयाः. The Prakritism in the form भारुकच्छ (Pāli भरुकच्छ) for the Sanskritic भृगुकच्छ is interesting to note. Bhṛgukaccha=Broach. Māheya=people of the Mahī valley. Both the tracts are in Gujarat.

2. Mk, Vmn, Vy *b*—सह सारस्वतैरपि; Mts—सह सारस्वतैस्तथा; Vy *acde*, Bmd—सहसा शाश्वतैरपि. Sārasvata=people of the Sarasvatī (Sarsuti) valley in the Eastern Punjab or in Gujarat (cf. Jarrett, *op. cit.*, p. 229).

3. Mk—काश्मीराश्च सुराष्ट्राश्च; Vy, Bmd—कच्छीयाश्च सुराष्ट्राश्च; Mts *a*—कच्छीकाश्चैव सौराष्ट्राः, *bc*—काच्छीकाश्चैव सौराष्ट्राः; Vmn—वात्सीयाश्च सुराष्ट्राश्च. Kacchīya=people of Kutch. Surāṣṭra=South Kathiawar.

4. Mk *acd*, Vmn—आवन्त्याश्चार्बुदैः सह, Mk *b*—अवन्त्याश्चार्बुदैः सह; Vy *acde* (v.l.)—अनर्त्ताश्चाबुर्दैः सह; Bmd, Vy *e*—आनर्त्ताश्चार्बुदैः सह; Mts—आनर्त्ता अर्बुदैः सहः; Vy *b*—अनर्त्ता हुद्बुदैः (sic) सह. Ānarta=district round Dvārakā. Arbuda=Mount Abu in the Sirohi District, Rajasthan.

5. Mk, Vy *e* (v.l.)—इत्येते ह्यपरान्ताश्च; Vy *acde*, Bmd—इत्येते सम्परीताश्च; Mts—इत्येते अपरान्तास्तु. Vmn has rather freely इत्येते पश्चिमामाशां स्थिता जानपदा जनाः.

6. Mk *acd*—शृणु विन्ध्यनिवासिनः, *b*—शृणु विन्धनिवासिनः; Vy *acde*, Bmd—शृणुध्वं विन्ध्यवासिनः; Mts—शृणु ये विन्ध्यवासिनः.

7. Vy *b* omits several lines and makes the *janapadas* mentioned in the following two Sections as those belonging to the West. See p. 40, note 3.

8. Mk *a*—सरजाश्च करुषाश्च,*bcd*—सरजाश्च करूषाश्च; Vy *acde*,Bmd,Mts—मालवाश्च करूषाश्च; Vmn—कारूषाश्चैकलव्याश्च; Vy *b*—मलदाश्च करूषाश्च. See above, p. 37, note 3. The Mālavas (Malloi of the Greeks) lived on the lower Irāvatī (Rāvī) in the Punjab in the fourth century B.C. Later they are found settled in Rajasthan and ultimately they gave their name to modern Malwa in Central India. Kārūṣa, which is also called Bṛhadgṛha, is the same as the present Shahabad District of Bihar. The Maladas may have lived in the Malda, District of West Bengal. Lama Tāranātha who locates a Mālava country near Prayāga seems to have confused the Maladas with the Mālavas of Western India on the strength of erroneous Puranic passages quoted in the present note. Cf. Bhandarkar's List, No. 1068.

9. Mk—केरलाश्चोत्कलैः सह; Vy *ae*—रोकलाश्चोत्कलैः सह; Bmd,

[1]उत्तमर्णा(का ?) दशार्णश्च[2] भोजाः किष्किन्धकैः सह[3] ॥
तोशलाः कोशलाश्चैव[4] त्रैपुरा वैदिशास्तथा[5] ।

Mts, Vmn, Vy *bcd*—मेकलाश्चोत्कलैः सह. The name Mekala is still preserved in that of the present Maikal range which is the connecting link between the Vindhyas and the Satpuras and stretches from the Khairagarh area in M.P. to the Rewah region (Raychaudhuri, *Stud. Ind. Ant.*, p. 134). The Mekalas lived near the Amarkantak hills. Utkala was the western fringe of the Midnapur District and the adjoining parts of Orissa.

1. The following line is wrongly placed in Section V in Mk *d*.

2. Mk, Bmd, Vmn, Vy *d*—उत्तमर्णा दशार्णश्च; Vy *ace*—उत्तमार्णा दशार्णश्च, *b*—उत्तमर्णा बशार्णश्च; Mts—औण्ड्रा माषा दशार्णश्च. Daśārṇa formed parts of modern East Malwa (cf. the river of the same name in Ch. IV, Sec. III) and the adjoining regions. It was also known as Ākara and had its capital at Vidiśā. The Uttamakas may be the Uttamabhadras of the Pushkar region.

3. Mk—भोज्याः किष्किन्धकैः सह; Vy, Bmd, Mts—भोजाः किश्किन्धकैः सह; Vmn—गोप्ता किकरवैः सह. The Bhojas originally lived in Berar, but later founded a kingdom in the Goa region. The Kiṣkindhakas may be the people of Kiṣkindhā identified with modern Kalyāṇpur in the south of the Udaipur division, Rajasthan, and not the well-known Kiṣkindhā in Mysore State. See Sircar, *The Guhilas of Kiṣkindhā*, p. 34.

4. Mk, Vmn *b*—तोशलाः कोशलाश्चैव; Vy *acde*—तोसलाः कोसलाश्चैव; Bmd—तोसलाः कौशलाश्चैव; Vmn *a*—तोशला शोकलाश्चैव; Mts *ac*—स्तोशलाः कोसलाश्चैव, *b*—स्तोशलाः कौसलाश्चैव; Vy *b*—recognises कोसल only. Both the names could be spelt either with the dental or with the palatal sibilant. Tośala was no doubt the district round the city of Tośalī (modern Dhauli, near Bhubaneswar in the Puri District, Orissa). In the early medieval period, the Tośala *janapada* is known to have been divided into Uttara-Tośala and Dakṣiṇa-Tośala. Kośala here is no doubt Dakṣiṇa-Kośala, identified with the modern Raipur-Bilaspur-Sambalpur region of M. P. and Orissa. See Raychaudhuri, *PHAI*, pp. 252, 452.

5. Mk *a*—तैपुरा वैदिशस्तथा, *bcd*—त्रैपुरा वैदिशस्तथा; Vy—त्रपुरा वैदिकास्तथा; Bmd, Mts—त्रैपुरा वैदिशास्तथा; Vmn—त्रैपुराः खेल्लिशास्तथा. The Traipuras were the people of Tripura or Tripurī, modern Tewar in the Jabalpur District. The Vaidiśas were the people of Vidiśā, modern Besnagar near Bhilsa in Madhya Pradesh.

तुमुरास्तुम्बुराश्चैव[1] पटवो निषधैः सह[2] ॥
[3]अनूपास्तुण्डिकेराश्च[4] वीतिहोत्रा अवन्तयः[5] ।

1. Mk *abc*—तुम्बुरास्तुम्बुलाश्चैव, *d*—तुम्बरास्तुम्बुलाश्चैव; Vy *acde*, Bmd—तुमुरास्तुम्बुराश्चैव; Mts *a*—तुमुरास्तम्बराश्चैव, *bc*—तुमुवास्तुम्बराश्चैव; Vmn—तुरगास्तुम्बराश्चैव; Vy *b*—थर्पुरास्तुम्बुराश्चैव. One of the names appears to be a mistake for तुम्बवन, modern Tumain in the Guna Dist., M. P. The other locality may be Tummāna (=Tumān), 45 miles north of Ratanpur in the Bilaspur Dist., M. P.

2. Mk *a*—पटवो निषधैः सह, *bc*—पटवो नैषधैः सह, *d*—पटवी नैषधैः सह; Vy *acde*, Bmd—षट्सुरा निषधैः सह; Vy *e* (v.l.)—षट्सुरा नैषधैः सह; Mts *ab*—पद्गमा निषधैः सह, *c*—पद्गमा नैषधैः सह; Vmn *a*—वहेला निषधैः सह, *b*—वहेला नैषधैः सह; Vy *b*—षत्तुमानाः पधैः सह. The Niṣadhas are associated with the Pāriyātra and identified with the Bhils of Rajasthan. They may have also lived about Nalapura, modern Narwar in the Shivapuri Dist., M. P. The other people's name and location are uncertain.

3. The following four lines are omitted in Vy *b*.

4. Mk *abcd*—अभ्रजास्तुष्टिकाराश्च, *c* (v.l.)—अनूपास्तुण्डिकेशाश्च; Vy *acde*, Bmd, Vmn—अनूपास्तुण्डिकेराश्च; Vy *e* (v.l.)—अणुपास्तुण्डिकेराश्च; Mts—अरूपाः शौण्डिकेराश्च. Omitted in Vy *b*. Anūpa was on the Narmadā with the city of Māhīṣmatī (modern Māndhātā in the Nimār District, M.P.) as its capital. The Haihaya king Arjuna, son of Kṛtavīrya, ruled at Māhiṣmatī. Like the Bhojas, Vītihotras and Avantis mentioned in the Section, the Tuṇḍikeras were a branch of the Haihayas. Acccrding to the Purāṇas (Mts, 43. 48-49; Vy, 94. 51-52; Bmh, 13. 203-04; Pdm, Sṛṣṭi-khaṇḍa, 12. 35-36; Hv, I. 33. 51-52), the Haihayas were subdivided into five clans, viz., Vītihotra (wrongly called वीरहोत्र in some texts), Bhoja, Avanti, Tuṇḍikera (called कुण्डकेर, तौण्डिकेर and तुण्डकेर in some texts) and तालजंघ. Pargiter (*AIHT*, pp. 98, etc.) refers to the name of the Haihaya clan शार्यात read in some texts as सुरताः, सञ्जाताः, असंख्याताः, सुजाताः, etc. The Haihayas were themselves a branch of the Yādavas. See, e.g., Vsn, IV. Ch. 11, Sections 1 ff.

5. Vy *acde*, Bmd—वीतिहोत्रा ह्यवन्तयः; Mk—वीरहोत्रा ह्यवन्तयः; Mts—वीतिहोत्रा अवन्तयः; Vmn—वीतिहोत्रास्त्ववन्तयः. Omitted in Vy *b*. The Avanti people lived in West Malwa and the Vītihotras probably in the same territory or on the Narmadā to the south of the Avantis.

एते जनपदाः सर्वे[1] विन्ध्यपृष्ठनिवासिनः[2] ॥

VII
Parvata (Himālaya)

अतो देशान् प्रवक्ष्यामि[3] पर्वताश्रयिणश्च ये[4] ।
निराहारा हंसमार्गाः[5] कुरवस्तङ्गणाः खसाः[6] ॥
कर्णप्रावरणाश्चैव[7] हूणा दार्वाः स-हूहुकाः[8] ।

1. Mk, Vy, Bmd—एते जनपदाः सर्वे; Mts—एते जनपदाः ख्याताः. Vmn has rather freely—सुकेशा विन्ध्यमूलस्थास्त्विमे जनपदा स्थि([b]—स्मृ)ताः.

2. Mk, Vy, Bmd, Mts—विन्ध्यपृष्ठनिवासिनः.

3. Mk, Vy *acde*, Bmd, Mts—अतो देशान् प्रवक्ष्यामि; Vmn—आद्यान् देशान् प्रवक्ष्यामः. The reference below is to the Himalayan peoples.

4. Mk,Vy *acde*, Bmd, Mts—पर्वताश्रयिणश्च ये; Vmn—पर्वताश्रयिणस्तु ये.

5. Mk—नीहारा हंसमार्गाश्च; Vy *acde*, Bmd—निगर्हरा हंसमार्गाः; Mts—निराहाराः सर्वगाश्च; Vmn *a*—निराहारं हंसमार्गाः, *b*—निराहारा हंसमार्गाः. The Haṁsamārgas of the Himalayan region are also known from other passages; cf. below, Chapter IV, Section viii; also above, p. 35, note 8. They had a settlement in North-West Kashmir. The name and location of the other people are uncertain.

6. Mk—कुरवो गुर्गणाः खसाः; Vy *acd*—क्षुपणास्तङ्गणाः खसाः, *e* (v.l.)—कुणपास्तङ्गणाः खसाः, क्षुपणास्तङ्गणास्त्रसाः, क्षुपणास्तङ्गणाः स्वसाः; Bmd—कुपथास्तङ्गणाः खसाः; Mts—कुपथा अपथास्तथा; Vmn—कुपथास्तङ्गणाः खशाः. These Kurus are doubtless the celebrated Uttara-Kurus living in the Himalayan region. The Khasas (identified with the modern Khakkas) lived about Kashmir. Mts, Vy and Bmd refer to a Himalayan locality called कुपथ or अपथ (cf. Chapter IV, Section X-6). For the Taṅgaṇas, see above p. 35, note 9.

7. Mk. कुन्तप्रावरणाश्चैव; Vy *acde*—कुशप्रावरणाश्चैव; Bmd, Vy *b*—कर्णप्रावरणाश्चैव; Mts. Vmn, Vy *e* (v.l.)—कुथप्रावरणाश्चैव. Both कुथप्रावरण and कर्णप्रावरण situated in the Himalayan region are elsewhere mentioned in Mts, Vy and Bmd (cf. Chapter IV, Section X—G and H). Rām., IV. 40. 25 locates कर्णप्रावरण in the eastern direction, probably in the eastern Himalayas.

8. Mk *a*—उर्णा दार्वाः सकृत्रकाः; *bcd*—ऊर्णा दार्वाः सकृत्रकाः; Vy *acde*—हूणा दर्वाः सहूदकाः; Bmd—हूणदर्वाः बहूवकाः Mts *a*—उर्णा दर्वाः समुद्गकाः, *bc*—ऊर्णा दर्वा समुद्गकाः; Vmn—ऊर्णाः प्लुष्टाः सहूहुकाः; Vy *b*—हूणा दर्वाः सहूहकाः; *e* (vl.)—हूणा दर्वाः सुहूहुकाः, हूणा दर्वाः सहूहुकाः. For the Hūṇas and Dārvas, see p. 36, note 2 above. For the Hūhukas of the Himalayan region, see Chapter IV, Section X—H.

त्रिगर्त्ता मालवाश्चैव[1] किरातास्तामरैः (स्तोमरैः ?) सह[2] ॥

VIII

Condensed List

[3]तास्त्विमे कुरु-पाञ्चाला[4] मध्यदेशादयो जनाः[5] ।
पूर्वदेशादिकाश्चैव[6] कामरूपनिवासिनः[7] ॥
पुण्ड्राः (पाण्ड्याः) कलिङ्गा मगधा (माहिषा)[8] दाक्षिणात्याश्च कृत्स्नशः[9] ।
तथापरान्ताः सौराष्ट्राः[10] शूद्राभीरास्तथार्बुदाः [11] ॥

1. Mk *ac*, Vy—त्रिगर्त्ता मालवाश्चैवः ; Mk *d*—त्रिगर्त्ता गालवाश्चैव; Bmd—त्रिगर्त्ता मालयाश्चैव; Mts—त्रिगर्त्ता मण्डलाश्चैव. Vmn has rather freely—त्रिगर्त्ताश्च किराताश्च तोमराः शशिखाद्रिकाः. Trigarta=modern Jalandhar region. For the *Sapta-Mālava* or seven localities called Mālava, see Raychaudhuri, *PHAI*, p. 492, note. Mālava in our text may be the same as the country of that name identified with the cis-Satlaj Districts of the Punjab together with some Himalayan territory.

2. Mk, Vy *acde*, Bmd—किरातास्तामसैः सह; Vy *b*—किरातास्तामरैः सह; Mts—किराताश्चामरैः सह. Kirātas=Himalayan mountaineers: cf. the name of modern Kirānti, etc. See above, p. 31, note 1. The reading तोमर (for तामर) is supported by some passages; cf. Chapter IV, Section X—H. See also p. 35, note 8 above.

3. This later list is very carelessly compiled and probably had some mistakes even in the original draft.

4. Vsn, Krm—तास्त्विमे कुरुपाञ्चालाः; Bmd—तास्त्विमे कुरुपञ्चालाः. See above, p. 30, note 2.

5. Vsn, Krm, Bmh—मध्यदेशादयो जनाः. Madhyadeśa lay between the East Punjab and Eastern U.P., and between the East Punjab and the Oxus valley lay Udīcya which was believed to have originally comprised the Himalayan region including Prāgjyotiṣa (Assam).

6. Vsn, Krm, Bmh—पूर्वदेशादिकाश्चैव. Originally Prācya or Pūrvadeśa included the lands lying to the east of Eastern U.P.

7. Vsn, Krm, Bmh—कामरूपनिवासिनः. Kāmarūpa (also called Prāgjyotiṣa)=Assam.

8. Vsn, Krm—पुण्ड्राः कलिङ्गा मगधाः; Bmh—पौण्ड्राः कलिङ्गा मगधाः. See above, pp. 38-39.

9. Vsn, Krm—दाक्षिणात्याश्च कृत्स्नशः; Bmh—दाक्षिणात्याश्च सर्वशः. Dākṣiṇātya or Dakṣiṇāpatha lay to the south of the Vindhya or the Narmadā.

10. Vsn—तथापरान्ताः सौराष्ट्राः; Bmh—तथापरान्त्याः सौराष्ट्राः; Krm—तथापरान्ताः सौराष्ट्राः. Aparānta lay to the west of Malwa. See p. 42, note 3.

11. Vsn, Bmh, Krm *ab* (v.l.)—शूद्राभीरास्तथार्बुदाः; Krm *bc*—शूद्रा हीनास्तथार्बुदाः. See p. 42, note 4; for Śūdra and Ābhīra in Udīcya, cf. p. 32 notes 7-8.

मा(म)रुका मालवाश्चैव[1] पारियात्रनिवासिनः[2] ।
सौवीराः सैन्धवा हूणाः[3] शाल्वाः शाकलवासिनः[4] ॥
मद्रारामा (द्रकारा?)स्तथाम्बष्ठाः[5] पारसीकादयस्तथा[6] ॥

1. Vsn—कारूषा मालवाश्चैवः; Krm *a*—मालका मालवाश्चैव, *b*—मालाका मलपाश्चैव, v.l.—मालका मलयाश्चैव, *c*—मालका मलपाश्चैव; Bmh—मारुका मालवाश्चैव. Maruka=Maru or Marwar (above, p. 35, note 2). For Mālava, see below, Chapter XII.

2. Vsn, Krm *a*—पारिपात्रनिवासिनः; Krm *b*—परियात्रानिवासिनः; Krm *c*, Bmh—पारियात्रनिवासिनः.

3. Vsn, Krm—सौवीराः सैन्धवा हूणाः; Krm *b* (v.l.)—सौवीराः सैन्धवाकूणाः, *c* (v.l.)—सौवीराः सैन्धवाकृष्णाः; Bmh—सोवीराः सैन्धवापस्नाः. See above, p. 33, notes 2 and 4.

4. Vsn, Bmh—शाल्वाः शाकलवासिनः; Krm *a*—शाल्वाः कान्यनिवासिनः, *b*—माल्या वाल्यानिवासिनः, v.l., *c* (v.l.)—शाल्वाः कल्पनिबासिनः. See above, p. 30, note 3; p. 33, note 2.

5. Vsn, Bmh—मद्रारामास्तथाम्बष्ठाः; Krm *a*—मद्रारामास्तथैवान्ध्राः, *bc*—माद्रारामास्तथैवान्ध्राः. See above, p. 30, note 4. For th eAmbaṣṭhas living on the lower Chenab, see Raychaudhuri, *PHAI*, pp. 206-07; Sircar, *Stud. Soc. Adm. Anc. Med. Ind.*, Vol. I, pp. 106ff.

6. Krm—पारसीकास्तथैव च; Vsn, Bmh—पारसीकादयस्तथा. The Pārasīkas (Persians) are not mentioned in the bigger list or in any work earlier than Mbh, VI. 9. 65-66, and Kālidāsa's *Raghuvaṁśa*, IV. 60. They are also mentioned in the *Pādatāḍitaka-bhāṇa* (5th century), the *Mudrārākṣasa* (c. 6th century) and the *Gaüḍavaha* (8th century). See *Suc. Sāt.*, p. 326; *JRAS*, 1946, p. 51.

CHAPTER IV

PURANIC LIST OF RIVERS

For introductory remarks and list of abbreviations, see above, pp. 26 ff.; Sircar, *Cosm.*, etc., p. 53, p. 55, note 88.

TEXT

Bmd, Ch. 49, vv. 28-42; Krm, I, Ch. 46 (Ch. 47 in some versions), vv. 28-39; Mk, Ch 57 (Ch. 54 in some versions), vv. 17-30 ; Mts, Ch. 114, vv. 20-32; Vmn, Ch. 13, vv. 20-33; Vy, Ch. 45 (of Book I according to some versions; cf. also Sachau, *Alb. Ind.*, I, pp. 357-39), vv. 94-107. Cf. Kirfel, *Bhāratavarṣa*, pp. 38-42; also *Vsn Dhm*, I. 10. 3-9.; Sircar, *Cosm.*, etc., pp. 83 ff.

I

Himavat

[1]पीयन्ते यैरिमा नद्यो[2] गङ्गा सिन्धुः सरस्वती[3] ।

1. The following line is omitted in Krm. A little independence as regards the order of enumeration of the rivers issuing from the Himavat (Himālaya) is noticed in Mk and, to a greater degree with deliberate omissions and commissions, in Vmn. Mk has गङ्गा सरस्वती सिन्धुश्चन्द्रभागा तथा परा। यमुना च शतद्रुश्च वितस्तेरावती कुहुः॥ गोमती धूतपापा (*c*–धृत०) च बाहुदा (*d*–ब०) सदृशद्वती (*d*–च०) । विपाशा (*b*–॰सा) देविका रङ्क्षुर्निश्चीरा गण्डकी तथा ॥ कौशिकी चापगा विप्र हिमवत्पादनिःसृताः ॥ Vmn has—सरस्वती पञ्चरूपा कालिन्दी च हिरण्वती। शतद्रुश्चन्द्रिका नीला वितस्तेरावती कुहुः॥ मधुरा हारावी च उशीरा धातुकी (*b*–०त०) रसा। गोमती धूतपापा च बाहुदा सा दृषद्वती ॥ निःस्वरा गण्डकी चित्रा कौशिकी तु वधूसरा। सरयूश्च सलोहित्या (*b*–०लौ०) हिमवत्पादनिःसृताः ॥ Most of the Himalayan rivers are noticed by Al-Bīrūnī. Vide *infra*, p. 51, note 3. Though he speaks of both Vy and Mts, possibly the names were quoted by him from the latter work with an amount of misunderstanding and confusion.

2. Vy, Bmd—[तैर्विमिश्रा जनपदा आर्यम्लेच्छाश्च (Vy *c*—र्यां०) नित्यशः।] पीयन्ते यैरिमा नद्यः; Mts—[तैर्विमिश्रा जनपदा (*bc*–जा०) आर्या म्लेच्छाश्च सर्वतः।] पिबन्ति बहुला नद्यः. Cf. p. 67, note. 4; p. 70, note. 9.

3. Vy, Bmd, Mts—गङ्गा सिन्धुः सरस्वती; Mk—गङ्गा सरस्वती सिन्धुः. Omitted in Krm. Some Mts versions read the names of the rivers in the second case-ending. It is well known that the Gaṅgā is the Ganges

शतद्रुश्चन्द्रभागा च[1] यमुना सरयूस्तथा[2] ॥

इरावती वितस्ता च[3] विपाशा देविका कुहुः[4] ।

and the Sindhu the Indus. The Sarasvatī rises in the Sirmur hills of the Siwalik range in the Himalayas and emerges into the plains at Ād-Badrī in the Ambala District, Punjab. It disappears once at Chalaur, but reappears at Bhavanipur ; then it disappears at Balchappar, but again appears at Bara Khera ; next it is joined by the Mārkaṇḍa at Urnali near Pehoa and the united stream, still called Sarasvatī, ultimately falls into the Ghaggar which is believed to have borne the name Sarasvatī in ancient times. In the early Vedic period, the Sarasvatī probably flowed into the Arabian Sea. Later literature speaks of its disappearance at Vinaśana (near modern Sirsa) in Kurukṣetra or Brahmāvarta (in the Eastern Punjab) which is described as the land between the Sarasvatī and the Dṛṣadvatī. Cf. दक्षिणेन सरस्वत्या उत्तरेण दृषद्वतीम् । ये वसन्ति कुरुक्षेत्रे ते वसन्ति त्रिविष्टपे ॥ (*Mbh*, III. 83. 204-05); सरस्वतीदृषद्वत्योर्देवनद्योर्यदन्तरम् । तं देवनिर्मितं देशं ब्रह्मावर्तं प्रचक्षते ॥ (*Manu*, II. 17). Vide *infra*, p. 50, note 2.

1. Vy, Bmd, Krm, Mts—शतद्रुश्चन्द्रभागा च; cf. Vmn—शतद्रुश्चन्द्रिका नीला. The Śatadru is the modern Satlaj. It is the Vedic Śutudrī and the Zaradros or Hesydrus of the Classical writers. The Candrabhāgā is the modern Chenab which is the same as Asiknī of the Vedic literature and the Acesines of the Classical authors. Ptolemy calls it Sandabala or Sandabal. The Nīlā referred to by Vmn reminds us of the name Nīlāb applied by the early Muslim writers to a part of the river Indus (cf. Ray, *DHNI*, Vol. I, p. 84).

2. Vy *a*, Mts *a*—यमुना सरयुस्तथा; Bmd, Vy *cde*, Mts *b*—यमुना सरयूस्तथा; Krm *a*—सरयू यमुना तथा; *bc*—सरयूर्यमुना तथा; Mts *c*, Vy *e* (v.l.)—यमुना शरयूस्तथा. The Yamunā is the modern Jumnā (Yamunā) joining the Ganges near Allahabad. The Sarayū is the modern Ghagra or Gogra on which the ancient city of Ayodhyā (near Fyzabad) is situated. It rises in the Kumaon hills and is called Sarayū, Ghagra or Devā (after its junction with the Kālīnadī).

3. Vy, Bmd, Krm, Mts *a*—इरावती वितस्ता च; Mts *bc*—ऐरावती वितस्ता च. The Irāvatī or Airāvatī is the modern Rāvī which was known to the Classical writers as Hydraotes. The Vitastā is the modern Jhelam, the upper course of which is still known by the old name. It was also known to the Classical writers as the Hydaspes or Bidaspes.

4. Vy *a*, Bmd—विपाशा देविका कुहुः; Krm, Mts *a*, Vy *cde*—विपाशा देविका कुहूः; Krm *b*(v. l.)—विपाशा द्रविका कुहुः; Mts *bc*—विशाला देविका कुहूः. The Vipāśā is the modern Beas, called Vipāś in earlier literature and Hyphasis or Bipasis in the works of the Classical writers. The Devikā

गोमती घृतपापा च[1] बाहुदा च दृषद्वती[2] ॥
[3]कौशिकी च तृतीया तु[4] निश्चीरा गण्डकी तथा[5] ।

is the modern Deeg, a tributary of the Rāvī. Another identification suggested by some writers is that with the Devā, a name applied to the lower course of the Sarayū. The Kuhū has been identified with the modern Kabul river, called Kubhā in the *Ṛgveda* and Kophen, Kophes or Koa by the Classical writers.

1. Mk *abd*, Vy *abc*, Bmd, Krm, Vmn—गोमती घृतपापा च; Vy *de*—गोमती धुतपापा च; Mts—गोमती धौतपापा च. Mk *c*— गोमती धूतपापा च. The river Gomatī has to be identified with the Ṛgvedic Gomatī which is the modern Gomal, a western tributary of the Indus, or with the modern Goomti which joins the Ganges below Varanasi. There is another river of this name in the Kangra District, Punjab. The Dhūtapāpā was a tributary of the Ganges in the Varanasi region. The name has been associated with Ahopāp on the Goomti, 18 miles south-west of Sultanpur in U. P.

2. Mk *abc*—बाहुदा सदृषद्वती, *d*—बाहुदा च दृषद्वती; Vy, Bmd, Krm, Mts—बाहुदा च दृशद्वती; Krm *b* (v. l.), Vy *e* (v. l.)—बहुदा च दृशद्वती; Vmn—बाहुदा च दृषद्वती. The Bāhudā has been identified by some with the Dhavalā (modern Dhumelā or Buḍhī-Rāptī), a tributary of the Rāptī, and by others with the Rāmgaṅgā that joins the Ganges near Kanauj. The river Dṛṣadvatī is usually identified with the Citang, Citrang or Cautang which runs parallel to the Sarasvatī ; but some writers prefer its identification with the Rakshi that flows by the south-east of Thaneswar. The view of certain earlier writers identifying the Dṛṣadvatī with the Ghaggar seems to be unjustifiable.

3. The following two lines are contracted into one in Krm—कौशिकी लोहिनी (*b* [v. l.], *c* [v. l.]—लोहिता) चेति हिमवत्पादनिःसृताः. Vmn and Mk arrange the names rather independently.

4. Vy, Bmd—कौशिकी च तृतीया तु; Mts—कौशिकी तु तृतीया च. Pargiter suggests the emendation—कौशिकी करतोया तु or कौशिकी च त्रिस्रोतास्तु. The Kauśikī is no doubt the modern Kosi which runs through Nepal and Tirhut and joins the Ganges below Patna; but originally the river seems to have passed through North Bengal to join the Brahmaputra. There is another Kosi (Kauśikī) which flows by Almora and Ramnagar in North-Western U. P. *Tṛtīyā* may be a mistake for Karatoyā the celebrated river of North Bengal. *Niścīrā* seems also to be a mistake for *Trisrotas*, the Sankrit name of the Teestā in North Bengal. Agn (Ch. 116. 7) seems to locate the rivers Tṛtīyā and Niścīrā in the Gayā region ; but the evidence is doubtful, and, moreover, in that case, they cannot be associated with the Himalayas. Cf. *infra*, note 5.

5. Mk, Vy, Bmd—निश्चीरा गण्डकी तथा; Mts—निश्चला गण्डकी

इक्षुर्लौहित्य इत्येता[1] हिमवत्पादनिःसृता[2] ॥[3]

तथा; Vmn—निःस्वरा गण्डकी चित्रा. For *Niścirā*, see remarks above, p. 50, note 4. The Caṇḍakī (modern Gaṇḍak) is the famous tributary of the Ganges and joins the river near Sonepur (Hariharakṣetra) in the Muzaffarpur District, Bihar.

1. Vy, Bmd—इक्षुर्लोहित इत्येताः; Mts *a*—इक्षुर्लोहितमित्येताः, *bc*—इक्षुर्लौहित्यमित्येताः; *c* (v. l.) — चङ्कुं लोहितामित्येताः. The name Ikṣu may indicate the Ikṣumatī (modern Kālīnadī, a tributary of the Ganges) ; but the form *Caṅku* as well as *Raṅkṣu* (or *Vaṅkṣu*) in Mk probably suggests that it is no other than the Vakṣu or Oxus (Amu Darya) ; cf. Kālidāsa's *Raghuvaṁśa*, IV. 67, in Vallabha's commentary. The Lohitya or Lauhitya is the same as the Brahmaputra.

2. Mk, Vy, Bmd, Krm, Vmn—हिमवत्पादनिःसृताः ; Mts—हिमवत्पार्श्वनिःसृताः, *c* (v. l.)—दिवमाप्नोति याः स्मरन्.

3. The following rivers are mentioned by Al-Bīrūnī (Sachau, *Alb Ind.*, I, Ch. 25) as issuing from the Himavat : 1. Sindh (Sindhu or Indus) or the river of Vaihand (ancient Udabhāṇḍa or Udahāṇḍa, modern Und near Attock), 2. Biyatta (Vitastā) or Jailam (Jhelam), 3. Candrabhāgā or Candrāha, 4. Biyāha (Vipāśā) to the west (sic—east) of Lahore, 5. Irāvatī to the east (sic—west) of Lahore, 6. Śatarudra or Śataldar (Śatadru or Satlaj), 7. Sarsat (Sarasvatī) flowing to the country of Sarsat (Sārasvata), 8. Jaun (Yamunā), 9. Gaṅgā, 10. Sarayū or Sarwa, 11. Devikā, 12. Kuhū, 13. Gomatī, 14. Dhūtapāpā, 15. Viśālā (cf. Mts *bc*, *supra*, p. 49, note 4), 16. Bāhudāsa (sic—Bāhudā, with *sa* prefixed to the next name in the text consulted ; cf. *supra*, p. 50, note 2), 17. Kauśikī, 18. Niścirā, 19. Gaṇḍakī, 20. Lohitā, 21. Dṛṣadvatī. Other names of this list appear to have been wrongly taken mostly from that of the rivers issuing from the Pāriyātra (vide pp. 52-54). They are : 22. Tāmrā Aruṇā (Tāmravarṇā ?), 23 Parṇāśā, 24. Vedasmṛti, 25. Vedāsinī or Vidāsinī (cf. *infra*, p. 52, note 1), 26. Candanā, 27. Kāwanā (same as Kawinī, tributary of the Sarwa ?), 28. Parā, 29. Carmaṇvatī, 30. Vidiśā, 31. Veṇumatī, 32. Śiprā that rises in the Pāriyātra and passes by Ujain (Ujjayinī), 33. Karatoyā, 34. Shmāhina (cf. *infra*, p. 53, note 2). Al-Bīrūnī further says (*op. cit.*, pp. 259 ff.), "In the mountains bordering on the kingdom of Kāyabish (Kāpiśa), i.e. Kābul, rises a river which is called the *Ghorvand* on account of its many branches. It is joined by several affluents : 1. the river of the pass of Ghūzak; 2. the river of the gorge of Panchīr (Panjshir falling into the Ghorvand), below the town of Parvān (about 8 miles to the north of Chārikar) ; 3-4. the river Sharvat and the river Sāva, which latter flows through the town of Lambagā (Lampāka), i.e. Lamghan ; they join the Ghorvand at the fortress of Drūta ; 5-6. the rivers Nūr and Kīrāt.—Swelled by these affluents, the Ghorvand (Kābul) is a great river opposite the town of Purshāvar (modern Peshāwar) being

II

Pāriyātra

वेदस्मृतिर्वेदवती[1] वृत्रघ्नी सिन्धुरेव च[2] ।

there called *the ford*, from a ford near the village of Mahanāra on the eastern bank of the river, and it falls into the river Sindh (Indus) near the castle of Bītūr, below the capital of al-Kandahār (Gandhāra), i.e. Vaihand (Und near Attock).—The river Biyatta (Vitastā), known as Jailam (Jhelam), from the city of this name on its western bank, and the river Candarāha (Candrahāgā) join each other nearly fifty miles above Jahrāvār (which together with Mūltān comprised the ancient Sauvīra country; cf. *loc. cit.*, pp. 300, 302), and pass along west of Mūltān (which was originally known as Kaśyapapura, next as Haṁsapura, then as Vegapura, afterwards as Sāmbapura and ultimately as Mūlasthāna ; cf. the views of Utpala ; *loc cit.*, p. 298).—The river Biyāh (Vipāśā) flows east of Mūltān, and joins afterwards the Biyatta (Vitastā) and Candarāha (Candrabhāgā).—The river Irāva (Irāvatī) is joined by the river Kaj which rises in Nagarkot in the mountains of Bhātul. Thereupon follows as the fifth, the river Śataladar (Śatadru).—After these five rivers have united below Mūltān at a place called *Pañcanada*, i.e. the meeting place of the five rivers, they form an enormous watercourse......The Muslims call the river, after it has passed the Sindhī city Aror, as a united stream, the *river of Mihran*. Thus it extends......enclosing in its course places like islands until it reaches al-Mansūra, situated between several of its arms, and flows into the ocean at two places, near the city of Loharānī, and more eastward in the province of Kacch at a place called *Sindhusāgara*, i.e. *the Sindh Sea*.—As the name *union of five rivers* (Pañcanada) occurs in this part of the world (the Punjab), we observe that a similar name is used also to the north of the above-mentioned mountain chains (i.e. the mountain bordering on the kingdom of Kāyabish), for the rivers which flow thence towards the north after having united near Tirmidh and having formed the river of Balkh, are called the union of *seven rivers* (cf. *hapta-hindu* of the ancient Iranians). The Zoroastrians of Sogdiana (Bukhara region) have confounded these two things; for they say that the whole of the seven rivers is *Sindh*, and its upper course Barīdīsh ......The river Sarsati (Prabhāsa-Sarasvatī, mod. Raunakshi) falls into the sea at the distance of a bow-shot east of Somnāth.—The river Jaun (Yamunā) joins the Ganges below Kanoj, which lies west of it. The united stream falls into the great ocean near Gaṅgāsāgara.—Between the mouths of the rivers Sarsati and Ganges is the mouth of the river Narmadā which descends from the eastern mountains, takes its course in a south-western direction and falls into the sea near the town of Bahroj (Broach) nearly sixty *yojana* east of Somnāth.—Behind the Ganges flow the rivers Rahab (modern Rāmgaṅgā) and Kavīnī which join the river Sarva near the city of Bārī (to the east of the Ganges at a distance of three to four days' march from Kanoj; cf. p. 199)."

1. Mk, Vy, Bmd, Krm—वेदस्मृतिर्वेदवती; Krm *b* (v. l.), *c* (v. l.)—वेदस्मृतिर्वेदरता; Mts—वेदस्मृतिर्वेत्रवती. The Vedasmṛti has been identified with the modern Besulā in Malwa. For the Vetravatī, vide *infra*, p. 53, note 5.

2. Mk, Vy *acde*, Bmd, Mts, Vmn—वृत्रघ्नी सिन्धुरेव च ; Krm—

पर्णाशा चन्दना चैव[1] सदानीरा मही तथा[2] ॥
[3]पारा चर्मण्वती लूपी(?)[4] विदिशा वेत्रवत्यपि[5] ।

व्रतघ्नी त्रिदिवा तथा; Vy *b* recognises Vṛtraghnī (cf. Vārtaghnī in Km.; modern Vātrak, a tributary of the Sabarmatī in Gujarāt) only. The Sindhu is no doubt the modern Kālīsindh, a tributary of the Jumna between the Chambal and the Betwa.

1. Mk. *abc*—वेण्वा सा नन्दनी चैव, *d*—वेण्वा सा नन्दना चैव ; Vy *a*, Krm *ab* (v. l.) *c* (v.l.)—पर्णाशा चन्दना चैव; Bmd, Vy *cde*, Krm *bc*—वर्णासा चन्दना चैव; Mts—पर्णाशा नर्मदा चैव; Vmn *a* पर्णाशा नन्दिनी चैव *b*—पर्णासा नन्दिनी चैव; Vy *b*—पर्णाशा नन्दना चैव; Krm *b* (v. l.) *c*(v. l.)—पर्णासा बन्धना चैव, *b*(v. l.)— पर्णाशा बन्धना चैव. The Parṇāśā is no other than the modern Banās, a tributary of the Chambal, in Rajasthan. *Candanā* is believed to have been another name of the Sābarmatī.

2. Mk, Bmd, Vy *d* (v. l.)—सदानीरा मही तथा; Vy *acde*—सतीरा महती तथा, *b*—सद्दानारा मदी तथा (sic), *c* (v. l.)—सदा तीरा मही तथा; Krm *ab*(v.l.) –सदानीरा मनोरमा, *bc*—सचर्मण्यवती सुरा; Mts—कावेरी महती तथा; Vmn—पावनी च मही तथा. The Mahī is no doubt the river of that name rising in Malwa and draining itself into the Gulf of Cambay. The Sadānīrā cannot be satisfactorily identified.

3. The following two lines have been contracted into one in Krm *bc*—विदिशा वेत्रवत्यपि पारियात्राश्रयाः स्मृताः, *b* (v. l.)– चर्मण्वती तथा सूर्या विदिशा वेत्रवत्यपि.

4. Mk *ac*–पारा चर्मण्वती तापी, *bd*—पारा चर्मण्वती लूपी; Bmd, Vy *acde*—पारा चर्मण्वती चैव; Vy *b*—परा चर्मण्वती लूपा (?), *e* (v.l.)—परा चर्मण्वती चैव; Krm—चर्मण्वती तथा दूर्या; *b*(v. l.)—चर्मण्वती तथा सूर्या, Mts—पारा च धन्वती रूपा, *c* (v.l.)—पारा चर्मण्वती रूपा. The Pārā may be the same as the Pārvatī which rises in Bhopal and is a tributary of the Chambal, while Carmaṇvatī is no other than the Chambal itself. The Chambal is a well-known tributary of the Jumna. The reading of the third name and its identification are uncertain.

5. Mk, Vy *acde*, Bmd, Krm—विदिशा वेत्रवत्यपि; Mts—विदुषा वेणुमत्यपि; Vmn—विदिशा वेणुमत्यपि; Vy *b* recognises Vidiśā only. The river Vidiśā is the modern Bes or Besali which falls into the Betwa near Besnagar (ancient Vidiśānagarī). The Vetravatī is the modern Betwa which is a tributary of the Jumna.

[1]शिप्रा ह्यवन्ती कुन्ती च[2] पारियात्राश्रयाः स्मृताः[3] ॥

III

Ṛkṣa

[4]शोणो महानदी चैव[5] नर्मदा सुरसा क्रिया[6] ।

1. This line is omitted in Krm *a* and Vy *b*.

2. Mk *abc*—शिप्रा ह्यवर्णी च तथा; Vy *acde*, Bmd—शिप्रा ह्यवन्ती च तथा; Vy *e* (v. l.), Mk *d*—क्षिप्रा ह्यवन्ती च तथा; Mts—शिप्रा ह्यवन्ती कुन्ती च; Vmn—चित्रा ह्योघवती रम्या. The Śiprā is mentioned in literature as the river on which the celebrated city of Ujjayinī (modern Ujjain in Madhya Pradesh) was situated. The river Avantī rises near Mhow and flows into the Chambal. The Kuntī, otherwise called Aśvanadī or Aśvarathanadī, is a small tributary of the Chambal (see Dey, *Geog. Dict.*, s. v. *Kuntī-Bhoja*).

3. Mk *abc*, Bmd—पारिपात्राश्रयाः स्मृताः; Vy, Mk *d*—परियात्राश्रयाः स्मृताः; Mts—पारियात्राश्रिताः स्मृताः; Vmn—पारियात्रोद्भवाः स्मृताः. Pāriyātra or Pāripātra was the name applied to the Western Vindhyas together with the Aravalli range.

4. This line is omitted in Mts *ab*. For the first three lines, Krm reads rather freely—नर्मदा सुरसा शोणो दशार्णा च महानदी । मन्दाकिनी चित्रकूटा तामसी च पिशाचिका ॥

5. Vy, Bmd, Mts *c*—शोणो महानदश्चैव; Vmn—शोणो महानदी चैव. The Śoṇa is the celebrated Sone which rises in the Amarkantak range and drains itself into the Ganges not far from Patna. The ancient city of Pāṭaliputra stood at the junction of the Śoṇa and the Gaṅgā. The Mahānadī also rises from the same range and flows through Orissa into the Bay of Bengal.

6. Mk—नर्मदा सुरथाद्रिजा; Vy *acde*—नर्मदा सुमहा द्रुमा, *b*—नर्मदा सुरसा किर्वा (sic—क्रिया), *c*(v. l.) *e* (v. l.)—नर्मदा सुरहा द्रुमा; Bmd—नर्मदा सुवहा द्रुमा; Vmn—नर्मदा सुरसा क्रिया; Mts *c*—नन्दना सुकृशा क्षमा; cf. सुरसा शिवा in Grd, I. 55. 8. The Narmadā is the famous Narbadā (Nerbudda) which rises in the Amarkantak range and falls into the Arabian Sea at Broach. The reading of the other names and their identification are uncertain.

मन्दाकिनी दशार्णा च[1] चित्रकूटा तथैव च[2] ॥
[3]तमसा पिप्पलश्रोणी[4] करमोदा पिशाचिका[5] ।
चित्रोत्पला विपाशा च[6] मञ्जुला बालुवाहिनी[7] ॥

1. Mk, Vy, Bmd, Mts, Vmn—मन्दाकिनी दशार्णा च. The Mandākinī (now called Mandākin) flows into the Paisuni near the Citrakūṭa hill, while the Daśārṇā is the modern Dhasan flowing past Sagar betwceen the Betwa and the Ken.

2. Mk—चित्रकूटा तथापरा; Vy, Bmd, Mts—चित्रकूटा तथैव च; Vmn—चित्रकूटा हि देविका. The Citrakūṭā is the same as the Mandākin or, more probably, a part of it.

3. This line is omitted in Krm. A little freedom is noticed in some texts as regards the order of the names in the following two lines. Mk has—चित्रोत्पला सतमसा करमोदा पिशाचिका । तथान्या पिप्पलिश्रोणिर्विपाशा (*d*—पिप्पल०) वञ्जुला नदी ॥ Mts has—तमसा पिप्पली श्येनी तथा चित्रोत्पलापि च । विमला चञ्चला चैव तथा च धूतवाहिनी ॥ Vmn has—चित्रोत्पला वै तमसा करमोदा (*b*—०तोया) पिशाचिका । तथान्या पिप्पलश्रेणी विपाशा वञ्जुलावती ॥

4. Vy *ade*, Bmd—तमसा पिप्पला श्रोणी; Mts—तमसा पिप्पली श्येणी; Vy *bc*—तमसा पिप्यला श्रोणी. The Tamasā is the modern Tons flowing into the Ganges below Allahabad. The Pippalaśroṇi may be the modern Paisuni or Parsaroni which is a tributary of the Jumna running between the Ken and the Tons.

5. Mk—करमोदा पिशाचिका; Vy *acde*, Bmd, Vmn—करतोया पिशाचिका; Vy *b*—करमोदा पिशाबिका (sic—पिशाचिका). Pargiter suggests the identification of the Karamodā with the modern Karamnāśā flowing into the Ganges above the Sone. The Piśācikā cannot be identified.

6. Vy *acde*, Bmd—नीलोत्पला विपाशा च; Krm—चित्रोत्पला विशाला च, *b* (v. l.) *c* (v. l.)—चित्रोत्पला विपाशा च; Vy *b*—चित्रपला (sic—चित्रोत्पला) महावेगा, *c* (v. l.)—चितोत्पला विपाशा च. The Citrotpalā is a branch of the Mahānadī in Orissa or the Mahānadī itself below its junction with the Pyri. The other river cannot be identified.

7. Vy *acde*—जम्बुला बालुवाहिनी, *bc* (v. l.)—वञ्जुला बालुवाहिनी; Bmd—जम्बूला बालुवाहिनी; Krm—मञ्जुला बालुवाहिनी, *b* (v. l.) *c* (v. l.)—मञ्जुला रत्नवाहिनी.

[1]शुनी लज्जा शुक्तिमती[2] शकुली त्रिदिवा क्रमुः[3] ।
ऋक्षपादात् प्रसूतास्ता[4] नद्यो मणिनिभोदकाः[5] ॥[6]

1. The following line is omitted in Krm.

2. Mk—सुमेरुजा शुक्तिमती; Vy *acde*, Bmd—सितेरजा शुक्तिमती; Vy *b* has Śuktimatī only, *c* (v. l.)—सितेरजा मुक्तिमती; Mts *a*— शुक्तिमती शुनी लज्जा, *bc*—शुक्तिमन्ती शुनी लज्जा; Vmn—सत्सन्तजा शुक्तिमती. The Śuktimatī is the modern Ken, a tributary of the Jumna, flowing through Bundelkhand. On its bank stood a city of the same name (Pali *Sotthivatī*) which was the ancient capital of the Cedi people. The reading of the other name and its identification are uncertain.

3. Mk *ab*—शकुली त्रिदिवा क्रमुः, *c*—शकुली त्रिदिवा क्रमः, *d*—सकुली त्रिदिवा क्रमुः; Vy *acde*—मक्रुणा त्रिदिवा क्रमात्, *c* (v. l.)—मक्षणा त्रिदिवा क्रमात्, *e* (v.l.)—मक्रणा त्रिदिवा क्रमा; Bmd—मक्षुणा त्रिदिवा क्रमात् ; Mts *ab*—मुकुटा ह्लदिकापि च, *e*—मुकुटा ह्लादिकापि च; Vmn—चक्रिणी त्रिदिवा वसुः ; Vy *b*—शक्रुणा त्रिदिवा क्रमात्. The Śakulī (Śakrulī ?) may be the Sakri which flows into the Seonath, a tributary of the Mahānadī. The reading of the other names and their identification are uncertain.

4. Mk *ac* (v. l.) *d*—ऋक्षपादप्रसूता वै; *bc*—स्कन्धपादप्रसूता वै; Vy, Bmd—ऋक्षपादात् प्रसूतास्ता;, Krm—ऋक्षवत्पादजा नद्यः; Mts *ac*—ऋक्षवन्त-प्रसूतास्ताः, *b*—ऋष्यवन्तप्रसूतास्ताः; Vmn—ऋक्षपादप्रसूता च. For the Ṛkṣavat, see below, note 6.

5. Mk—तथान्या वेगवाहिनी; Vy, Bmd—नद्यो मणिनिभोदकाः; Krm—सर्वपापहरा नृणाम्; Mts—नद्योमलजलाः शुभाः ; Vmn *a*—तथान्या बलवाहिनी, *b*—तथान्या वल्गुवाहिनी (फल्गु॰?).

6. It is to be noted that the short list in Section IX below (pp. 63-64) wrongly speaks of some of these rivers as flowing not from the Ṛkṣavat but from the Vindhya. The name Ṛkṣavat was applied to a particular section of the range that was in a general sense called the Vindhya. While the name Vindhya was loosely applied to the whole chain of hills running from Gujarat (cf. the title 'lord of the Vindhya' claimed by the medieval rulers of Rajpipla in the Broach District in *Ep. Ind.*, Vol. XXXIV, p. 140) to the Gayā region (cf. the Nagarjuni hill mentioned as Vindhya in *CII*, Vol. III, p. 227) and lying on both sides of the Narmadā, that of the Ṛkṣavat, when especially mentioned in literature, is always associated with the central

IV

Vindhya

[1]तापी पयोष्णी निर्विन्ध्या[2] शिप्रा च निषधावती[3] ।

part of the Narmadā valley, of which Māhiṣmatī (Māndhātā in the Nimār District of Madhya Pradesh, Maheśvar in the former Indore State according to some) was the most important city and Daśārṇā (see p. 55, ncte 1) a notable river. The Vindhya, when distinguished from the Ṛkṣavat, denotes the chain lying to the south of the Narmadā as suggested by Nīlakaṇṭha who explains the passage *Vindhy-Arkṣavantāv=abhitaḥ* (i.e., 'lying between the Vindhya and the Ṛkṣavat') in the *Hv*, II. 38. 7 (in respect of the location of Māhiṣmatī and Purikā which were on the Ṛkṣavat and also between the two Vindhyas, i.e. the Vindhya and the Ṛkṣa) as *Vindhyasy=ottarata Ṛkṣavato dakṣiṇata ity=arthaḥ* (i.e., 'lying to the north of the Vindhya and to the south of the Ṛkṣavat'). Cf. Raychaudhuri, *Stud. Ind. Ant.*, p. 128. The *Rāmāyaṇa* (VI.27.9—*Ṛkṣavantaṁ giri-śreṣṭham=adhyāste Narmadāṁ piban*) associates the Ṛkṣavat with the banks of the Narmadā, while the *Mahābhārata* (III.61.22—*ete gacchanti panthāno bahavo Dakṣiṇāpatham · Avantiṁ samatikramya Ṛkṣavantaṁ ca parvatam*) locates the Ṛkṣa to the south of West Malwa. The *Skanda Purāṇa* (Revā, V.51) refers to the Ṛkṣa as the source of the Narmadā and seems to include the Amarkantak and Maikal in the said range. The Puranic texts quoted above (cf. pp. 54 ff.) would also suggest that the Ṛkṣavat was that part of the Vindhyan range which lies to the north of the Narmadā and runs from the Malwa region right up to the sources of that river as well as of the Mahānadī.

1. In some texts slight independence is noticed in the order of the names contained in this line. Mk has—शिप्रा (*d*—क्षिप्रा) पयोष्णी निर्विन्ध्या तापी सनिषधावती (*d*—च निषधावती); Vmn has—शिवा पयोष्णी निर्विन्ध्या तापी सनिषधावती.

2. Vy *abde* (v. l.), Bmd, Krm, Mts—तापी पयोष्णी निर्विन्ध्या; Vy *ce*—तापी पयोष्णी निर्वन्ध्या. The Tāpī, otherwise called Tāptī, is the celebrated river that rises near Multai in the Betul District, Madhya Pradesh, and flows into the Gulf of Cambay (Arabian Sea) near Surat. The Payoṣṇī has been identified by some scholars with the Pain or Paingaṅgā which is a tributary of the Godāvarī. Cf. also the name of the modern Paisani, the small tributary of the Jumna in Bundelkhand. The Nirvindhyā is the modern Newuj, a tributary of the Chambal flowing between the Betwa and the Kālīsindh.

3. Vy *acde*, Bmd—मद्रा च निषधा नदी; Vy *b*—सिर्वा (sic—सिप्रा) च निषधा नदी; Krm—शीघ्रोदा च महानदी; Mts *a*—क्षिप्ता च ऋषभा नदी, *bc*—क्षिप्रा च ऋषभा नदी. Mk has *Śiprā* (*d*—*Kṣiprā*) and Vmn *Śivā* for *Madrā* (Bmd, and Vy with the exception of *b*), *Śīghrodā* (Krm) and *Kṣiptā* or *Kṣiprā* (Mts). The Śiprā passed by the ancient city of Ujjayinī. The reading of the other name and its identification are uncertain.

वेण्वा वैतरणी चैव[1] शिनिबाहुः कुमुद्वती[2] ॥
ब्रह्माणी च महागौरी[3] दुर्गा चान्तःशिला तथा[4] ।
विन्ध्यपादप्रसूतास्ता[5] नद्यः पुण्यजलाः शुभाः[6] ॥[7]

1. Mk *ac*, Vy *abce* (v. l.), Bmd, Krm *a*—वेण्वा वैतरणी चैव; Vy *te*—वेन्वा वैतरणी चैव; Krm *bc*—विन्ना वैतरणी चैव, *b*(v. l.), *c* (v. l.)—वेण्या वैतरणी चैव, चिन्ता वैतरणी चैव; Mts, Vmn *b*—वेणा वैतरणी चैव; Vmn *a*—वणा (sic— वेणा) वैतरणी चैव; Mk *bd*— वेण्या वैतरणी चैव. The Veṇvā may be the Waingaṅgā, a tributary of the Godāvarī. Pargiter thinks that Veṇvā is the Waingaṅgā together with the Prāṇhitā called Praṇītā in the Manthani inscription of 1199 A.D. (*Ep. Ind.*, Vol. XXXIV, p. 67). The Vaitaraṇī seems to be the celebrated river of that name flowing through Orissa into the Bay of Bengal.

2. Mk *abc*—सिनीबाली कुमुद्वती, *d*—सिनीवाली कुमुद्वती; Vy *acde*, Bmd—शितिबाहु कुमुद्वती; Krm—वलाका च कुमुद्वती; Mts—विश्वमाला कुमुद्वती; Vmn—सिनीबाहुः कुमुद्वती; Vy *b*—सिनिहाहुः कुमुद्वती, *c* (v. l.)—शिनिबाहुः कुमुद्वती. The identification of the Kumudvatī and the reading of the other name are uncertain.

3. Mk—करतोया महागौरी; Vy *a*— तोण्वा चैरमहागौरी; Vy *cde*, Bmd, Krm *a*, Mts—तोया चैव महागौरी; Vy *b*—तोबा (sic—तोया) चैव महागौरी; Krm *bc*—तथा चैव महागौरी; Vmn—तोया रेवा महागौरी; cf. ब्रह्माणीं च महागौरीं दुर्गामपि च भारत in *Mbh*, VI. 9.33. The Brahmāṇī seems to be the same as the Brāhmaṇī river flowing through Orissa. The other river cannot be identified.

4. Mk *abc*—दुर्गा चान्तःशिरा तथा, *d*—दुर्गा चान्तःशिवा तथा; Vy, Bmd, Krm—दुर्गा चान्तःशिला तथा; Mts—दुर्गमा तु शिला तथा; Vmn—दुर्गन्धा वा शिला तथा; cf. दुर्गां चित्रशिलां चैव in *Mbh*, VI. 9. 30. It is to be noted that *Mbh* mentions Durgā once in relation to Mahāgaurī (cf. note 3 above) and again in connection with Citraśilā (sic—*c*=*Antaḥśilā*) probably due to confusion. The rivers cannot be identified. The Citraśilā was in upper Assam according to the Paschimbhag plate of Śrīcandra.

5. Mk, Krm *a*, Mts—विन्ध्यपादप्रसूतास्ताः; Vy, Bmd, Vmn—विन्ध्यपादप्रसूताश्च; Krm *bc*—विन्ध्यपादप्रसूतास्तु. For the Vindhya, cf. note 3 above and p. 56, note 6.

6. Mk, Vy, Bmd, Vmn—नद्यः पुण्यजलाः शुभाः; Krm—सद्यः पापहरा नृणाम्, *b* (v. l.) *c* (v. l.)—सद्यः पुण्यजलाः शुभाः; Mts—सर्वाः शीतजलाः शुभाः; *c* (v. l.)—नद्यः शीतजलाः शुभाः.

7. Note that the short list in Section IX below (pp. 63-64)

V

Sahya

गोदावरी भीमरथी[1] कृष्णा वेण्वा च वञ्जुला[2] ।
तुङ्गभद्रा सुप्रयोगा[3] बाह्या कावेरी चैव तु[4] ॥

wrongly speaks of some of these rivers as flowing from the Ṛkṣavat and not from the Vindhya. Vide p. 56. note 6 above.

1. Mk *abc*—गोदावरी भीमरथा; Vy, Bmd, Mts, Vmn, Mk *d*, Krm *c*—गोदावरी भीमरथी; Krm *ab*—गोदावरी भीमरक्षी. The Godāvarī, still known by its ancient name, rises in the Western Ghats and flows through the Deccan into the Bay of Bengal. The Bhīmarathī or Bhīmarathā is the modern Bhīmā which is a tributary of the Kṛṣṇā.

2. Mk *ac*—कृष्णवेण्वा तथा परा, *bd*— कृष्णा वेण्या तथा परा; Vy *acde*, Bmd—कृष्णा वैण्यथ वञ्जुला; Krm *a*— कृष्णा वेणा च वैश्यता, *bc*—कृष्णा वेणा च वश्यता, *b* (v. l.)—कृष्णा वेणी च पासरी; Mts—कृष्णवेणी च वञ्जुला; Vmn *a*—कृष्णवेण्या सरिद्वती, *b*—कृष्णा वेण्या सरिद्वती; Vy *b*—कृष्णा वैण्या सवञ्जुला. The Kṛṣṇā is still known by its ancient name and flows from the Western Ghats through the Deccan into the Bay of Bengal. The Veṇvā is the modern Varnā (running between the Satara and Kolhapur Districts) which is a tributary of the Kṛṣṇā, the combined stream of the two rivers being often mentioned as Kṛṣṇaveṇā or Kṛṣṇaveṇī. The Vañjulā is no other than the modern Mañjirā which is a southern tributary of the Godāvarī.

3. Mk, Vy, Bmd, Mts—तुङ्गभद्रा सुप्रयोगा; Vmn—विशमद्री सुप्रयोगा. The Tuṅgabhadrā is a well-known tributary of the Kṛṣṇā, while the Suprayogā is a small river of the Nellore-Guntur region.

4. Mk—बाह्या कावेर्यथापगा; Vy *acde*, Bmd—कावेरी च तथापगा; Krm—कावेरी च द्विजोत्तमा:; Mts—बाह्या कावेरी चैव तु; Vmn—बाह्या कावेरीरेव च; Vy *b*—पाजय (sic—बाह्या) कावेरी चैव तु. The Kāverī is the celebrated holy river of the southernmost part of India. It is mentioned by the Greek geographer Ptolemy as Khaberos rising from the Adeisathron mountain range. The other river cannot be identified.

दक्षिणापथनद्यस्तु[1] सह्यपादाद्विनिःसृताः[2] ॥

VI

Malaya

कृतमाला ताम्रपर्णी[3] पुष्पजा चोत्पलावती[4] ।
[5]मलयाभिजाता नद्यः[6] सर्वाः शीतजलाः शुभाः[7] ॥

1. Mk *abc*—विन्ध्य (sic—सह्य०)पादविनिष्क्रान्ता, *d* (v. l.)—सह्यपादविनिष्क्रान्ता; Vy, Bmd, Mts *a*, Krm—दक्षिणापथनद्यस्तु; Mts *bc*—दक्षिणापथनद्यस्ताः; Vmn- एताश्चापि महानद्यः, *b*—दुग्धोदा नलिनी चैव वारिसेना कलस्वना। एताश्चापि महानद्यः. Sahya is the name applied to that part of the Western Ghats which lies to the north of the Travancore hills. For the spurious addition of a line in Vmn *b*, cf. *JRASB*, Letters, Vol. XIV, p. 44, note 3; p. 45, note 5, etc.

2. Mk—इत्येताः सरिदुत्तमाः; Vy, Bmd, Mts, Krm *bc*—सह्यपादाद्विनिःसृता; Krm *a*—सह्यपादविनिःसृताः; Vmn—सह्यपादविनिर्गताः.

3. Mk, Vy *a*, Krm , Mts, Vmn—कृतमाला ताम्रपर्णी; Krm *bc*—ऋतुमाला ताम्रपर्णी; Bmd, Vy *bcde* (v. l.) —कृतमाला ताम्रवर्णा; Vy *e*—कृतमाला ताम्रवर्णी. The Kṛtamālā is no other than the modern Vaigai running past Madurai (cf. *Caitanyacaritāmṛta*, Ch. IX). The Tāmraparṇī is now known as Tāmbravari which combines its stream with the Chittar in the Tirunelveli District of Madras.

4. Mk—पुष्पजा सूत्पलावती; Vy, Bmd, Krm *a*—पुष्पजात्युत्पलावती; Mts—पुष्पजा ह्युत्पलावती; Vmn—वञ्जुला चोत्पलावती; Krm *b*—पुष्पवत्युत्पलावती, *c*— पुष्पवत्युत्पलावती. These rivers cannot be identified.

5. The following line is omitted in Vmn together with the two lines in Section VII following. Thus the Vmn text would make the rivers mentioned in Section VI flow from the Śuktimat mountain (cf. Section VIII below) apparently through mistake.

6. Mk—मलयाद्रिसमुद्भूताः, Vy—मलयाभिजातास्ता नद्यः; Bmd—मलयाभिजाता नद्यः; Krm—मलयान्निःसृता नद्यः; Mts *a*—मलयप्रसूता नद्यः; *bc*—मलयप्रसूता नद्यस्ताः. Malaya (derived from the Dravidian word *malai* meaning 'hill') was the name applied to the Travancore hills and the southernmost part of the Western Ghāṭs.

7. Mk—नद्यः शीतजलास्त्विमाः; Vy, Bmd, Mts—सर्वाः शीतजलाः शुभा; Krm—सर्वाः शीतजलाः स्मृताः.

VII
Mahendra

[1]त्रिसामा ऋषिकुल्या च[2] इक्षुला त्रिदिवा च या[3] ।
[4]लाङ्गुलिनी वंशधरा[5] महेन्द्रतनयाः स्मृताः[6] ॥

1. The following two lines, as already indicated above (p. 60, note 5), are omitted in Vmn, and are represented in Krm by one line reading ऋषिकुल्या त्रिसामा च गन्धमादनगामिनी without any reference to the range from which the rivers spring. Thus the Krm text would wrongly represent the rivers mentioned in Section VII as issuing from the Śuktimat (see Section VIII).

2. Mk—पितृसोमर्षिकुल्या च; Vy *acd*—त्रिसामा ऋतुकुल्या च, *b*—त्रिसागा ऋषिकुल्या च, *c*— त्रिसामा ऋतुकूल्या च; Bmd—त्रिसामा ऋषिकुल्या च; Krm—ऋषिकुल्या त्रिसामा च; Mts—त्रिभागा ऋषिकुल्या च. The Ṛṣikulyā, still bearing its ancient name, flows into the Bay of Bengal near Berhampur on the South-Eastern Railway in the Ganjam District of Orissa. The other river cannot be satisfactorily identified.

3. Mk—इक्षुका त्रिदिवा च या; Vy *acde*, Bmd—इक्षुला त्रिदिवा च या ; Mts—इक्षुदा त्रिदिवाचला ; Vy *b*—इक्षुला त्रिदिवायना. Cf. त्रिदिवामिक्षुलां क्रिमिम् in *Mbh*, VI.9.17.

4. Mts reads for this line rather freely—ताम्रपर्णी तथा मूली शवरा (*c*—शबरा) विमला तथा । महेन्द्रतनयाः सर्वाः प्रख्याताः शुभगामिनीः ॥ It may be noted that *śubha-gāminīḥ* does not suit the context. Cf. Pargiter, *op. cit.*, pp. 78 ff.

5. Mk *ac*—लाङ्गलिनी वंशकरा, *bd*—लाङ्गुलिनी वंशकरा; Vy *abd*, Bmd—लाङ्गुलिनी वंशधरा; Vy *ce*—लाङ्गूलिनी वंशधराः The Lāṅgulinī is the modern Lāṅguliyā running past Śrīkākuḷam in the District of that name in Andhra Pradesh. The Vaṁśadharā (°dhārā), which is still known by the old name, runs past Kalingapatam near Śrīkākuḷam.

6. Mk—महेन्द्रप्रभवाः स्मृताः; Vy, Bmd—महेन्द्रतनयाः स्मृताः. Mahendra was the name applied to the Eastern Ghāṭs. A small river passing by Parlakimedi is called Mahendratanayā, though the said expression in our text really describes the rivers as sprung from the Mahendra.

VIII
Śuktimat

[1]ऋषिकुल्या कुमारी च[2] मन्दगा मन्दवाहिनी[3] ।
कूपा पलाशिनी चैव[4] शुक्तिमत्प्रभवाः स्मृताः[5] ॥

1. Vmn contracts the two following lines into one—शुनी चैव सुदामा च शुक्तिमत्प्रभवास्त्विमाः (*b*—शक्तिमत्०) while Krm reads rather freely—क्षिप्रा (*b* [v.l.]—रूपा) पलासिनी चैव ऋषीका वंशधारिणी (*b* [v. l.], *c* [v. l.]—बलकारिणी, रसकारिणी) । शुक्तिमत्पादसञ्जाताः सर्वपापहरा नृणाम् ॥

2. Mk—ऋषिकुल्या कुमारी च; Vy *acde*—ऋषीका सुकुमारी च, *b*—ऋषीका सकुमारी च; Bmd—ऋषिका सुकुमारी च; Mts—काशिका सुकुमारी च. Cf. कुमारीमृषिकुल्याञ्च in Mbh, VI. 9. 36, and note that in this case at least Mk seems to preserve the original reading (cf. p. 64, note 8 below). Raychaudhuri's identification of Kumārī with the river of the same name in the former Mānbhūm District in South Bihar seems to be quite probable in view of the plausibility of his identification of the *kula-parvata* called Śuktimat with the hills of Eastern India extending from Chhatisgarh to the Santal Parganas (see note 5 below). The Ṛṣikulyā may be no other than the Koel in Chhota Nagpur. Another Ṛṣikulyā is no doubt the Kiul, a tributary of the Ganges.

3. Mk, Vy *acde*, Bmd, Mts—मन्दगा मन्दवाहिनी; Vy *b*—बालुका मन्दवाहिनी; cf. Krm—गन्धमादनगामिनी and Mbh, VI. 9. 33—मन्दगां मन्दवाहिनीम्. The rivers cannot be identified.

4. Mk *abc*, Vy *e* (v.l.) —कूपा पलाशिनी चैव; Mk *c* (v.l.)—कृशा पलाशिनी चैव, *d*—कुशा पलाशिनी चैव; Vy *acde*—कूपा पलाशिनी चैव *b*—किर्पा (sic—कूपा) पलासिनी चैव; Bmd—कुपा पलाशिनी चैव; Krm—क्षिप्रा पलाशिनी चैव; Mts—कूपा च पाशिनी चैव. Raychaudhuri identifies the Kūpā with the modern Kopā, a tributary of the Bāblā in Eastern India, and the Palāśinī with the modern Parās, a tributary of the Koel in Chota Nagpur. The other river cannot be identified.

5. Mk *abc*, Vy, Bmd—शुक्तिमत्प्रभवाः स्मृताः; Mk *b*—शुचिमत्प्रभवाः स्मृताः; Mts—शुक्तिमन्तात्मजास्तु ताः The name of the Śuktima is preserved in that of the Śakti hills in Raigarh, Madhya Pradesh, and possibly in that of the Suktel river which joins the Mahānadī near Sonepur in Orissa. The name Śuktimat was probably applied to the chain of hills that extends from Sakti in Raigarh, M. P., to the Dalma hills in the old Mānbhūm District drained by the Kumārī and perhaps even to the hills in the Santal Parganas washed by the affluents of the Bāblā (Raychaudhuri, *op. cit.* p. 120).

IX

Condensed List

Vsn, II, Ch. 3, vv. 9-13 ; Bmh, Ch. 19, vv. 10-14 ; cf. Sv, Dharmasaṁhitā, Ch. 34, vv. 9-12 ; Agn, Ch. 118, vv. 6-8.

[1]शतद्रुश्चन्द्रभागाद्या[2] हिमवत्पादनिर्गताः[3] ।
वेदस्मृतिमुखाद्याश्च[4] पारियात्रोद्भवा मुने[5] ॥
[6]नर्मदासुरसाद्याश्च[7] नद्यो विन्ध्याद्रिनिर्गताः[8] ।
[9]तापीपयोष्णीनिर्विन्ध्याप्रमुखा ऋक्षसम्भवाः[10] ॥

1. This line is omitted in Sv and Agn. See *Vsn Dhm*, I. 10. 3-9.

2. Vsn, Bmh—शतद्रुचन्द्रभागाद्याः (see p. 49, note 1 above).

3. Vsn—हिमवत्पादनिर्गताः ; Bmh—हिमवत्पादनिःसृताः (see p. 51, note 2 above).

4. Vsn—वेदस्मृतिमुखाद्याश्च; Bmh—वेदस्मृतिमुखाश्चान्याः; Sv, Agn—वेदस्मृतिमुखा नद्यः (see p. 52, note 1 above).

5. Vsn— पारिपात्रोद्भवा मुने; Bmh, Sv—पारियात्रोद्भवा मुने; Agn—पारिपात्रोद्भवास्तथा (see p. 54, note 3 above).

6. For the following line, Sv has—नर्मदासुरसाद्याश्च सन्त्यन्याश्च सहस्रशः । विन्ध्योद्भवा महानद्यः सर्वपापहराः शुभाः ॥ Agn has—विन्ध्याच्च नर्मदाद्याः स्युः सह्यात् (sic—ऋक्षात्) तापी पयोष्णिका.

7. Vsn—नर्मदासुरसाद्याश्च; Bmh—नर्मदासुरमाद्याश्च (see p. 54, note 6 above).

8. Vsn—नद्यो विन्ध्याद्रिनिर्गताः ; Bmh—नद्यो विन्ध्यविनिःसृताः. Note that the source of the Narmadā group of rivers is actually the Ṛkṣavat and not the Vindhya ; vide p. 56, note 6 above.

9. For the following line, Bmh has— तापीपयोष्णीनिर्विन्ध्या-कावेरीप्रमुखा नदीः । ऋक्षपादोद्भवा ह्येताः श्रुताः पापं हरन्ति याः ॥ For the wrong *vibhakti* in *nadīḥ*, see p. 61, note 4 above. Note that the Tāpī and Kāverī are wrongly grouped together as rivers issuing from the Ṛkṣavat (see p. 57, note 2; p. 58, note 5; p. 60, note 2 above). This line is omitted in Sv.

10. Vsn—तापीपयोष्णीनिर्विन्ध्याप्रमुखा ऋक्षसम्भवाः. Note that these rivers actually flow from the Vindhya and not from the Ṛkṣavat. See p. 56, note 6; p. 58, note 5 above.

[1]गोदावरी भीमरथी कृष्णवेण्यादिकास्तथा[2] ।
सह्यपादोद्भवा नद्यः स्मृताः पापभयापहाः[3] ॥
[4]कृतमालाताम्रपर्णीप्रमुखा मलयोद्भवाः ।[5]
त्रिसामा[6] ऋषिकुल्याद्या महेन्द्रप्रभवाः स्मृताः[7] ॥
ऋषिकुल्याकुमार्याद्याः[8] शुक्तिमत्पादसम्भवाः[9] ॥[10]

1. For the following two lines, Sv has—गोदावरीभीमरथीतापीप्रमुखतोयगाः । विन्ध्याद्रिनिर्गताः पुण्याः स्नानात् पापभयापहाः ॥ सह्यपादोद्भवा नद्यः कृष्णवेण्यादिकास्तथा ॥

2. Vsn, Bmh—गोदावरी भीमरथी कृष्णवेण्यादिकास्तथा; Agn—गोदावरी भीमरथी कृष्णवेण्वादिकास्तथा.

3. Vsn, Bmh—सह्यपादोद्भवा नद्यः स्मृताः पापभयापहाः.

4. Agn has rather freely—मलयात् कृतमालाद्यास्त्रिसामाद्या महेन्द्रजाः । कुमाराद्याः शुक्तिमतो हिमाद्रेश्चन्द्रभागका ॥

5. Vsn, Bmh—कृतमालाताम्रपर्णीप्रमुखा मलयोद्भवाः.

6. Vsn—त्रिसामा चार्ष्यकुल्याद्याः; Bmh—त्रिसान्ध्य ऋषिकुल्याद्याः; Sv—त्रियामा ऋ षिकुल्याद्याः (see p. 61, note 2 above).

7. Vsn, Bmh, Sv—महेन्द्रप्रभवाः स्मृताः (see p. 61, note 6 above).

8. Vsn, Sv—ऋषिकुल्याकुमार्याद्याः; Bmh—ऋषिकुल्याकुमाराद्याः (see p. 62, note 2 above).

9. Vsn, Sv, Bmd—शुक्तिमत्पादसम्भवाः (see p. 62, note 5 above).

10. Grd (I, Ch. 56, 8-10) reads rather freely—वेदस्मृतिर्नर्मदा च वरदा सुरसा शिवा । तापी पयोष्णी सरयूः कावेरी गोमती तथा ॥ गोदावरी भीमरथी कृष्णवेणी महानदी । केतुमाला ताम्रपर्णी चन्द्रभागा सरस्वती ॥ ऋषिकुल्या च कावेरी मत्त्यगङ्गा पयस्विनी । विदर्भा च शतद्रुश्च नद्यः पापहराः शुभाः ॥ Bgvt (V, Ch. 19, Section 17) has in prose—चन्द्रवशा ताम्रपर्णा अवटोदा कृतताला वैहायसी कावेरी वेण्वा पयस्विनी शर्करावर्त्ता तुङ्गभद्रा कृष्णवेण्वा भीमरथी गोदावरी निर्विन्ध्या पयोष्णी तापी रेवा सुरसा नर्मदा चर्मण्वती अन्धः शोणश्च नदौ महानदी वेदस्मृति ऋषिकुल्या त्रिसामा कौशिकी मन्दाकिनी यमुना सरस्वती दृशद्वती गोमती सरयुरोघवती षष्ठवती सप्तवती सुषोमा शतद्रु-

X

Gaṅgā

Bmd, Ch. 51, vv. 40-60; Mts, Ch. 121, vv. 39-51 ; Vy, Ch. 47, vv. 38-48; cf. Sachau, *Alb. Ind.*, I, pp. 24-62.

A
Seven Streams

नद्याः स्रोतस्तु गङ्गायाः[1] प्रत्यपद्यत सप्तधा[2] ।
नलिनी ह्लादिनी चैव[3] पावनी चैव प्रागगताः[4] ॥
सीता चक्षुश्च सिन्धुश्च[5] प्रतीचीं दिशमाश्रिताः[6] ।
सप्तमी त्वनुगा तासां [7]दक्षिणेन भगीरथम्[8] ॥

श्चन्द्रभागा मरुद्वृधा वितस्ता असिक्नी विश्वेति महानद्यः ॥ Cf. Km. (pp. 93-94): (a) पूर्वदेश—शोणलौहित्यौ नदौ गङ्गाकरतोयाकपिशाद्याश्च नद्यः । (b) दक्षिणापथ—नर्मदातापीपयोष्णीगोदावरीकावेरीभैमरथीवेण्वाकृष्णवेण्वा-वञ्जुरातुङ्गभद्राताम्रपर्ण्युत्पलावतीरावणगङ्गाद्या नद्यः । (c) पश्चाद्देश—सरस्वतीश्वभ्रवतीवार्तघ्नीमहीहिडिबाद्या नद्यः। (d) उत्तरापथ—गङ्गासिन्धु-सरस्वतीशतद्रुचन्द्रभागायमुनेरावतीवितस्ताविपाशाकुहूदेविकाद्या नद्यः । The Kapiśā is the modern Kaśāi running through the Midnapur District., West Bengal. Śvabhravatī=mod. Śābarmatī. For longer lists of rivers, see Mbh, VI, Ch. 9, 14-36; *Varāha Purāṇa*, Ch. 85, etc. *Kaśāi* is pronounced by some as *Kānsāi*.

1. Vy *adc*, Bmd—नद्याः स्रोतस्तु गङ्गायाः, c—नद्याः श्रोतस्तु गङ्गायाः; Mts—स्रोतांसि त्रिपथायास्तु. Note that Al-Birūnī utilised, in regard to this Section, the text of Mts; therefore here we shall have to speak of Mts *d* instead of Vy *b*. Thus Vy, when unspecified, would indicate here all the versions excepting Vy *b*.

2. Vy, Bmd—प्रत्यपद्यत सप्तधा; Mts—प्रत्यपद्यन्त सप्तधा.

3. Vy, Bmd, Mts *d*—नलिनी ह्लादिनी चैवः; Mts *abc*—नलिनी ह्लादिनी चैव.

4. Vy *a*—पावनी चैव प्रगता; Vy *cde*—पावनी चव प्रागगता. Mts—पावनी चैव प्राच्यगा.

5. Vy, Mts, Bmd—सीता चक्षुश्च सिन्धुश्च.

6. Vy, Bmd—प्रतीचीं दिशमाश्रिताः; Mts—तिस्रस्ता वै प्रतीच्यगाः.

7. Vy, Mts—सप्तमी त्वनुगा तासां; Bmd—सप्तमी हि समानीता.

8. Vy *acd*—दक्षिणेन भागीरथी; Vy *e*, Mts—दक्षिणेन भगीरथम्; Bmd—भगीरथमहात्मना.

तस्माद्भागीरथी या सा[1] प्रविष्टा दक्षिणोदधिम्[2] ॥
सप्तैता भावयन्तीह[3] हिमाह्वं वर्षमेव तु[4] ।

1. Vy *a*—अस्माद् भागीरथी या सा, *cde*, Bmd—तस्माद् भागीरथी या सा; Mts—तस्माद् भागीरथी सा वै. For these seven rivers, cf. *Rāmāyāṇa* (I. 43. 11-14 ; 44 6): विससर्ज ततो गङ्गा हरो विन्दुसरः प्रति । तस्यां विसृज्यमानायां सप्तस्रोतांसि जज्ञिरे ॥ ह्लादिनी पावनी चैव नलिनी च तथैव च । तिस्रः प्राचीं दिशं जग्मुर्गङ्गाः शिवजलाः शुभाः । सुचक्षुश्चैव सीता च सिन्धुश्चैव महानदी । तिस्रश्चैता दिशं जग्मुः प्रतीचीन्तु दिशं शुभाः ॥ सप्तमी चान्वगात्तासां भगीरथरथन्तदा ....त्रीन् पथो भावन्तीति तस्मात् त्रिपथगा स्मृता ॥ Of the seven streams, there is no difficulty about the identification of the Gaṅgā or Bhāgīrathī, which is no other than the Ganges, and of the Sindhu which is the Indus. The western stream Cakṣu is also apparently the same as Vakṣu or Vaṁkṣu, i. e. the Oxus or Amu Darya, while the second western river called Sītā may be the Tagdumbash branch of the Yarkand. As the eastern stream Hlādinī or Hrādinī is said to have run through the country of the Kirātas, who were hill-men inhabiting the eastern Himalayas, this river may be tentatively identified with the upper Brahmaputra. Indradvīpa, through which the other eastern river called Pāvanī passed, is identified by some scholars with Burma. Cf. Subsection G below. See also the note in the same Subsection on Vetrapatha and Śaṅkupatha (in Suvarṇabhūmi) through which the Pāvanī is said to have passed. This river may thus be tentatively identified with the Irawadi. The third eastern stream called Nalinī cannot be identified; but it is said to have entered into the sea in the Śaimi *maṇḍala* which reminds us of Siam. The name of Siam (Syāma-raṭṭha), which is unconnected with Sanskrit *śyāma*, is derived from that of a tribe called in Chinese *Sien*, Burmese *Shan*, Malayan *Syam*, Annamite *Xiem* and Cham *Syāṁ*. The Nalinī may thus be tentatively identified with the Salween or the Mekong. It must be remembered that the theory attributing the origin of all the seven rivers to the same lake in the Himalayas has to be regarded as a mere flight of fancy. The lists of countries washed by the rivers do not appear to be always accurate.

2. Vy, Mts *a*, Bmd—प्रविष्टा लवणोदधिम् ; Mts *bc*—प्रविष्टा दक्षिणोदधिम्.

3. Vy, Bmd—सप्तैता भावयन्तीह; Mts—सप्त चैता प्लावयन्ति.

4. Vy, Bmd—हिमाह्वं वर्षमेव तु; Mts *a*—वर्षन्ति हिमसाह्वयम्, *bc*—वर्षं तु हिमसाह्वयम्. The Hima *varṣa* was otherwise called the Haimavata or Bhārata *varṣa* (Bmd, I. 33. 55, etc.) lying between the Himalayas and the sea.

प्रसूताः सप्त नद्यस्ता:[1] शुभा विन्दुसरोद्भवाः[2] ॥
नानादेशान् भावयन्त्यो[3] म्लेच्छप्रायांश्च सर्वशः[4] ।
[5]उपगच्छन्ति ताः सर्वा[6] यतो वर्षति वासवः[7] ।

B

Sitā

सिरीन्ध्रान् कुकुरांश्चीनान्[8] वर्वरान् यवनान् शकान्[9] ।
रुषाणांश्च कुणिन्दांश्च[10] अङ्गलोकवरांश्च यान्[11] ॥

1. Vy, Bmd—प्रसूताः सप्त नद्यस्ताः; Mts—प्रसूताः सप्त नद्यस्तु.
2. Vy, Bmd, Mts—शुभा विन्दुसरोद्भवाः.
3. Vy *a*—नानादेशान् भावयन्तो, *d*—नानादेशान् भावयन्यो; Bmd, Vy *ce*—नानादेशान् भावयन्त्यो; Mts—तान् देशान् प्लावयन्ति स्म.
4. Vy, Mts, Bmd—म्लेच्छप्रायांश्च सर्वशः; cf. p. 70, note 9 below, and p. 48, note 2 above.
5. The following line is omitted in Mts.
6. Vy, Bmd—उपगच्छन्ति ताः सर्वाः.
7. Vy, Bmd—यतो वर्षति वासवः.
8. Vy—सिरीन्ध्रान् कुन्तलाश्चीनान्; Mts *abc*—सशैलान् कुकुरान् रोध्रान्; Bmd—सिरीन्ध्रान् कुकुरांश्चीनान्; Mts *d*—सलिलान् कस्तुंबांश्चीनान्. For the Cīnas, see above p. 34, note 6; below, p. 68, note 3.
9. Vy—वर्वरान् यवसान् द्रुहान्; Bmd—वर्वरान् यवनान् द्रुहान्; Mts *ab*—वर्वरान् यवसान् खसान्, *c*—वर्वरान् यवनान् खसान्, v. l.—वर्वरान् यवनाञ्छकान्, *d*—वर्वरान् यवसान् ब्रहान्. For the Varvaras and Yavanas, see above, p. 33, note 1, p. 34, note 5. The original home of the Śakas was the valleys of the Jaxartes and Oxus. The Yavanas or Greeks ruled once in the Oxus valley.
10. Vy, Bmd—रुषाणांश्च कुणिन्दांश्च; Mts *abc*—पुलिकांश्च कुलत्थांश्च, पुष्करांश्च कुलतांश्च. For the Kuṇindas, see above, p. 33, note 3, although the people indicated here must have lived in Central Asia. The Ruṣāṇas were the people of Roshan in the Tajik Republic (U.S.S.R.).
11. Vy, Bmd—अङ्गलोकवराश्च ये; Mts *abc*—अङ्गलोक्यान् वरांश्च यान्, *d*—मङ्गलकवरांश्च यान्. See above, p. 34, note 5.

कृत्वा द्विधा हिमवन्तं[1] सीतागात् पश्चिमोदधिम्[2] ॥

C

Cakṣu

अथ चीनमरूंश्चैव[3] तङ्गनान् सर्वशूलिकान्[4] ।
साघ्रांस्तुषारान् लम्पाकान्[5] पह्लवान् पारदान् शकान्[6] ॥
एतान् जनपदांश्चक्षु:[7] प्लावयन्ती गतोदधिम्[8] ॥

D

Sindhu

दरदांश्च सकाश्मीरान्[9] गान्धारान् औरसान् कुहून्[10] ।

1. Vy—कृत्वा द्विधा सिन्धुमरुं; Bmd—कृत्वा द्विधा सिन्धुमरुं; Mts *abc*—कृत्वा द्विधा हिमवन्तं, *d*—कृत्वा द्विधा सङ्गवन्तं.

2. Vy, Bmd, Mts *d*—सीतागात् पश्चिमोदधिम्; Mts *abc*—प्रविष्टा दक्षिणोदधिम्.

3. Vy, Bmd, Mts *ad*—अथ चीनमरूंश्चैव; Mts *bc*—अथ वीरमरूंश्चव. For the Cīna desert, cf. above, p. 34, note 6; p. 67, note 8.

4. Vy *ac*, Bmd—तङ्गणान् सर्वमूलिकान्, *cd*—नङ्गणान् सर्वमूलिकान्, *c* (v. l.)—तङ्गणान् सर्वशूलिकान्; Mts *a*—कालिकांश्चैव चूलकान्, *bc*—कालिकांश्चैव शूलिकान्, *d*—कालिकांश्चैव घूलिकान्. For the Taṅgaṇas and Cūlikas, see above, p. 35, notes 4 and 9. But these Taṅgaṇas may have lived near about the Western Himalayas. The correct reading of the last name may be *Cūlika*.

5. Vy *ace*—सान्घ्रांस्तुषारांस्तम्पाकान्, *d*—साघ्रांस्तुषारांस्तम्पकान्; Mts *abc*—तुषारान् ववंराकारान्, *d*—तुखारान् ववंराकारान्; Bmd—साघ्रांस्तुषारान् लम्पाकान्. For the Tuṣāras or Tukhāras and Lampākas, see above, p. 34, note 6; p. 35, note 3.

6. Vy *a*—पह्लवान् दरदान् शकान्; Bmd, Vy *cde*—पह्लवान् दरदान् शकान्; Mts *abc*—पह्लवान् पारदान् शकान्, *d*—बारवाञ्चतान् (sic—पारदाञ्छकान्). See above, p. 32, note 9; p. 33, note 4; p. 67, note 9.

7. Vy, Bmd, Mts—एतान् जनपदांश्चक्षु:.

8. Vy *ac* (v. l.) *e*—प्लावयन्ती गतोदधिम्, *cde* (v. l.), Bmd—स्रावयन्ती गतोदधिम्; Mts—प्लावयित्वोदधिं गता.

9. Vy, Bmd—दरदांश्च सकाश्मीरान्; Vy *e* (v. l.)—मरदांश्च सकाश्मीरान्; Mts *abc*—दरदोर्ज्जगुडांश्चैव, *d*—दरदजिन्दुतुन्दांश्च. See above, p. 34, note 4; p. 35, note 9.

10. Vy, Bmd—गान्धारान् वरपान् ह्लदान्; Mts *abc*—गन्धारान्

शिवपौरान् इन्द्रमरून्[1] वसातींश्च विसर्जयान्[2] ॥
सैन्धवान् रन्ध्रकरकान्[3] भ्रमराभीररोमकान्[4] ।
शुनामुखांश्चोर्ध्वमरून्[5] सिन्धुरेताम्निषेवते[6] ॥

E
Gaṅgā

गन्धर्वान् किन्नरान् यक्षान्[7] रक्षोविद्याधरोरगान्[8] ।

औरसान् कुहून्, *d*—गान्धारान् रूरसान् (sic—औरसान्) क्रूरान्. See above, p. 33, note 1; p. 35, note 6. The Kuhus appear to have been the people inhabiting the valley of the Kuhu or Kabul river.

1. Vy, Bmd—शिवपौरान् इन्द्रहासान् ; Mts—शिवपौरान् इन्द्रमरून्. The Śivapauras must have been the inhabitants of Śivapura, i.e. modern Shorkot in the Jhang District of West Pakistan. The other people cannot be identified.

2. Vy— वदातींश्च विसर्जयान् ; Bmd—वसातींश्च विसर्जयान् ; Mts *abc*—वसतीन् समतेजसम्, *d*—सबातीन् (sic—वसातीन्)...The Vasātis (Ossadioi of the Greeks) appear to have occupied parts of the territory drained by the lower Chenab and situated between the confluences of that river respectively with the Rāvī and the Indus.

3. Vy, Bmd—सैन्धवान् रन्ध्रकरकान् ; Vy *a* (v. l.)— सैन्धवान् रन्ध्रवरकान् ; Mts *abc*—सैन्धवानुर्वसान् वर्वान्, *d*—सैन्धवान् कुबतान् (sic—कुपथान्)... Other names in *Alb. Ind.* are Bahīmarvara, Mara, Mrūṇa and Supurda which are apparently due to mistakes in the Arabic manuscripts of Al-Bīrūni's work consulted by Sachau. For the Saindhavas, see above, p. 33, note 2, and for the variant readings of the other name, p. 33, note 5.

4. Vy—भ्रमराभीररोहकान्, *e* (v. l.)—भ्रमटाभीररोहकान्; Bmd—भ्रमराभीररोमकान् ; Mts *abc*—कुपथान् भीमरोमकान्. Above, p. 32, note 7 would suggest आभीरान् कालतोयकान्. But the Romakas may be the people of Rumā lying probably near the Salt Range.

5. Vy, Bmd—शुनामुखांश्चोर्द्धमनून्; Mts *abc*—शुनामुखांश्चोर्द्ध्वमरून्.

6. Vy—सिद्धचारणसेवितान्, *c* (v. l.), Mts *abc*, Bmd—सिन्धुरेताम्निषेवते.

7. Vy, Mts, Bmd—गन्धर्वान् किन्नरान् यक्षान्. Al-Bīrūnī explains the Gandharvas as musicians. These names originally indicated certain tribal peoples, although later they came to imply classes of mythical beings. The Kinnaras and Yakṣas were probably names applied to some Himalayan tribes, while Gandharva appears to have been the original name of the people later called Gandhāra (cf. *Rām.*, VII. 101. 11 : तक्षं तक्षशिलायां तु पुष्कलं पुष्कलावते। गन्धर्वदेशे रुचिरे गान्धारविषये च सः॥ etc.). The names however seem to have been used in the present context in the sense of mythical tribes.

8. Vy, Bmd, Mts—रक्षोविद्याधरोरगान्. Al-Bīrūnī regards the

कलापग्रामकांश्चैव[1] तथा किंपुरुषान् खसान्[2] ॥
किरातांश्च पुलिन्दांश्च[3] कुरून् सभरतानपि[4] ।
पञ्चालकाशिमत्स्यांश्च[5] मगधाङ्गांस्तथैव च[6] ॥
ब्रह्मोत्तरांश्च वङ्गांश्च[7] ताम्रलिप्तांस्तथैव च ।
एतान् जनपदानार्यान्[9] गङ्गा भावयते शुभा ॥[10]

Uragas as those who creep on their breasts, i.e. the serpents. These three names, like those mentioned above, p. 69, notes 5ff., originally indicated certain non-Aryan tribal peoples but were later used to indicate mythica tribes. They appear to be used in the present context in the mythical sense.

1. Vy, Bmd, Mts—कलापग्रामकांश्चैव. All-Bīrūni explains Kalāpagrāma as the city of the most virtuous people.

2. Vy, Bmd—पारदान् सीगनान् खसान्, *e* (v. l.) —पारदांस्ताङ्गणान् खसान्; Mts *abc*—तथा किंपुरुषान् नरान्, *d*—तथा किंपुरुषान् खसान्. Al-Bırūnī explains the Khasas as mountaineers. The name Kimpuruṣa was originally applied to a Himalayan people but was later used to indicate a mythical tribe. The Khasas were a Himalayan tribe now represented by the Khakkas of Kashmir. See above p. 36, note 6.

3. Vy, Bmd, Mts—किरातांश्च पुलिदांश्च. Al-Bīrūmī explains the Pulindas as hunters of the plains or robbers. For the Kirātas, see above p. 31, note 1. The Pulindas were an aboriginal people inhabiting the Vindhyan region ; but the name was later applied to Vindhyan mountaineers in general and still later to any aboriginal people. The meaning of *Kirāta*, originally a Himalayan tribe, was also similarly modified. Cf. p. 39, note 5.

4. Vy, Bmd—कुरून् सभरतानपि ; Mts—कुरून् वै भरतानपि. According to the epic and Puranic traditions, the Kurus and the Bharatas belonged to the same clan. For the Kurus, see above, p. 30, note 2.

5. Vy, Bmd—पञ्चालकाशिमत्स्यांश्च; Mts *a*—पञ्चालकौशिकान् मत्स्यान्, *bc*—पाञ्चालान् कौशिकान् मत्स्यान्, *d*—पाञ्चालान् कौषकान् मत्स्यान्. See above, p. 30, note 2; p. 31, notes 1 and 2. But *Matsya* may be a mistake for *Vatsa* which was the name of the people inhabiting the Allahabad region.

6. Vy, Bmd, Mts *abc*—मगधाङ्गांस्तथैव च. Mts *d* omits *Aṅga*. See above, p. 36, note 5; p. 38, note 1.

7. Vy, Bmd, Mts *abc*—ब्रह्मोत्तरांश्च वङ्गांश्च. Mts *d* omits *Vaṅga*. See above, p. 36, note 5; p. 37, note 5.

8. Vy, Bmd, Mts—ताम्रलिप्तांस्तथैव च. See above, p. 37, note 8.

9. Vy, Bmd, Mts—एतान् जनपदान् आर्यान्. Cf. the reference here to *ārya* with *mleccha-prāyāṁś=ca sarvaśaḥ* at p. 67, note 4 above.

10 Vy, Bmd—गङ्गा भावयते शुभान्; Mts, Vy *e* (v. l.)—गङ्गा भावयते शुभा ।

ततः प्रतिहता विन्ध्ये[1] प्रविष्टा दक्षिणोदधिम्[2] ॥

F

Hlādinī

ततश्च ह्लादिनी पुण्या[3] प्राचीनाभिमुखं ययौ[4] ।
प्लावयन्त्युपकानांश्च[5] निषादानपि सर्वशः ॥[6]
धीवरानृषिकांश्चैव[7] तथा नीलमुखानपि[8] ॥
केकरानुष्ट्रकर्णांश्च[9] किरातानपि चैव हि ॥[10]
कालोदरान् विवर्णांश्च[11] कुशिकान् स्वर्ग(र्ण)भूमिकान्[12] ।

1. Vy, Bmd, Mts— ततः प्रतिहता विन्ध्ये.

2. Vy, Bmd, Mts—प्रविष्टा दक्षिणोदधिम्.

3. Vy—ततश्चाह्लादिनी पुण्या; Bmd—ततश्च ह्लादिनी पुण्या; Mts—ततस्तु ह्लादिनी पुण्या.

4. Vy *ac*—प्राचीनाभिमुखं ययौ; Mts *be*, Vy *de*—प्राचीनाभिमुखी ययौ; Vy *c* (v. l.), Mts *a*—प्राचीनाभिमुखा ययौ; Bmd—प्राचीमाभिमुखी ययौ.

5. Vy, Bmd—प्लावयन्त्युपभोगांश्च; Mts *abc*—प्लावयन्त्युपकांश्चैव, *d*—प्लावयन्त्युपकानांश्च.

6. Vy, Bmd— निषादानाञ्च जातयः; Mts *abc*—निषादानपि सर्वशः, *d*—निषावानपि (sic) सर्वशः.

7. Vy, Mts *abc*—धीवरानृषिकांश्चैव; Bmd—धीवरानृषकांश्चैव; Mts *d*—धीवरान् प्रिषकांश्चैव.

8. Vy, Bmd, Mts—तथा नीलमुखानपि.

9. Vy, Bmd—केरलानुष्ट्रकर्णांश्च; Vy *e* (v. l.)—केरलानोष्ट्रकर्णांश्च; Mts *abc*—केकरान् एककर्णांश्च, *c* (v. l.) *d*— केकरान् उष्ट्रकर्णांश्च. Al-Bīrūnī explains *Uṣṭrakarṇa* as a people whose lips are turned like their ears possibly through a confusion of the word *uṣṭra* with *oṣṭha*.

10. Vy, Bmd,Mts—किरातानपि चैव हि.

11. Vy,Bmd—कालोवरान् विनर्णांश्च; Mts *abc*—कालञ्जरान् विवर्णांश्च, *c* (v. l.)—कालिन्दगतिकांश्चैव, *d*—कलीदरान् विवर्णांश्च. Al-Bīrūnī explains *Vivarṇa* as the colourless people so called on account of their intensely dark complexion.

12. Vy, Bmd—कुमारान् स्वर्णभूषितान्; Vy *e* (v. l.)—कुमारीस्वर्णभूषितान्;

सा मण्डले समुद्रस्य[1] तिरोभूतानुपूर्वतः[2] ॥

G

Pāvanī

ततस्तु पावनी चैव[3] प्राचीमेव दिशं गता[4] ।
कुपथान् प्लावयन्तीह[5] इन्द्रद्युम्नसरांस्यपि[6] ॥
तथा खरपथांश्चैव[7] वेत्रशंकुपथानपि[8] ।

Mts *abc*—कुशिकान् स्वर्गभौमकान्, *c* (v. l.) *d*—कुशिकान् स्वर्गभूमिकान्, कुशिकान् पर्णभूमिकान्. Al-Bīrūnī explains *Svargabhūmi* as a country resembling paradise. But the reference is to Survarṇabhūmi. Cf. above, p. 66, note 1; below, notes 7-8.

1. Vy, Bmd, Mts—सा मण्डले समुद्रस्य.

2. Vy, Bmd—तिरोभूतानुपूर्वतः ; Mts *abc*—तीरे भुत्वा तु सर्वशः, (v. l.)—तीरे भूत्वा चतुर्दश.

3. Vy, Bmd, Mts *d*—ततस्तु पावनी चैव; Mts *abc*—ततस्तु नलिनी चापि. Note that there is difference between the extant versions of Mts and those of Vy and Bmd, but that the Mts manuscript consulted by Al-Bīrūnī supports Vy and Bmd. See also above, p. 65, notes 3 and 4; p. 66, note 1.

4. Vy, Bmd—प्राचीमेव दिशं गता; Mts—प्राचीमेव दिशं ययौ.

5. Vy *acde* (v. l.), Bmd—अपथान् भावयन्तीह; Vy *e*—अपथान् प्लावयन्तीह; Mts—कुपथान् प्लावयन्ती सा. Al-Bīrūnī explains Kupatha as a people who are far from sin. For names ending in *patha* or *mārga*, cf. Siddha-patha (*Rājatar.*, VIII. 557), the old name of the Sidau or Budil pass (14000 ft.) in Kashmir. See also note 8 below. Cf. p. 35, note 8; p. 45, note 5; p. 74, note 2.

6. Vy, Bmd—इन्द्रद्युम्नसरोपि च; Mts—इन्द्रद्युम्नसरांस्यपि. Al-Bīrūnī explains the passage as 'the cisterns of king Indradyumna'.

7. Vy, Bmd—तथा खरपथांश्चैव ; Mts— तथा खरपथान् देशान्. Kharapatha reminds us of Ajapatha and Verāpatha, mentioned in the Buddhist *Niddesa* commentary, the latter also by Ptolemy as Berabai which was not far from Takkola about the present Isthmus of Kra (cf. Majumdar, *Suvarṇadvīpa*, Vol. I, pp. 56-60 ; Lévi, *Etudes Asiatiques*, Vol. II, pp. 1-55). See note 8 below.

8. Vy, Bmd—इन्द्रशंकुपथानपि; Mts—वेत्रशंकुपथानपि. Śaṅkupatha is mentioned in the *Niddesa* commentary (cf. note 7 above) along with Verāpatha, Jaṇṇupatha, Ajapatha, Meṇḍhapatha, Chatrapatha, Vaṁśa-

मध्येनोद्यानकमरून्[1] कुथप्रावरणान् ययौ[2] ।
इन्द्रद्वीपसमुद्रे तु[3] प्रविष्टा लवणोदधिम्[4] ॥

H

Nalini

ततश्च नलिनी चागात्[5] प्राचीमाशां जवेन तु[6] ।

patha, Śakuṇapatha, Mūṣikapatha and Darīpatha. Vetrapatha (possibly called Vettādhāra or Vettācāra in the *Niddesa*) is mentioned in connection with Suvarṇabhūmi (the land beyond the eastern sea or the Bay of Bengal) in the *Bṛhatkathāślokasaṁgraha*. These extraordinary routes (passes ?) are also referred to in other early Indian works such as the *Vimānavatthu*, the *Tittira Jātaka*, the *Milindapañha*, Patañjali's comment on Pāṇini, V. 1. 77, and the *Gaṇapāṭha*. It has been suggested that the knowledge of the Far East exhibited by the *Niddesa* did not exist in India before the first century A.D., but that it is earlier than the third century A.D. The *Niddesa* list, with which the Purāṇic section under discussion may be contemporaneous, has been assigned to a date between the end of the first and the beginning of the third century A.D. Incidentally it may be pointed out that the knowledge of the Far East exhibited by the *Mahābhāṣya* points to a late date of the work in its present form as suggested in *IHQ*, Vol. XV, pp. 933 ff.

1. Vy *a*—मध्येनोद्यानमकरारान् (sic), *cde* (v. l.), Bmd—मध्येनोद्यानमस्कारान् ; Vy *e*—मध्येनोद्यानमकरान्; Mts *abc*— मध्येनोज्जानकमरून्, *d*—मध्येनोद्यानमरूरान्.

2. Vy, Bmd, Mts *abc*—कुथप्रावरणान् ययौ; Mts *d*—कुशप्रावरणान् ययौ. See above, p. 45. note 7.

3. Vy, Bmd, Mts *a*—इन्द्रद्वीपसमुद्रे तु; Mts *bc*— इन्द्रद्वीपसमीपे तु, *c* (v. l.)—इन्द्रद्वीपसमुद्रान्ते. For the identification of Indradvipa with Burma, see Cunningham, *Anc. Geog. Ind.*, ed. Majumdar Sastri, pp. 751-52; Sircar, *Cosm. Geog. E. Ind. Lit.*, p. 56.

4. Vy, Bmd, Mts—प्रविष्टा लवणोदधिम्.

5. Vy, Bmd, Mts *d*—ततश्च नलिनी चागात् ; Mts *abc*—ततस्तु पावनी प्रायात्. Note the difference between the two versions; cf. above, p. 72, note 3.

6. Vy, Bmd, Mts—प्राचीमाशां जवेन तु.

तोमरान् भावयन्तीह्[1] हंसमार्गान् सहूहुकान्[2] ॥
पूर्वान् देशांश्च सेवन्ती[3] भित्त्वा सा बहुधा गिरीन्[4] ।
कर्णप्रावरणांश्चैव[5] प्राप्य चाश्वमुखानपि[6] ॥
सिक्त्वा पर्वतमरून् सा[7] गत्वा विद्याधरानपि[8] ।
शैमिण्डलकोष्ठे तु[9] प्रविष्टा सा महोदधिम्[10] ॥

1. Vy *acde* (v. l.), Bmd— तोमरान् भावयन्तीह; Vy *c*— तोमरान् प्लावयन्तीह; Mts *abc*—तोमरान् प्लावयन्ती च, *d*—तामरान् प्लावयन्ती च. See above, p. 35, note 8; p. 46, note 2.

2. Vy *acd*—हंसमार्गान् सहूहुकान्, *c* (v. l.),Bmd—हंसमार्गान् बहूदकान्; Vy *c*— हंसमार्गान् सद्दूहुकान् (sic); Mts *abc*— हंसमार्गान् समूहकान्, *d*—हंसमार्गान् समूहुकान्. See above p. 35, note 8: p. 45, note 5.

3. Vy, Bmd, Mts *abc*—पूर्वान् देशांश्च सेवन्ती; Mts *d*—पूर्णान् देशांश्च सेवन्ती.

4. Vy, Bmd, Mts *d*—भित्त्वा सा बहुधा गिरीन् ; Mts *abc*—भित्त्वा सा बहुधा गिरिम्.

5. Vy, Bmd—कर्णप्रावरणांश्चैव ; Mts— कर्णप्रावरणान् प्राप्य. Al-Bīrūnī explains *Karṇaprāvaraṇa* as a people whose ears used to fall down on their shoulders. Cf. above, p. 45, note 7.

6. Vy, Bmd—प्राप्य चाश्वमुखानपि; Mts— गता साश्वमुखानपि. Al-Bīrūnī explains *Aśvamukha* as a people with horse's face.

7. Vy, Bmd— सिकतापर्वतमरून् ; Mts *abc*—सिक्त्वा पर्वतमेरुं सा, *d*—सिक्त्वा पर्वतमरुं सा. Al-Bīrūnī explains *parvata-maru* as mountainous steppes.

8. Vy, Bmd—गत्वा विद्याधरान् ययौ; Mts—गत्वा विद्याधरानपि.

9. Vy—नेमिमण्डलकोष्ठे तु; Bmd—नेमिमण्डलमध्येन; Mts *ac*—शैमिमण्डलकोष्ठं तु, *b*—शौमिण्डलकोष्ठं तु, *d*—रूमीमण्डल कोष्ठं तु. *Kosthe* may be a mistake for *kacche*. See above, p. 66, note 1.

10. Vy, Bmd, Mts *d*—प्रविष्टा सा महोदधिम्; Mts *abc*— सा प्रविष्टा महत्सरः.

CHAPTER V

ACCOUNT OF FIFTY-SIX COUNTRIES

There is the manuscript of a small work entitled *Ṣaṭpañcāśaddeśavibhāga* (G. 9660 ; foll. 1 B-3 B) in the library of the Asiatic Society, Calcutta. While entering it in the *Descriptive Catalogue of Sanskrit Manuscripts in the Government Collection under the Care of the Asiatic Society of Bengal,*[1] Pandit H. P. Sastri remarked that it may 'be part of some Tantra'; but he did not try to find out the Tantra to which it belongs. The *Ṣaṭpañcāśaddeśavibhāga* is, however, actually the seventh Paṭala of Book III of the *Śaktisaṅgama Tantra.* There are four fragmentary manuscripts of this Tantra in the Society's library ; but only one of them[2] contains the section in question. A few years ago, Book I (Kālī-khaṇḍa) of the *Śaktisaṅgama Tantra,* edited by B. Bhattacharya, was published in the Gaekwad Oriental Series ; but the remaining Books remained in manuscripts. Bhattacharya, then Director of the Oriental Institute, Baroda, kindly sent me, at my request, a copy of the *Śaktisaṅgama Tantra,* Book III, Paṭala vii. His letter to me dated the 2nd April, 1941, reads, "This reading is based on four manuscripts...........We have in our library only one manuscript of the *Śaktisaṅgama Tantra*; but the readings found in it contain omissions and errors which are corrected with the help of other manuscripts. I feel that the accompanying reading is fuller and free from errors as far as the manuscript material permits."[3] In my edition of the *Ṣaṭpañcāśaddeśavibhāga* published here, I have therefore utilised the following three manuscripts :—

A—Manuscript of the *Ṣaṭpañcāśaddeśavibhāga* in the library of the Asiatic Society (No. 9660).

B—Manuscript of the *Śaktisaṅgama Tantra* in the library of the Asiatic Society (No. 323).

C—Copy of the manuscript of the *Śaktisaṅgama Tantra,* Book III, Paṭala vii, kindly supplied by B. Bhattacharya.

1. Vol. IV, 1923, pp. 35 f.
2. 323-5524, foll. 209A-212A.
3. Volumes II (Tārākhaṇḍa) and III (Sundarīkhaṇḍa) were published respectively in 1941 and 1947. For the *Ṣaṭpañcāśaddeśavibhāga,* see Vol. III, pp. 66 ff.

The *Śaktisaṅgama Tantra*, to which the *Ṣaṭpañcāśaddeśavibhāga* belongs, is a late work that is roughly assignable to the seventeenth century. In his introduction to the *Kālīkhaṇḍa* (Book I of the *Śaktisaṅgama Tantra*), B. Bhattacharya assigns the work to the period between 1555 and 1607 A.D. or roughly to *circa* 1581 A.D.[1] This date is, however, based on the supposition that 'in the *Tantrasāra* by Kṛṣṇānanda Āgamavāgīśa (written sometime after 1577 A.D.),[2] the *Śaktisaṅgama Tantra* is largely quoted.'[3] As a matter of fact, there is not a single quotation from the *Śaktisaṅgama* in the *Tantrasāra*. The *Ṣaṭpañcāśaddeśavibhāga* (verse 11), on the other hand, refers to the inclusion of Ujjayinī in the country of Mahārāṣṭra and this seems to point to the Marāṭhā occupation of Malwa under Bajī Rāo I (1720-40 A.D.). Verse 16 again speaks of Karṇāṭa in the sense of the 'Carnatic' and not of the Kannaḍa-speaking country. We know that the successors of Veṅkaṭa II (1630-42) of the Āravīḍu dynasty of Vijayanagara kings, even though they ruled a small tract in the east coast and had little or no authority over the Kannaḍa country, still called themselves 'Lords of Karṇāṭaka'. The verse seems to point to a date when the name Karṇāṭaka had become stereotyped as the designation of a part of the east coast. This was done possibly about the time when Zulfiqār 'Alī Khān (1692-1703 A.D.) was created the 'Nawab of the Karṇāṭaka' by the Mughal emperor Aurangzīb.[4] If these suggestions are correct, the *Ṣaṭpañcāśaddeśavibhāga* cannot be assigned to a period earlier than the first quarter of the 18th century. The manuscript of the Tantra, here marked B, contains the date *Śaka* 1674=1752 A.D. when apparently the manuscript was copied. But that the *Śaktisaṅgama Tantra* was composed sometime before 1744 A.D. is indicated by the fact that two of its stanzas are quoted by

1. *Ibid.*, p. vii.

2. As we have shown elsewhere, the *Tantrasāra* was composd in the first half of the seventeenth century. See *The Śākta Pīṭhas*, p. 80. Reference has been made to a manuscript of this work bearing the date 'Saṁvat 1586' (Gode, *Studies*, Vol. I, pp. 154 ff., Poleman, *Census of Indic Manuscripts in U. S. A.*, etc.). But *Saṁvat* in this case seems to be a misreading for *Śaka*.

3. Cf. p. vi.

4. See *infra*.

Harikṛṣṇa Miśra in his *Vaidikavaiṣṇavasadācāra* composed between 1713 and 1744 A.D.[1]

The *Ṣaṭpañcāśaddeśavibhāga* belongs to that class of works which is called the gazetteer literature in Sanskrit and on which Pandit H.P. Sastri wrote an interesting but short paper in *JBORS*, Vol. IV, pp. 14-25. The work gives an account of fifty-six countries lying in and on the borders of India. The list is drawn from the pilgrim's point of view and most of the localities mentioned are holy places. The majority of the localities are again Śaiva and Śākta Tīrthas and this fact indicates Tantric influence. The importance of the number 56 is, however, not determinable although it appears to be conventionally used to mean 'all' as the numbers, 18, 36, etc. This number is found adhered to in some other works of the class. The earliest work containing a list of fifty-six countries seems to be the *Candragarbhasūtra* or *Candragarbhavaipulya* (translated into Chinese by Narendrayaśas in 566 A.D.)[2] mentioning them in connection with the Buddha's manifestations in Jambudvīpa. All of them cannot be identified; but 'apparently less than half are within India proper'.[3] It has been suggested that the Sūtra was composed or re-edited in Central Asia; it is therefore possible that the importance attached to the number fifty-six is essentially foreign. Similar lists are found in some other medieval Tantric texts; *e.g.*, the *Sammoha Tantra* (composed before 1450 A.D. according to Gode)[4] gives two lists of fifty-six countries. These two lists are quoted below since they can be compared fruitfully with the list of the *Ṣaṭpañcāśaddeśavibhāga* under study.

I

कादौ षट्पंचाशद्देशाश्च(द्देशा) हादावपि तथा प्रिये ।
तद्देशशक्तिसंयोगस्त(न्तद् or गात्तद्)राज्यमिति कीर्त्तनम् ॥१
तत्संगमजपेणै(नै)व तद्देशराज्यमाक्नुयात् (?) ।
तत्रादौ कादिदेशाञ्च(श्च) कथ्यते(न्ते) परमेश्वरि ॥ २

1. See *Poona Orientlist*, Vol. XXI, 1956, pp. 4 ff.
2. See Bagchi, *Le Canon Buddhique*, Tome I, p. 270.
3. See Elliot, *Hinduism and Buddhism*, Vol. III, p. 215; Lévi, *BEFEO*, Tome V, pp. 261 f.
4. *ABORI*, Vol. XIX, pp. 184 f.

अंगवंगौ कलिंगश्च कालिंग(गः) केरलस्तथा ।
काश्मीर(रः) कामरूपश्च महाराष्ट्रोऽथ एव च ॥ ३
सौराष्ट्रो द्रविडश्चैवं(व) तैलिंगमलयाद्र(द्रि)कौ ।
कर्णाटावन्त्यवैदर्भसर्पा(प्ता?)भीर(राः) समालवा(वाः) ॥ ४
चौ(चो)लयां(पां)चालकांबोज(जाः) वैराद्र(ट्)पाम्यां(ण्ड्यौ) विदेहभूः ।
वाह्लीकश्च किरातश्च कैकटा(कर्वटा)वन्तकौ(?) तथा ॥ ५
ऐराकभोटांतचीन(ना) महाचीनस्तथैव च ।
नेपाल(लः) शीलहट्टश्च गौडकोशलमा(म)गवा(धाः) ॥ ६
श्नो(उ)त्कलाः कुंतलहूनाः(णौ) कोंकणः केकयस्तथा ।
शूरसेनः कौरवाश्च सिंहलाश्च पुलिंदका(काः) ॥ ७
कच्छसेवनमद्राश्च सौवीरलाटवर्वराः ।
मत्स्या(त्स्याः) सैंधव इत्येत(ते) षट्पंचांश(शत्) प्रकीर्त्तिता(ताः) ॥ ८

II

कादिदेशाः समाख्याता हादिदेशाः(शान्) शृणु प्रिये ।
अंगवंगकलिंग(गाः) स्यात् कालिंगः स्यात् सु(सौ)वीरक(कः) ॥ १
काश्मीरश्चैव काम्बोजा(जाः) सौराष्ट्रो मगधस्तथा ।
महाराष्ट्रो मालवस्तु नेपालः केरलस्तथा ॥ २
चोलयां(पां)चालगौडाश्च मलयांढ(याल)श्च सिंहलः ।
व्योंकविडो व्योन्तश्चैव (?) का(क)र्णाटो लाट एव च ॥ ३
मलाटश्चैव पानाटाः पावाद्यान्धक(?)पुलिन्दकः(काः) ।
हु(हू)णकौरवगान्धारविदर्भाः सविदेहका(काः) ॥ ४
वाह्लीक(को) वर्वरो देवि केकयः कोशलोऽपि च ।
कुन्तलश्च किला(रा)तश्च शूरसेनश्च सेवर(नः) ॥ ५
वनाटः टंक(ग)णश्चैव कोंकणा मत्स्यमद्रकौ ।
मैड(मरु)सैन्धवसन्धथाः(?)पाश्वंकीकौ(पारसीकस्)ततः स्मृतौ(तः) ॥६
ड्योर्जलि(?)यवनौ देवि जल(?)जालन्ध(न्ध्र)साल्वला(काः) ।
सिन्धुश्च वंद्यदेशाश्च हादिपर्यायवाधका(चकाः) ॥ ७[1]

1. Quoted by P. C. Bagchi, *Studies in the Tanras*, pp. 97-99, from a late Newari manuscript in the Durbar Library, Nepal. The text is faulty; but it can be partially corrected with the help of other materials and may also be used in correcting the latter. It is, however, possible that some of the mistakes belong to the original. The *Kādi* and *Hādi* sets of Tantric groupings are given 9 subdivisions each. There is also a fourfold division: 1. *Kerala* extending from Aṅga to Mālava, 2. *Kāśmīra* from Madra-deśa to Nepāla, 3. *Gauḍa* from Śilahaṭṭa to Saindhava, and 4. *Vilāsa* prevalent everywhere.

The geographical information derived from the *Ṣaṭpañcāśaddeśavibhāga* is very interesting, and, in many cases, it throws welcome light on some problems.[1] But the difficulties in dealing with the text are also numerous. The sources of the author's information are works like the epics, the Purāṇas and the Tantras and, above all, travellers' or pilgrims' tales. The author might have himself visited some of the places on pilgrimage. The manuscripts he relied on may have misled him in some cases with their wrong readings and many such mistakes may be also due to the copyists of his work.[2] Moreover the author, like many of the Tantric writers of his time, seems to have only an insufficient knowledge of Sanskrit. Partly due to this fact and partly to the inadvertence of the copyists, a great confusion has been created with reference to the actual position of many countries and others bordering on them. As will be indicated below in the notes on the countries, many of these mistakes can be quite easily corrected.[3] It must, however, be admitted that we cannot expect in all cases accurate geographical knowledge from our author who had to depend on a pilgrim's knowledge and had no opportunity of consulting any scientifically compiled text on geography or a scientifically prepared map. In some cases, therefore, we find that the position of one country given in relation to another is only partially true or even wrong.[4] Owing to the same reason, some countries are given as abutting on each other, though they are actually separated by other countries.[5] The author generally mentions places, usually those sacred to Śiva and Pārvatī, as boundaries of the countries. Some of these places are difficult to identify, and, in many cases, the name of a god or goddess in a particular temple actually indicates the whole district round the shrine.[6] This difficulty is enhanced by the fact that, while in some cases the boundary mentioned is to be included in the country, in others

1. See notes on verses 3, 21, 39, 46, 49, etc.
2. Cf. verses 5, 43, 47, etc.
3. Cf. verses 19, 47, etc.
4. *E. g.*, verse 18 wherein 'east' seems to be used for 'south-east'.
5. Cf. verses 13, 25, 33, 57, etc.
6. Cf. verses 12, 36, 39, etc.

it is to be excluded from it.[1] In many cases again, more than one country are mentioned in the same area and they overlap one another.[2] In view of the above observations, it will be seen that the *Ṣaṭpañcāśaddeśavibhāga* only attempts a rough estimate of the extent of countries, and sometimes the extent recorded is wider or narrower than it actually is.

The 8th Paṭala of the *Śaktisaṅgama Tantra*, Book III, is also very interesting to the students of early Indian geography. It gives an account of a fivefold division of India. The Chinese authors generally refer to the Five Indies, *viz.*, the Northern, Western, Central, Eastern and Southern, and the *Bhuvanakośa* section of the Purāṇas divide the country into Madhyadeśa, Udīcya, Prācya, Dakṣiṇāpatha and Aparānta, although sometimes the Himalayan and Vindhyan regions are added to them. The *Kāvyamīmāṁsā* of Rājaśekhara also refers to this fivefold division when it speaks of Pūrva-deśa beyond Vārāṇasī, Dakṣiṇāpatha beyond Māhiṣmatī, Paścād-deśa beyond Devasabhā and Uttarāpatha beyond Pṛthūdaka. The five divisions of India as given in the *Śaktisaṅgama Tantra* from the Tantric point of view are different. They are Indra-prastha, Yama-prastha, Varuṇa-prastha, Kūrma-prastha and Deva-prastha. We have appended this interesting section as a supplement to the text of the *Ṣaṭpañcāśaddeśvibhāga*.

षट्पंचाशद्देशविभागः

(*Śaktisaṅgama Tantra*, Book III,
Chapter VII)

[3]ॐ श्रीगणेशाय नमः ॥

[4]शंकर उवाच ।

[5]रहस्यातिरहस्यं हि[6] प्रोक्तं देवि तवाग्रतः ।
देशव्यवस्था देवेशि कथ्यते शृणु यत्नतः[7] ॥ १

1. Cf. verses 9, 11, 14, etc.
2. Cf. verses 4, 12, 14, 42, etc.
3. Not found in B and C.
4. Not found in B and C.
5. Not found in A.
6. C—च.
7. B—तत्परः ।

वैद्यनाथं समारभ्य भूवनेशान्तगं(गः) शिवे ।
तावद्वंगाभिधो[1] देशो यात्रायां नहि दुष्यति[2]॥ २
रत्नाकर[3] समारभ्य ब्रह्मपुत्रान्तगं(गः) शिवे[4] ।
वंगदेशो मया प्रोक्तः सर्वसिद्धिप्रदायकः[5] ॥ ३
जगन्नाथात् पूर्वभागात्[6] कृष्णातीरान्तगं(गः) शिवे ।
कलिंगदेशः संप्रोक्तो वाममार्गपरायणः[7] ॥ ४
कलिंगदेशमारभ्य पंचाष्टयोजनं शिवे ।
दक्षिणस्यां महेशानि कालिंगः परिकीर्त्तितः ॥ ५
सुब्रह्मण्यं समारभ्य यावद्देवो जनार्दनः ।
तावत्केरलदेशः स्यात्तन्मध्ये सिद्धिकेरलः[8] ॥ ६
रामेश्वरो वेंकटेशो[9] हंसकेरलबोधकः ॥ ७
अनन्तशैलमारभ्य यावत्स्यादुडुपं पुरम्[10] ।
तावत्सर्वेशनामा तु[11] केरलः परिकीर्त्तितः ॥ ८
शारदामठमारभ्य कुंकुमाद्रितटान्तकम्[12] ।
तावत्काश्मीरदेशः स्यात्पंचाशद्योजनान्तकः[13] ॥ ९
कालेश्वरश्वेतगिरिं[14] त्रिपुरान्नीलपर्वतम् ।
कामरूपाभिधो देवि गणेशगिरिमर्द्धनि ॥ १०
त्रयंबकं[15] समारभ्य मध्ये चोज्जयिनी शिवे ।

1. B—°विधो.
2. C—दुष्यते.
3. A—वज्राकार; C—वज्राकर.
4. C—परे.
5. AB—°प्रदर्शकः.
6. Read जगन्नाथपूर्व°.
7. B—राम°.
8. A—सिद्ध°.
9. A—विकुंठेशो; B—वैकंटेशो. Read रामेश्वराद्वेंकटेशं.
10. AB—परे.
11. C—सर्वे समानास्तु.
12. B—°माधि°
13. BC—°न्तकम्.
14. B—केरलेश्वरे; C—कोलेश्वरं. Read °श्वराच्छ्वेत°.
15. BC—त्रिपंचकं due possibly to the spelling त्रियंबकं. Better read त्र्यंबकेशं.

मार्जारतीर्थं[1] राजेन्द्रे[2] कोलापुरनिवासिनी ॥
तावद्देशो महाराष्ट्रः कर्णाटस्याभिगोचरः[3] ॥ ११
जगन्नाथाद्वूर्द्ध्वभागे[4] अर्वाक् श्री भ्रमरांबिका ।
तावदंध्राभिधो देशः सौराष्ट्रं शृणु साम्प्रतम्[5] ॥ १२
कोंकणात्पश्चिमं[6] तीर्त्वा समुद्रप्रान्तगोचरः ।
हिंगुलाजान्तको[7] देवि शतयोजनमाश्रितः[8] ॥
सौराष्ट्रो देशो देवेशि नाम्ना तु गुर्जराभिधः[9] ॥ १३
श्रीशैलन्तु समारम्य चोलेशान्मध्यभागतः ।
तैलंगदेशो देवेशि ध्यानाध्ययनतत्परः ॥ १४
मूकांबिकां समारम्य मलयाद्रचन्तगं[10] शिवे ।
मलयालाभिधो[11] देशो मन्त्रसिद्धिप्रवर्त्तकः ॥ १५
रामनाथं समारम्य श्रीरंगान्तं वरेश्वरि[12] ।
कर्णाटदेशो देवेशि साम्राज्यभोगदायकः[13] ॥ १६
ताम्रपर्णीं समारम्य[14] शैलार्द्धे रुतरार्द्धतः[15] ।
अवन्तीसंज्ञको[16] देशः कालिका तत्र तिष्ठति ॥ १७
भद्रकाली महत्पूर्वं[17] रामदुर्गाच्च[18] पश्चिमे ।
श्रीविदर्भाभिधो देशो वैदर्भी तत्र[19] तिष्ठति ॥ १८

1. B—°तीर.
2. B—राजेन्द्र; C—राजेन्द्रः.
3. B—कपाटस्वाभि°.
4. Read जगन्नाथ ऊर्द्ध्व°.
5. AC—सादरं.
6. A—कौंकणासम्मिंम.
7. C—°न्तकं.
8. B—जन्तुयोजनने शतः; C—जन्तवो जनलेशतः.
9. A—षण्मातुरजराभिधः.
10. B—मनपाड्यान्तगं. Read °न्तगः or °न्तकं.
11. AB—मलपाला°.
12. BC—बिलेश्वरि.
13. B—भोम°.
14. B—समासाद्य.
15. B—शैलार्द्धेषूत्तरोर्द्धतः; C—शैलद्वयेत्यत्तरतोर्द्धतः. Read शैलार्षादु.°
16. Usually अवन्ति.
17. A—मरुत्पूर्वा. Possibly स्थानात्पूर्वे. Read भद्रकालीदक्षपूर्वे. Cf. below. p. 83, note 12.
18. A—दुर्गा च.
19. A—वैदर्भा तत्र; C—वेदभास्तत्र.

गुर्जरात्पूर्वभागे[1] तु द्वारकातो हि[2] दक्षिणे ।
मरुदेशो महेशानि उष्ट्रोत्पत्तिपरायणः ॥ १९
श्रीकोंकनादधोभागे[3] तापीतः पश्चिमोत्तरे[4] ।
आभीरदेशो देवेशि विंध्यशैले व्यवस्थितः ॥ २०
अवन्तीतः पूर्वभागे गोदावर्यास्तथोत्तरे ।
मालवाख्यो महादेशो धनधान्यपरायणः[5] ॥ २१
द्राविडतैलंगयोर्मध्ये[6] चोलदेशः प्रकीर्त्तितः ।
लम्बकर्णाश्च ते प्रोक्ता भेदोह्यवान्तरो[7] भवेत् ॥ २२
कुरुक्षेत्रात् पश्चिमे तु तथा चोत्तरभागतः[8] ।
इन्द्रप्रस्थान्महेशानि दशत्रियोजनोत्तरम्[9] ।
पाञ्चालदेशो देवेशि सौन्दर्यगर्वभूषितः ॥ २३
पाञ्चालदेशमारभ्य म्लेच्छाद्दक्षिणपूर्वतः ।
काम्बोजदेशो देवेशि वाजिराशिपरायणः[10] ॥ २४
वैदर्भदेशादूर्द्ध्वं च इन्द्रप्रस्थाच्च दक्षिणे ।
मरुदेशात्पूर्वभागे विराटः परिकीर्त्तितः[11] ॥ २५
काम्बोजाद्दक्षभागे[12] तु इन्द्रप्रस्थाच्च[13] पश्चिमे ।
पाण्डुदेशो[14] महेशानि महाशूरत्वकारकः ॥ २६

1. Read गुर्जरपूर्वभागात्तु.

2. Read द्वारका यस्य.

3. A—श्रीकोंक°. Read श्रीकोंकणस्त्वधोभागे.

4. AB—पश्चिमे वरे. Read तापी च पश्चिमोत्तरे.

5. A—धनधाम.

6. B—तिलंग. Read तैलंगद्रविडान्तरे. Note that the rules of metre are often disregarded.

7. B—ह्यपरान्तरे.

8. B—भारतः

9. A—योजनाद्वयः; B—योजनद्वये.

10. Read वाजिराजि.

11. AC—वैराट्देशः प्रकीर्त्तितः.

12. A—भोजा°. Note that the word *dakṣa* is used in our text as well as in many other Tantra works in the sense of *dakṣiṇa*.

13. AB—इन्द्रदेशा°.

14. B—पाण्डयदेशो; C—याम्यदेशो corrected to पाण्डयदेशो.

गंडकीतीरमारम्य[1] चम्पारण्यान्तकं शिवे ।
विदेहभूः समाख्याता तैरभुक्त्यभिधः[2] स तु ॥ २७
काम्बोजदेशमारम्य महाम्लेच्छातु पूर्वके ।
बाह्लीकदेशो[3] देवेशि अश्वोत्पत्तिपरायणः ॥ २८
तप्तकुंडं[4] समारम्य रामक्षेत्रान्तकं शिवे ।
किरातदेशो देवेशि विन्ध्यशैले च तिष्ठति ॥ २९
करतोयां समारम्य हिंगुलाजान्तकं शिवे ।
वक्रान्तदेशो[5] देवेशि महाम्लेच्छपरायणः ॥ ३०
हिंगुपीठं समारम्य मत्केशान्तं[6] महेश्वरि ।
खुराशनाभिधो देशो म्लेच्छमार्गपरायणः ॥ ३१
तन्मध्ये चोत्तरे देवि ऐराकः[8] परिकीर्त्तितः ॥ ३२
काश्मीरन्तु समारम्य कामरूपाच्च पश्चिमे ।
भोटान्तदेशो देवेशि मानसेशाच्च दक्षिणे ॥ ३३
मानशेषा (सेशा) द्दक्षपूर्वे चीनदेशः प्रकीर्त्तितः ॥ ३४
कैलासगिरिमारम्य[9] सरयूयोनितः परे ।
आमोंगान्तं[10] महेशानि महाचीनाभिधो[11] भवेत् ॥ ३५
जटेश्वरं समारम्य योगिन्यन्तं[12] महेश्वरि ।
नेपालदेशो देवेशि शिलहट्टं शृणु प्रिये ॥ ३६
गणेश्वरं समारम्य महोदध्यन्तगं (गः) शिवे ।
शिलाहट्टाभिधो देशो पर्वते तिष्ठति प्रिये ॥ ३७[13]

1. B—नीर°.
2. B—तिरभुक्ता°. Read भिधा सा.
3. A—वाहिक; B—वाहीक.
4. BC—तत्तत्कुण्डं.
5. B—वभक्तान (बदक्षाणः ?).
6 A—मत्कोशान्तं; B—मत्कशान्तं. Read मक्केशान्तं.
7. B—षुरासन°; C—खुरासना°. A reads मत्कोशान्तं महेश्वरि । after करतोयां समारम्य and thus omits the stanza referring to Vakrānta-deśa.
8. B—ऐकीरः
9. B—कैलानरं समारम्य; C—कैलासतीरमारम्य.
10. A—आमोरंगं; B—आमारागां.
11. B—चीनाधिपो.
12. B—योगिनिन्तं.
13. Not found in A.

वंगदेशं समारम्य भुवनेशान्तगं(गः) शिवे[1] ।
गौडदेशः समाख्यातः सर्वविद्याविशारदः ॥ ३८
गोकर्णशात्पूर्वभागे[2] आर्यावर्त्तात्तु चोत्तरे ।
तंरभुक्तात्[3] पश्चिमे तु महापुर्यश्च पूर्वतः[4] ।
महाकोशलदेशश्च[5] सूर्यवंशपरायणः ॥ ३९
व्यासेश्वरं[6] समारम्य तप्तकुंडान्तकं शिवे ।
मगधाख्यो[7] महादेशो यात्रायां नहि दुष्यति ।
दक्षोत्तरक्रमेणैव क्रमात्कीकटमा(म)गधौ ॥ ४०
चरणाद्रिं समारम्य गृध्रकूटान्तकं[8] शिवे ।
तावत्कीकटदेशः स्यादुत्तरो[9] मा(म)गधो मतः ॥ ४१
जगन्नाथप्रान्तदेश[10]श्चोत्कलः[11] परिकीर्तितः ॥ ४२
कामगिरिं समारम्य द्वारकान्तं महेश्वरि ।
श्रीकुन्तलाभिधो देशो हूणं शृणु महेश्वरि ॥ ४३
कामगिरेर्दक्षभागे मरुदेशात्तथोत्तरे ।
हूणदेशः समाख्यातः शूरास्तत्र वसन्ति हि[12] ॥ ४४
अथ घट्टं[13] समारम्य कोटीशस्य तु मध्यगः[14] ।
समुद्रप्रान्तदेशो हि कोंकणः[15] परिकीर्त्तितः ॥ ४५
ब्रह्मपुत्रात्कामरूपान्मध्यभागे[16] तु कैकयः ॥ ४६

1. A—भुवनेशं महेश्वरि; B—भुवनेशान्तं महेश्वरि.
2. Read गोकर्णशाद्दक्षभागे.
3. A—तैर्भुक्तात्.
4. AB—सर्वतः.
5. C—देशः स्यात्.
6. B— कालेश्वरं.
7. B—मघदा°.
8. BC—°न्तगं.
9. A—स्यात्तदूर्द्ध्वः; B—स्यदूर्द्ध्वस्तु.
10. B—प्रोतदेश°; C—प्रोक्तदेश°.
11. C—°श्चौत्कलः.
12. B—च.
13. A—अथेरर्द्धं; B—अयतयदं.
14. B— कोटीशश्च वसकगे; C—मध्यगे.
15. A— कौंकणः.
16. B— °गत.

मगधाद्दक्षभागे[1] तु विन्ध्यात्पश्चिमतः शिवे ।
सौरसेनाभिधो[2] देशः सूर्य (यदु) वंशप्रकाशकः ॥ ४७
हस्तिनापुरमारभ्य कुरुक्षेत्राच्च दक्षिणे ।
पाञ्चालपूर्वभागे तु[3] कुरुदेशः प्रकीर्त्तितः ॥ ४८
मरुदेशात्पूर्वभागे कामाद्रेर्दक्षिणे शिवे ।
सिंहलाख्यो महादेशः सर्वदेशोत्तमोत्तमः ॥ ४९
शिलहट्टात्[4] पूर्वभागे कामरूपात्तथोत्तरे ।
पुलिन्ददेशो[5] देवेशि नरनारायणपरः ॥ ५०
गणेश्वरात्पूर्वभागे समुद्रादुत्तरे शिवे ।
कच्छदेशः समाख्यातः सेवनं[6] शृणु सादरम् ॥ ५१
पुलिन्दादुत्तरे भागे कच्छाच्च पश्चिमे शिवे ।
मत्स्यदेशः समाख्यातः मत्स्यबाहुल्यकारकः ॥ ५२
वैराटपाण्डययोर्मध्ये[7] पूर्वदक्षक्रमेण च ।
मद्रदेशः समाख्यातः माद्रीशास्तत्र[8] तिष्ठति ॥ ५३
शूरसेनात्[9] पूर्वभागे कण्ठकात्[10] पश्चिमे वरे ।
सौवीरदेशो देवेशि सर्वदेशाधमाधमः[11] ॥ ५४
अवन्तीतः पश्चिमे तु वैदर्भाद्दक्षिणोत्तरे[12] ।
लाटदेशः समाख्यातो वर्वरं शृणु पार्वति ॥ ५५
मायापुरं समारभ्य सप्तशृंगात्तथोत्तरे ।

1. Read मगधो दक्षपूर्वे.
2. Usually शूर° or शौर°.
3. A—पाञ्चालात्.
4. A—सिलात् evidently for सिंहलात्.
5. A—पुलिन्घ्र°; B—पुलिन्द्र°.
6. See below, note on Kaccha.
7. Better read पांडुविराटयोर्मध्ये.
8. A—माद्रीकास्तत्र; C—माद्रीतत्त्वाऽत्र. Read माद्रीशस्तत्र, मद्रेशस्तत्र or मद्रेशी तत्र.
9. Read शूरसेनः.
10. AB—कण्टकातः. Should we read कण्ठकः ?
11. A—देशः सर्वोत्तमोत्तमः; C—°शाधमोधमः.
12. Read पश्चिमोत्तरे.

वर्वराख्यो महादेशो सैन्धवं शृणु सादरम् ॥ ५६
लंकाप्रदेशमारभ्य मक्कान्तं[1] परमेश्वरि ।
सैन्धवाख्यो महादेशः पर्वते तिष्ठति प्रिये ॥ ५७
एते षट्पंचाशद्देशा मया प्रोक्ता तव प्रिये[2] ।
एतन्मध्येऽपि देवेशि देशभेदास्त्वनेकशः ।
कोटिशः सन्ति देवेशि एते मुख्याः प्रकीर्त्तिताः ॥ ५८
रहस्यातिरहस्यं च गोप्तव्यं पशुसंकटे ।
इति संक्षेपतः प्रोक्तं किमन्यच्छ्रोतुमिच्छसि ॥ ५९

Supplement[4]

(*Śaktisaṅgama Tantra*, Book III,
Chapter VIII)

श्रीदेव्यूवाच ।

देवेश श्रोतुमिच्छामि पंच प्रस्थान[5] साम्प्रतम् ।
इन्द्रप्रस्थं यमप्रस्थं वरुणप्रस्थमेव च ।
कूर्मप्रस्थं महादेव देवप्रस्थं च पञ्चमम् ॥ १
इन्द्रप्रस्थं महेशानि शृणु वक्षे[7] यथाक्रमम् ।
इन्द्रप्रस्थं महेशानि योगिनीसंयुतम्[8] ॥ २
मथुरागोकुलं पूर्व[9] भागे तस्य व्यवस्थितम् ।
वृन्दावनं कोलदेशे हस्तिनापुरमुत्तरे ॥ ३

1. A—मच्छान्तं.

2. C—द्देशान्मया प्रोक्ताद्य भक्तितः. In the place of this stanza, A has a concluding verse which reads as follows after minor corrections :

षट्पंचाशन्मिता देशा मया प्रोक्ता तवानघे ।
तेषां शास्ता भवेद्यो वै चक्रवर्ती स उच्यते ॥

3. B—गोप्तकं.

4. From B only.

5. Read प्रस्थानि.

6. Read श्रीभगवानुवाच before this stanza. Better read पंच प्रस्थानि देवेशि in place of इन्द्रप्रस्थं महेशानि

7. Read वक्ष्ये.

8. Read योगिनीपुरसंयुतम्.

9. Read पूर्वे.

द्वारका पश्चिमायां च गदावर्त्तस्तु दक्षिणे ।
इन्द्रप्रस्थं मध्यभागे कालिकामुखगोचरम् ॥ ४
द्रावाह्री[1]क्षेत्रमुद्दिष्टमिन्द्रप्रस्थं तदुच्यते ।
सिंहासनं चाश्वपतेस्तदेव परिकीर्त्तितम् ॥ ५
यमप्रस्थं महेशानि दक्षिणस्यां व्यवस्थितम् ।
सोमेश्वरः पश्चिमायां सप्तश्रृंगात्तु पूर्वगाः[2] ॥ ६
मायापुरं वामभागे व्यंकटेशस्तु दक्षिणे ।
यमप्रस्थं भवेन्मध्ये कमलामुखगोचरम् ॥ ७
शंखावर्त्तो भवेत्तत्र क्षेत्रं मायाभिधं भवेत् ।
सिंहासनं नरपतेस्तद्दक्षे परिकीर्त्तितः[3] ॥ ८
ततः पूर्वे जगन्नाथः प्रान्ते कविपति भवेत्[4] ।
वारुणाख्यं तथा देवि कथ्यते श्रृणु साम्प्रतम् ॥ ९
मक्केश्वर[5] पश्चिमायामुत्तरे हिंगुला भवेत् ।
त्रैलोक्यविजया देवि दक्षभागे प्रकीर्त्तिता ॥ १०
राजावर्त्तस्तत्र भवेत्तत्र शीतामुखं च तत् ।
तदन्ते सागराः सप्त समुद्राश्च तदन्तके ॥ ११
तावत् श्रीवारुणाख्यन्तु[6] प्रस्थं प्रोक्तं मया तव ।
कूर्मप्रस्थं महेशानि कथ्यते श्रृणु साम्प्रतम् ॥ १२
गोकर्णेश[7] दक्षभागे कामाख्या पूर्वगोचरः[8] ।
उत्तरे मानसेशः स्यात् पश्चिमे सारदा भवेत् ॥ १३
त्रिगुणा वर्त्तते तद्धि[9] वज्रेस्या मुखमीरितम्[10] ।

1. The intended reading may be वाराही°.

2. Possibly we have to suggest सप्तश्रृगन्तु पूर्वके.

3. Read °कीर्त्तितम्.

4. Read °पतिर्भवेत्. *Kavi-pati* may be an error for *kari-pati*, i.e. *gaja-pati*.

5. Read °श्वरः.

6. Read तावच्छ्री°. But the rules of *sandhi* and metre are often disregarded.

7. Read गोकर्णेशो.

8. Read °गोचरा.

9. Possibly त्रिगुणावर्तस्तत्र हि.

10. Possibly वज्रेशी°.

षट्पंचाछथाटि चामुंडागणैर्व्याप्तिन्तु[1] तद्भवेत् ॥ १४
पूर्वे वैरजनाथस्तूत्तरेऽमरकण्टकम्[2] ।
कांचीपुरं मध्यभागे मोहनावर्त्तमे(ए)व च ॥ १५
ललिताया मुखं तच्च विद्यागणविभूषितम् ।
कामाख्यात्र समुद्दिष्टं महामन्त्रविभूषितम् ॥ १६
नवलक्षमहाविद्या सगणे परिभूषितः[3] ।
पंच प्रस्थानि देवेशि कथितानि मया तव ॥ १७

II

I. *Aṅga* (v. 2).

Aṅga is said to have extended from Vaidyanātha upto Bhuvaneśa. Vaidyanātha is no doubt the same as Baidyanāthdhām in the Santal Parganas District, very near the southern frontiers of the Monghyr and Bhagalpur Districts, in Bihār. Bhuvaneśa reminds us of the celebrated Bhubaneswar in the Puri District of Orissa.

The ancient Aṅga country is supposed to have comprised the present Monghyr and Bhagalpur Districts of Bihar excluding the parts lying to the north of the Ganges. Its capital Campā is believed to have been situated at the confluence of the Ganges and the Campā (modern Cāndan) and the two villages Campānagar and Campāpur, near Bhāgalpur, are supposed to represent the actual site of the ancient capital of Aṅga. Since this view is supported by Hiuen-tsang and Al-Bīrūnī,[4] the city should not be located near the Lakhisarai railway station in the

1. Possibly षट्पंचाशद्भिश्चामुण्डागणैर्व्याप्तन्तु.

2. Possibly पूर्वे वैरजनाथस्तु उत्तरे or पूर्वे वै चन्द्रनाथस्तु उत्तरे.

3. Possibly महाविद्याः स्वगणैः परिभूषिताः.

4. Cf. Sircar, *Cosm. Geog. E. Ind. Lit.*, p. 35 and note, p. 36 note, pp. 77, 99, 104, 152, 155.

western fringe of the Monghyr District.[1] There are traditions regarding the expansion of the boundary of the Aṅga country especially in the south. The *Kathāsaritsāgara*,[2] *e.g.*, refers to the Aṅga city called Viṭaṅkapura which was situated on the sea-shore. The commentary on Vātsyāyana's *Kāmasūtra* (VI. 6) places Aṅga to the east of the Mahānadī.

It will be seen that Baidyanāthdhām is situated near the southern limits of the original Aṅga country; but Vaidyanātha may here indicate the district round the holy place, and the reference to Bhubaneswar may point to the country's extension towards the Bay of Bengal. The verse appears to place Aṅga to the south of the Ganges, and it must be admitted that there is no definite evidence indicating the extension of the Aṅga country to the north of that river. According to some manuscripts of the *Bhaviṣya Purāṇa* (see No. 37 below), Vaidyanātha-Mahādeva lay in the Jāṅgala-Jhārīkhaṇḍa country which was situated to the north of the Dārukeśvara river and to the west of the Bhāgīrathī, between Pañcakūṭa (the former Pañcakoṭ State in the Dhanbad District) and Kīkaṭa (the Gayā region in South Bihar).

2. *Vaṅga* (v. 3).

The country of Vaṅga is described as extending from the sea as far as the Brahmaputra. The sea is no doubt the Bay of Bengal in the south, and the Brahmaputra, the northern boundary, seems to indicate that portion of the river which bifurcates from the Jamuna. Vaṅga therefore included the eastern part of the Sundarbans in the south and half of the Mymensingh District in the north. The verse seems to exclude the region to the east of the Brahmaputra and the Meghnā (cf. v. 37) and agrees with medieval epigraphic evidence which places the heart of Vaṅga in the Vikramapura-bhāga comprising the Munshiganj and Madaripur Subdivisions of the Dacca and Faridpur Districts of East Pakistan. The original habitat of the Vaṅga people has been discussed in a separate chapter (Ch. XI) below.

1. *JBRS*, Vol. XLI, Part 2, p. 8.
2. 25.35; 26.116; 82.3-16.

3-4. *Kaliṅga* (v. 4) and *Kāliṅga* (v. 5).

Kaliṅga extended from the east of Jagannātha as far as the banks of the Kṛṣṇā. Jagannātha here evidently refers to the celebrated deity in the shrine of Purī in the Puri District of Orissa. It is interesting to note that the Tantric texts recognise Purī as one of the Pīṭha-sthānas where the Bhairava is Jagannātha and the Bhairavī Vimalā. That the Purī region was also called Utkala is known from v. 42.

The name Kaliṅga has been used here in a wide sense. The *Mahābhārata* (III. 114. 4) recognises the Vaitaraṇī river as the north-eastern boundary of Kaliṅga, and at the time of Aśoka and Khāravela it certainly comprised the Purī-Cuttack region in the east. At the time of the Eastern Gaṅga king Anantavarman Coḍagaṅga, the Kaliṅga kingdom extended from the Ganges to the Godāvarī. The statement in the commentary on the *Kāmasūtra* (VI. 6) that Kaliṅga lies to the south of the Gauḍa-viṣaya appears to indicate the extension of ancient Kaliṅga as far as the Ganges. The expansion of Kaliṅga from the Mahānadī to the river Kṛṣṇā in the south seems to be referred to in an inscription of the fifth century A.D.[1] The reference in our text may, however, be to the dominions of the Sūryavaṁśī Gajapatis of Orissa. Their empire originally included certain tracts lying to the south of the Kṛṣṇā, though they were later ousted from those areas by the kings of Vijayanagara.

On the southern confines of Kaliṅga was a country called Kāliṅga according to our text. But we do not know of any country of this name to the south of the Kṛṣṇā from any other source. The author probably refers to the Nellore-Guntur region which once formed parts of the dominions of the Sūryavaṁśī Gajapatis of Orissa.

5. *Kerala* including *Siddhikerala* (v. 6)

The Kerala country is said to have extended from Subrah-

1. Cf. *Ep. Ind.*, Vol. XXX, p. 114. Some allied topics have been discussed in a separate chapter (Ch. VII).

maṇya upto Janārdana and Siddhikerala to have formed a part of it. Roughly speaking, Kerala comprised the coastal region inhabited by the Malayālam-speaking people. But our text makes separate mention not only of two other Keralas called Haṁsa (v. 7) and Sarveśa (v. 8), but also of Malayāla (v. 15).

There are many temples of Subrahmaṇya (Kārttikeya) in South India and it is difficult to determine which of them is meant. The same is the case with Janārdana, as we can trace several Janārdana temples in the area in question. It is, however, possible that Subrahmaṇya here indicates the village of that name in the South Kanara District, which is famous for its temple and cattle fair, and that Janārdana is the temple at Varkala or Varkkalai between Quilon and Trivandrum.

6. *Haṁsa-Kerala* (v. 7).

The territory including Rāmeśvara and Veṅkaṭeśa is called Haṁsa-Kerala, a name otherwise unknown. The shrine of Rāmeśvara and the celebrated Veṅkaṭeśvara temple at Tirupati are quite well known; but it is difficult to determine whether they are actually intended. If, however, the verse refers to the country from Rāmeśvara to Tirupati, the name Haṁsa-Kerala would appear to indicate the same territory as the old Drāviḍa.

7. *Sarveśa-Kerala* (v. 8).

The Kerala country designated by the name Sarveśa extended from the Ananta-śaila upto the city called Uḍupa. The Ananta hill may refer to Trivandrum which is known as Ananta-śayana or Anantapura. It is possible that the place is referred to by Varadarāja (17th century) whose *Saṁskṛtamañjarī* speaks of the southern Tīrthas called Janārdana-kṣetra, Gokarṇa-kṣetra (in the North Kanara District), Anantasena (°śayana ?)-kṣetra, and Subrahmaṇya-kṣetra. Uḍupa-pura is possibly Uḍipi, headquarters of the Uḍipi Taluk in the South Kanara District. The celebrated Vaiṣṇava saint Madhva was born at Kalyāṇapura near Uḍipi.

8. *Kāśmīra* (v. 9).

Kāśmīra (Kashmir) is described as the country lying between the Śāradā-maṭha and the Kuṅkum-ādri and extending over 50 *yojanas*. The Kuṅkuma hill cannot be satisfactorily identified; but it may refer to the *Kuṅkuma* (saffron)-growing plateau above Pampur near Srinagar, capital of Kashmir.[1] Śāradā-maṭha is evidently modern Sardi near the confluence of the Kishenganga and Kankatori rivers in Kashmir. The shrine of the goddess Sāradā is mentioned in Kalhaṇa's *Rājataraṅgiṇī* (VIII. 2556, 2706). According to Stein, the old shrine is in complete ruins and it is substituted by the late Sāradā temple at Guṣa (ancient Ghoṣa) which is now visited by the pilgrims.

9. *Kāmarūpa* (v. 10).

The country of Kāmarūpa is said to have extended from Kāleśvara to the Śveta-giri and from Tripura to the Nīla-parvata. The 'Blue Mountain' seems to be the Nīl-ādri or Nīlakūṭa, the name of the Kāmakhyā hill according to the *Kālikā Purāṇa* (Ch. 79, v. 74). The Gaṇeśa-giri seems to be referred to as lying in the heart of Kāmarūpa. Tripura can be quite satisfactorily identified with Tripurā (Tipperah) which is now partly in East Pakistan. The heart of ancient Kāmarūpa was the Gauhati region of Assam. The celebrated temple of Kāmākhyā, called the *Yoni-pīṭha*, lies not far from the city of Gauhati. According to Chinese evidence, Kāmarūpa lay to the east of the Karatoyā, while the *Yoginī Tantra* includes in the country the Brahmaputra valley together with Rangpur and Cochbihar.

10. *Mahārāṣṭra* (v. 11).

Mahārāṣṭra (i. e. the Marāṭhī-speaking area) extended from Tryambaka to Karṇāṭa, and comprised Ujjayinī, Mārjāra-tīrtha and Kolāpura-nivāsinī. Tryambaka is certainly the celebrated Tryambakeśvara Śiva-liṅga near Nāsik and here

1. Cf. Watt, *The Commercial Products of India*, p. 429.

2. See *Rājatar.*, trans., Vol. II, pp. 279, 289.

indicates the whole of the Nasik area. Ujjain and Kolhāpur are also well-known localities. Kolāpura-nivāsinī appears to be the goddess Mahālakṣmī of the great shrine at Kolhāpur. As has already been pointed out, the inclusion of Ujjain appears to refer to the Marāṭhā conquest of Malwa during the rule of Peshwā Bājī Rāo I (1720-40 A.D.); and in that case, we have to assume that the *Śaktisaṅgama Tantra*, or at least the *Deśavibhāga* section of it, was composed not earlier than the first quarter of the 18th century. Karṇāṭa of this verse seems to be different from that of v. 16 and to indicate the original Karṇāṭa, i.e. the Kannaḍa-speaking area. It is interesting to note that the *Kāmasūtra* (VI. 5. 29) commentary (middle of the thirteenth century) locates Mahārāṣṭra between the Narmadā and Karṇāṭa.

11. *Andhra* (v.12).

Andhra is said to have been the country which has Jagannātha above and Bhramarāmbikā below. Jagannātha is the celebrated god of Purī; but here the district round the shrine seems to be indicated. Bhramarāmbikā is no doubt the famous Bhramarāmbā who is one of the 18 Śaktis in India and is enshrined on the Śrīśaila with the god Mallikārjuna, one of the 12 Jyotirliṅgas. It has to be noted that our author separates Andhra from Tailaṅga (v. 14) and also from Kaliṅga (vv. 4-5).

12. *Saurāṣṭra* or *Gurjara* (v. 13).

From the Koṅkaṇa up to Hiṅgulāja in the west lies, according to the verse, the Saurāṣṭra country covering a hundred *yojanas* on the coast. This country has also been called Gurjara. Koṅkaṇa is the strip of land between the Western Ghāṭs and the Arabian Sea, and Hiṅgulāja is the famous Tīrtha of Hiṅglāj near the Arabian Sea (between lat. 25° and 26° and long. 65° and 66°), more than 100 miles to the west of Karachi.

Surāṣṭra originally indicated the southern part of Kathiawar ; but in the later period, the name Gujarāt is found to be used in a wide sense to comprise the whole of Kathiawar and the adjoining regions. The verse in question ment ons the wide extent of the country. It must, however, be remembered that, from a pilgrim's point of view, Gujarāt may be roughly placed between the Northern Koṅkan (comprising the modern Thana District of Maharashtra State) and the district round Hiṅglāj. Pilgrims seem to have reached Somanātha in Kathiawar by boat from the Northern Koṅkan and then proceeded to Hiṅglāj again by boat.[1]

13. *Tailaṅga* (v. 14).

From the Śrī-śaila up to the middle of the distance between the same and Coleśa lies the Tailaṅga country. Coleśa may indicate the god of the celebrated Bṛhadīśvara temple at Tañjāvūr. Tailaṅga is the same as the name Telengana and is related to *Telugu.* But the origin of this geographical name is unknown. The author's separation of Tailaṅga from Andhra (v. 11) is comparable to that of Kāliṅga from Kaliṅga (vv.4-5). Really Tailaṅga means the Telugu-speaking area which is the same as Andhra.[2] An inscription[3] of 1358 A. D. gives the following boundaries of the Tiliṅga or Tailaṅga country :

पश्चात्पुरस्ताद्यस्य देशौ महाराष्ट्रकलिंगसंज्ञौ ।
अवागुदक् पांड्यककान्यकुब्जौ देशस्स तत्रास्ति तिलिंगनामा ॥

that is, the Telugu country is bounded by Mahārāṣṭra in the west, Kaliṅga in the east, Pāṇḍya in the south and Kānyakubja in the north. There is reference here to the old empire of the Kākatīyas. Of course, in this age, the Kānyakubja country was no longer a political unit ; but the reference may be to the empire of the Turkish Sulṭāns of Delhi.

14. *Malayāla* (v. 15).

The Malayāla country, which is separated from the tracts called Kerala (vv. 6-8), is described as lying between Mūkāmbikā

1. Cf. v. 57.

2. The description does not suit the present Telengana area of Andhra Pradesh.

3. *Ep. Ind.*, Vol. XIV, p. 90.

and the Malay-ādri. The Malay-ādri or Malaya range has been identified with the Travancore hills. The goddess is no doubt one of the same name worshipped at Kollūr or Barkalūr in the northern part of the South Kanara District, formerly in Madras but now in Mysore. This goddess is mentioned in inscriptions like C. P. Nos. 3-6A of *A.R.Ep.*, 1927-28, and in stone inscription No. 403 of the same year. See also the *Śivatattvaratnākara*, Kallola V, Ch. I, vv. 27 ff.

15. *Karṇāṭa* (v. 16).

The Karṇāṭa country is said to have extended from Rāmanātha upto Śrīraṅga which is either Śrīraṅgam opposite Tañjāvūr on the other side of the Kāverī or Śrīraṅgapattana (Seringapatam) near Mysore. Rāmanātha seems to be either the Rāmanāthapuram (former Ramnad) District, or the Rāmanātha Maṭha in the Madurai District, or the Rāmeśvara Tīrtha near the junction of the rivers Tuṅgā and Bhadrā.[1] The possible reference to Śrīraṅgam or Ramnad as a boundary of Karṇāṭa suggests that it is not the ancient Karṇāṭa country, the heart of which was in the Kannaḍa-speaking land, *Kannaḍa* being the same as Sanskrit *Karṇāṭa*. With the expansion of the empire of the Kanarese kings of Vijayanagara, the name Karṇāṭa extended over a large part of the Deccan. After the battle of Talikoṭa in 1665 A.D., the Vijayanagara kings withdrew first to Candragiri (Chittoor District) and then to Vellore (North Arcot District). But, as has already been indicated, even when their kingdom became confined to a very small area far away from ancient Karṇāṭa, they were known as the Rāyas of Karṇāṭaka. About the end of the seventeenth century, Zulfiqār 'Alī' Khān (*circa* 1692-1703 A.D.), the progenitor of the Nawābs of Arcot, was created the 'Nawāb of the Carnatic (Karṇāṭaka)'. It is not impossible that the verse actually refers to the kingdom of these Nawābs of the Carnatic.

16. *Avanti* (v. 17).

Avanti is described as extending from the Tāmraparṇī

1. See *Bomb. Gaz.*, Vol. I, Part ii, pp. 377,397. The place is identified with Rāmeśvara near Alampur in the Mahbubnagar Dist., A. P. (*EP. Ind.*, vol. 33, p. 332 n).

as far as the northern side of some hills and as the place of the goddess Kālikā. This Tāmraparṇī is an unknown river and not the celebrated rivulet of that name in the extreme south of India.[1] The hills referred to may be the Western Vindhyas.

Avanti is generally identified with the district round Ujjain (in the former Gwalior State) which was its capital.[2] Thus it roughly corresponded with modern West Malwa. Kālikā mentioned in the verse is evidently the Śakti of the celebrated Mahākāla of Ujjain.[3]

17. *Vidarbha* (v. 18).[4]

The country of Vidarbha is said to have been situated to the east of Bhadrakālī and to the west of Rāmadurga or with Rāmadurgā to its west. This Bhadrakālī appears to be the same goddess as Kālikā [of Ujjain] mentioned in the preceding verse. Rāmadurga reminds us of the former State of this name in the South Marāṭhā country lying between Belgaum and Bijapur, although it is far from Vidarbha, i.e. modern Berar. The Ujjain region lies to the north or north-west of Berar. Even if we accept an old tradition according to which Avanti included the Māhiṣmatī (modern Māndhātā in the Nimar District or Maheswar in the former Indore State) region on the Narmadā, it cannot be placed to its west. I have therefore suggested the reading *Bhadrakālī-dakṣapūrve*, 'to the south-east of Bhadrakālī.'

18. *Maru* (v. 19).

The Maru country seems to be described as lying to the north-west of the Gujarāt region and to the north of Dvārakā and also as famous for its camels. It is the great Indian desert called the Thar or Rājputānā desert. The land no doubt also includes the Mārwār (i.e. Maru) or Jodhpur area.

1. See above, p. 60, note 3.
2. Cf. Mālava below, No. 20.
3. For the separate mention of Avanti and Mālava, see below, No. 20 (p. 98, note 2); also Chapter XII.
4. For Vidarbha as the name of the Sadiya region of Assam, see *JAIH*, Vol. I, p. 19.

19. *Ābhīra* (v. 20).

The Ābhīra country appears to be placed on the Vindhya between the Koṅkaṇa in the south and the Tāpī or Tāptī in the north-west.

The Ābhīras originally lived in the desert; but they gradually pushed towards the south.[1] In the 1st and 2nd centuries A.D., Aberia or the Ābhīra country lay not far from Surāṣṭra or Southern Kathiawar. In the 3rd century A.D., the Ābhīras founded a kingdom comprising parts of the Northern Koṅkan and the Nāsik region. But it is difficult to believe that this area was called Ābhīra in the late medieval period. The reference seems to be to the Ahirwas fort in the former Indore State, lying at lat. 22° 31′, long. 76° 31′, though, in that case, we have to read *śrī-Koṅkaṇāt pūrva-bhāge*, since the area in question does not lie to the south of the Koṅkaṇ. The *Kāmasūtra* (VI. 4. 24) commentary applies the name Ābhīra to the tracts including Śrīkaṇṭha (Thanesar) and Kuru-kṣetra (cf. also *Mbh*, IX. 37. 1).

20. *Mālava* (v. 21)

The great country of Mālava is described as lying to the east of Avanti and to the north of the Godāvarī. The verse evidently indicates Eastern Malwa, the ancient name of which was Ākara or Daśārṇa having its capital at Vidiśā (modern Besnagar near Bhilsa in the former Gwalior State). It has to be noticed that the *Kāmasūtra* (VI. 5. 22 and 24) mentions Avanti and Mālava separately and that its commentary identifies Mālava with Pūrva-Mālava and gives the name Avanti to the Ujjain region.[2] Bāṇa's *Kādambarī* supports the same tradition when it associates Mālava ladies with the Vetravatī (modern Betwa) surrounding Vidiśā and speaks of Ujjayinī as a city of the Avanti country in the present West Malwa.[3]

1. See *The Age of Imperial Unity*, pp. 221 ff.

2. आवन्तिका उज्जयिनीदेशभवाः । ता एवापरमालव्यः । मालव्य इति पूर्वमालवभवाः ।

3. Cf. मज्जन्मालवविलासिनीकुचतटास्फालनजर्जरितोर्मिमालया...

21. *Cola* (v. 22).

The Cola country is said to have been situated between Drāviḍa and Tailaṅga (v. 14), and the three countries, which were very similar to one another, were also known generally as the [land of the] Lambakarṇas.

The Cola country proper was the Tañjāvūr-Tiruciräppalli region. The verse, however, appears to refer to the country of the Telugu Coḍas of the Anantapur-Cuddappa area.

22. *Pāñcāla* (v. 23).

Pāñcāla appears to have been described as lying at 13 or 30 *yojanas* from Indraprastha to the west and north of Kuru-kṣetra. Indraprastha is the modern Delhi region and Kuru-kṣetra was the country to the south of the Sarasvatī and to the north of the Dṛṣadvatī, in the Karnal-Ambala region of the Eastern Punjab.

Ancient Pāñcāla was divided into two divisions, viz., the northern and southern, and the former comprised roughly the modern Rohilkhand Division of U. P. Ahicchatrā, capital of North Pāñcāla, has been identified with modern Rāmnagar in the Bareily District, while Kāmpilya, which was the capital of South Pāñcāla, has been located at modern Kāmpil in the Farrukhabad District to the south of the Ganges. But this is not the country mentioned in the stanza. For its association with the Pir Panchal range, see Chapter XII, Section ii, below.

23. *Kāmboja* (v. 24).

The Kāmboja country is described as extending from Pāñcāla and lying to the south-east of the Mleccha country. This Mleccha is apparently the same as Mahā-Mleccha of v. 28 and indicates the Muslim countries on the north-western borders of medieval India.[1]

वेत्रवत्या परिगता विदिशाभिधाना नगरी राजधान्यासीत् and विजितामर-लोकद्युतिरवन्तिषूज्जयिनीनामनगरी in Siddhāntavāgīśa's ed., pp. 19, 183. See also below, Chapter XII, Section iii.

1. Cf. Nos. 25 and 27 below.

The ancient Kāmbojas may be supposed to have lived in various settlements in the wide area between the Punjab and Iran, to the south of Balkh (cf. No. 27). From a pilgrim's standpoint therefore the country may have been described as having Pāñcāla (the region near about the Pir Panchal range) in the east and the Muhammadan countries in the west. The horses of the Kāmboja country are often mentioned in literature and inscriptions. The Kāmbojas living in Aśoka's empire are mentioned in his inscriptions along with the Yavanas or Greeks of the area around Kandahar where an edict in two versions meant for these two peoples has been discovered. See below, Chapter XII.

24. *Virāṭa* (v. 25).

The country called Virāṭa or Vairāṭa is placed to the north of Vidarbha, to the south of Indraprastha and to the east of Maru-deśa.

Of course Vidarbha is far to the south ; but the ancient Virāṭa or Matsya country actually lay in the Alwar-Bharatpur-Jaipur region to the south of the Delhi area and to the east of the desert. Vidarbha was never its southern boundary ; but it may have been mentioned as a country lying in the south—whatever be the distance—of the Virāṭa country. One of the Śākta Pīṭhas is said to be in Virāṭa where the Devī is Ambikā and the Bhairava Amṛta. There was a Virāṭa country in Northern Bengal, according to some manuscripts of the *Bhaviṣya Purāṇa*.[1] The Nivṛti country (comprising Vardhanakūṭa = Bardhankoṭ, Kacchapa = Cochbihār or, more probably, Kachhar, and Rangpur) has there been described as :

निवृतिः सा च यत्रैव जनिष्यति तृणादिकम् ।
वरेन्द्रोत्तरभागे च विराटस्य च सन्निधौ ।
वंगस्य पश्चिमोत्तरे निवृतिर्विषयो मतः ॥

This Virāṭa is apparently connected with the 'Birāṭ-rājār Gaḍh' in the Rangpur District, or Bairāṭṭā in the Dinajpur District, in

1. See No. 37 below.

the northern part of Bengal and has nothing to do with the country of the same name mentioned in our text.

25. *Pāṇḍu* (v. 26).

The Pāṇḍu or Pāṇḍya country is placed to the south of Kāmboja and to the west of Indraprastha or the Delhi region.

It is interesting to note that there is evidence of the existence of an ancient Pāṇḍu country in the Punjab. Ptolemy's Geography[1] places the country of Pandoouoi (i. e. Pāṇḍava) about the Bidaspes = Vitastā = Jhelam and includes in it Sagala = Sākala = Sialkot. Poros (i. e. Puru or Paurava), contemporary of Alexander, evidently belonged to this Pāṇḍu country. The *Vedic Index* suggests that either the Jhelam region was the early home of the Purus where some remained while others migrated to the east, or that that region was later conquered by the Purus who had been settled in the east. But it is difficult to determine how this ancient tradition finds an echo in the late medieval text before us. See Chapter XII, Section ii.

26. *Videha* or *Tairabhukti* (v. 27).

The Videha country, also called Tīrabhukta or Tairabhukti, is described as extending from the banks of the Gaṇḍakī as far as the Camp-āraṇya. Gaṇḍakī-tīra here appears to indicate the southern boundary of the country and the place where the modern Gaṇḍak falls into the Ganges. Camp-āraṇya is the modern Campāraṇ which seems to be the northern boundary. The name Tīrabhukti still survives in the modern form *Tirhut*. It is the same as the Darbhanga-Muzaffarpur region of Upper Bihar to the north of the Ganges.

27. *Bāhlīka* (v. 28).

The Bāhlīka country is said to be bordering on Kāmboja and situated to the east of Mahā-Mleccha. Bāhlīka was Bactria (modern Balkh) in the north of Afghanistān; but Vāhīka was an old name of the Punjab. The reading Bāhlīka is better as the

1. VII. i. 46.

country is placed between Kāmboja and the great Mleccha world of the west. It is interesting to note that the Kāmbojas, elsewhere associated with Pañcāla (v. 24). are here connected with the northern part of Afghanistān.

28. *Kirāta* (v. 29).

The Kirāta country is described as extending from Tapta-kuṇḍa as far as Rāma-kṣetra and as lying on the Vindhyas. Tapta-kuṇḍa may indicate the hot springs at Rājgir in the Patna District, Bihar, or those near Monghyr. Rāma-kṣetra may be Rāma-giri or Rāmtek. Kirāta here seems to indicate some Vindhyan hill tribes, though in old literature they are usually connected with the Himalayan region. The fact is that such names as Pulinda and Kirāta really indicated particular hill tribes; but later their meaning expanded so as to signify any hill tribe.

29. *Vakrānta* (v. 30).

The Vakrānta country is said to have extended from the Karatoyā as far as Hiṅgulāja and abounded in a large number of Mlecchas. The Karatoyā river mentioned here is not easy to identify; but the reference to the Hiṅglāj area as one of the bordering regions suggests that Vakrānta-deśa is possibly not Wakhan, but very probably Makran between Baluchistan and Persia. Pargiter[1] notices three Karatoyās; the first a tributary of the Brahmaputra in North Bengal, the second belonging to the Himalayan region in North India and the third rising from the Vindhyan range. The second seems to be indicated in the verse; but the plurality of the name may also have been caused by wrong readings.

30. *Khurāśana* (v. 31).

The country called Khurāśana (Khorasan) is described as extending from the Hiṅgu-pīṭha, i.e. Hiṅgulāja, upto Makkeśa, 'lord of Mecca', which is an imaginary Śiva-liṅga at Mecca,

1. *Mārkaṇḍeya Purāṇa*, trans., p. 301.

the most sacred of the Muslim holy places. According to a popular Bengali tradition, apparently influenced by the Tantras, in an underground room underneath the floor of the Kaaba at Mecca there lies Śiva in deep sleep; pious Hindus are never allowed to approach him; if, however, a devout Hindu could place on his head a *bilva-patra* only once, the god would at once rise up and destroy all the Muhammadans of the world.

Khorasan is the north-eastern province of Persia, to the west of Afghanistān; but the name here seems to be employed to indicate the kingdom of Persia as known in the late medieval period.

31. *Airāka* (v. 31).

In the northern part of the country of Khurāśana, i.e. the Persian kingdom or Western Asia, is placed Airāka, i.e. Iraq.

32. *Bhoṭānta* (v. 33).

The Bhoṭānta country is said to have extended from Kāśmīra and lay to the west of Kāmarūpa and to the south of Mānaseśa. Bhoṭānta is evidently Bhoṭān and Mānaseśa refers to the god Śiva at the Mānasa lake (Mānsarovar) in the Himalayas. Modern Bhoṭān is a small state to the north of Bengal; but the Tibetan people are also known as Bhauṭṭa or Bhoṭa. Taken in this wider sense, Bhoṭānta may actually be placed between Kāśmīra in the west and Kāmarūpa in the east.

33. *Cīna* (v. 34).

Cīna-deśa is described as lying to the south-east of Mānaseśa. The country to the south-east of the Mān-sarovar is Tibet which appears to be indicated by the name Cīna in the verse. Tibet formed a part of the dominions of the Chinese emperors. Abul Fazl's '*Ain-i-Akbarī* refers to 'Pegu which former writers called Cheen, accounting this to be the capital city', and seems to include Burma in the Cīna country.[1] According to Pargiter, Cīna comprised the country of Tibet 'along

1. Jarrett's trans., Vol. II, ed. Sarkar, p. 132.

the whole range of the Himalayas.'[1] The Cīnas are sometimes linked in the *Mahābhārata* with the Kāmbojas of the north-west[2] and often mentioned among the retinue of the king of Prāg jyotiṣa or Assam in the east[3] and are also sometimes placed near the source of the Ganges in the country lying midway between the two regions.[4]

34. *Mahā-Cīna* (v. 35).

The Mahācīna country is said to have extended from the Kailāsa-giri and from the place of origin of the Sarayū as far as Moṅga. Mount Kailāsa is that part of the Himalayan range which lies to the north of the Mān-sarovar, and the origin of the Sarayū (Ghogra) is not far off. Moṅga and its variations appear to refer to Mongolia to the north of China. Mahācīna therefore refers to China proper.[5] The suggestion is supported by the fact that, in the seventh century A.D., the Chinese traveller Hiuen-tsang described his native country as 'the Mahācīna of the Indians.'[6]

35. *Nepāla* (v. 35)

The country of Nepāla is placed between Jaṭeśvara and Yoginī which seems to refer to Yoginīpura or Delhi. The Nepāla country to the north of Bihār and U.P. is well-known; but Jaṭeśvara cannot be satisfactorily identified. It is possible that the original reading was Jalpeśvara, the famous Śiva of the Jalpaiguri District in North Bengal, sometimes referred to in the later Puranic literature.

36. *Śilahaṭṭa* (v. 36).

The country called Śilahaṭṭa is described as extending

1. *Mārk. Pur.*, trans., p. 319, note.
2. VI. 9. 66.
3. V. 19. 15.
4. III. 177. 12.
5. For Cīna and Mahācīna, see Hobson-Jobson, s. v. *China* and *Macheen*. The evidence of our text is supported by the *'Ain-i-Akbari*, trans. Jarrell and Sarkar, Vol. II, p. 131.
6. Cf. Watters, *On Yuan Chwang's Travels in India*, Vol. I, p. 349.

from Gaṇeśvara as far as the sea. Gaṇeśvara appears to be identical with the Gaṇeśa-giri placed in the heart of the country of Kāmarūpa (v. 10). Śilahaṭṭa is no doubt modern Sylhet which, however, does not now extend up to the Bay of Bengal. The land lying to the east of the Meghna in South-East Bengal (now mostly in East Pakistan) seems to be included in Śilahaṭṭa.

37. *Gauḍa* (v. 38).

Gauḍa is said to have extended from the Vaṅga country upto Bhuvaneśa (i.e. Bhubaneswar in Orissa). In modern times, Gauḍa implies the whole area inhabited by the Bengali-speaking people. Originally, however, the country seems to have comprised the present Murshidabad District with the inclusion of the southernmost part of Malda. The Chinese traveller Hiuen-tsang describes Karṇasuvarṇa as the name of both the kingdom and capital of Śaśāṅka who has been called 'the lord of Gauḍa' in the *Harṣacarita*. According to the Chinese pilgrim, the Raktamṛttikā-vihāra lay in the suburbs of the Karṇasuvarṇa city and the country was about 730 or 750 miles in circuit. As the monastery referred to has been located at modern Rāṅgāmāṭi (Sanskrit *Rakta-mṛttikā*), of Karṇasuvarṇa or Gauḍa country can be identified with the Murshidabad District. This identification is again supported by epigraphic evidence and an interpolated section of the *Bharisya Purāṇa*,[1] which says:

पुण्ड्रदेशे सप्तदेशास्तेषां नामानि त्वं शृणु ।
गौडो वरेन्द्रो निवृत्तिः सुह्मदेशः प्रकीर्त्तितः ॥
जांगलो झारिखंडश्च वराहभूमिरेव च ।
वर्द्धमानो विंध्यपार्श्वे सप्तैते परिकीर्त्तिताः ॥
पद्मानद्या दक्षभागे वर्द्धमानस्य चोत्तरे ।
गौडदेशः स विज्ञेयो गौडेशी यत्र तिष्ठति ॥

It is said that Gauḍa (comprising Navadvīpa, Śāntipura, Maulapattana and Kaṇṭakapattana) lies between the river Padmā and the Vardhamāna district and forms a part of Puṇḍra, a name here used to indicate Bengal without its south-eastern part, but with some parts of Bihār and Orissa. The southern part of the Malda District comprising the site of the city of Gauḍa

1. See *Ep. Ind.*, Vol. XXXVII, pp. 25ff.; MS. No. 3582 of the Asiatic Society, Calcutta.

(Gaur) seems to have been to the south of the Ganges and included in the Gauḍa country in the early medieval period, and the capitals of some of the Pāla and Sena kings in Gauḍa were situated near the modern site of Gaur. Our work divides Bengal in two broad divisions, viz. East Bengal (Vaṅga) and West Bengal (Gauḍa). It is interesting in this connection to note that Muslim historians sometimes called this country 'Gaur-Bangāl'. Gauḍa offers one of the numerous instances of the expansion of the connotation of a geographical term.[1]

38. *Mahā-Kośala* (v. 38).

The Mahā-Kośala country is placed to the south of Gokarṇeśa, to the north of Āryāvarta, to the west of Tairabhukti and to the east of Mahāpurī. Gokarṇa is the name of a village on the Bāgmatī in Nepāl. Āryāvarta is not the same as that (=North India) defined by Manu, but the Tantric Āryāvarta defined as

विन्ध्योत्तरगतेनैव मगधाच्चापरेण तु ।
हिमाद्रेर्दक्षिणे भागे पाञ्चालात्पूर्वतस्तथा ॥
आर्यावर्त्तं इति ख्यातः.........[2]

Mahāpurī appears to refer to Delhi which was the capital of the Mughul emperors at the time of the composition of the work to which the verse belongs. Mahā-Kośala is no doubt the same as old Kośala ruled by the Sūrya-vaṁśī kings from their capital at Ayodhyā near modern Faizābād in Oudh (Ayodhyā). The name Mahā-Kośala thus appears to be wrongly applied now to indicate South Kośala, *i.e.* the Raipur-Sambalpur-Bilaspur region in Madhya Pradesh and Orissa.[3] If we prefer the reading *sarvataḥ* to *pūrvataḥ*, Mahāpurī would possibly indicate Ayodhyā, the ancient capital of the Kośala *janapada*.

1. Cf. Chapter V below.

2. Bagchi, *Studies in the Tantras*, p. 108. Here also the name Pāñcāla appears to be associated with the Pir Pantsal range.

3. See Pargiter, *Mārk. Pur.*, trans., pp. 308, 342.

39. *Magadha* (v. 39).

The great country of Magadha is described as extending from Kāleśvara or Vyāseśvara as far as the Tapta-kuṇḍa. Kāleśvara is evidently the celebrated Kāla-Bhairava of Varanasi and Vyāseśvara reminds us of Vyāsa-Kāśī. Tapta-kuṇḍa seems to refer to the hot springs (*e.g.*, Sītākuṇḍa) near Monghyr. The Magadha *deśa* is therefore placed between the Benares region and the Monghyr District. The ancient Magadha country comprised the present Patna and Gaya Districts of Bihar and had its capital originally at Girivaja-Rājagṛha (Rājgir) and then at Pāṭaliputra (near Pāṭnā), both in the present Patna District.

40. *Kīkaṭa* (v. 40).

The Kīkaṭa country is placed in the southern part of Magadha, the northern half of which is called Magadha proper. Kīkaṭa is described as extending from the Caraṇ-ādri upto Gṛidharakūṭa. The celebrated Gṛdhrakūṭa or 'Vulture Peak' (modern Śaila-giri) was situated near Rājagṛha, modern Rājgir in the Patna District. Caraṇ-ādri appears to indicate either the Viṣṇupada hill at Gayā, or Chūnār. In the Purāṇic literature, Kīkaṭa is essentially connected with the Gayā region. But the traditional identifications of Kīkaṭa with Gayā-pradeśa and with Magadha appear to be both right. The verse under study indicates that originally the Gayā region was called Kīkaṭa and the Pāṭnā region Magadha; but soon the former became a part of the latter.

41. *Utkala* (v. 41).

The country which had Jagannātha (the god in the shrine at Purī on the sea) on the boundary is described as Utkala. The original connotation of the name and its later expansion have been discussed in a separate chapter (Ch. X) below.

42. *Śrīkuntala* (v. 43).

The country called Śrīkuntala is placed between the Kāmagiri and Dvārakā. V. 44 shows that the Kāma-giri was far to the north of the desert. In this connection, it is interesting to note that an inscription of 1176 A.D.[1] represents a ruler of the Kāma country as a tributary to the king of the Sapādalakṣa Mountain (i.e. the Siwalik range). It is also not impossible that the name Kāma has something to do with that of Kumaon. The Śrīkuntala country thus appears to be placed between the Siwalik-Kumaon region in the north and Dvārakā in the south. Śrīkuntala is, however, otherwise unknown and the well-known Kuntala lies in the Kanarese area. Is Śrīkuntala a mistake for Śrikaṇṭhaka or the Thānesar region? The wide extent of the Śrīkaṇṭha country can, however, only be accepted if we refer to the time of Harṣavardhana (606-47 A.D.), king of Thānesar and Kanauj.

Pargiter[2] notices three Kuntalas, one in the Deccan, the second near Chūnār to the south of Varanasi and the third 'in the west'. This third Kuntala may be indicated in the verse. It is, however, probable that this plurality of Kuntala as accepted by Pargiter is due to wrong readings which might have misled even the author of the verse in question.

43. *Hūṇa* (v. 44).

Hūṇa-deśa is placed to the south of the Kāma-giri (see No. 42) and to the north of Maru-deśa, i.e. the Desert, and is called the land of heroes. The Hūṇas had a place in Indian politics even long after the disintegration of the extensive empire of Toramāṇa and Mihirakula. The *Harṣacarita* places the Hūṇa country in the Punjab region practically suggesting the same area as the verse under study. Hūṇa royal houses are mentioned in the records of the Pratihāras, Kalacuris and others. The Hūṇas are ultimately known to have been regarded as one of the 36 clans of the Rajputs.

44. *Koṅkaṇa* (v. 45).

Koṅkaṇa has been described as lying on the sea-coast. It seems to have extended from the Ghaṭṭa as far as the middle of

1. *Ind. Ant.*, Vol. X, p. 342. Sapādalakṣa may also be the Sāmbhar region and Kāma, modern Kaman near Bharatpur (cf. Ch. XVIII below).

2. *Mārkaṇḍeya Purāṇa*, trans., p. 308.

the district round Koṭīśa. The Koṅkaṇ, as we know, is the strip of land between the Western Ghāts and the Arabian Sea. Ghaṭṭa therefore appears to indicate the Western Ghāṭs, while Koṭīśa reminds us of the Koṭeśvara Mahādeva near the sea, not far from Bhuj in the Gujarat State, and Koṭeśvara near Udipi in the South Kanara District of Mysore, the first of the two being out of question in the present context. Our Koṭīśa may be the same as Saptakoṭīśvara of the inscriptions and coins of the Kadambas of Goa.[1]

45. *Kaikaya* (v. 46).

Kaikaya is placed between the Brahmaputra and Kāmarūpa. The ancient Kekaya country lay in the Punjab to the east of Gandhāra (the Peshawar-Rawalpindi region). But a Kekaya kingdom is known to have existed in the northern part of Mysore in the fifth century A.D.[2] The Kaikaya country mentioned in our verse seems to be a place in North-East Bengal, which was possibly connected by local traditions with the ancient Kekayas of the Punjab. Kaikaya here may even indicate the land of the Kukis in Assam and Maṇipur. If this identification is accepted, it shows how such tribal names were given a classical colour in the late medieval works.

46. *Śaurasena* (v. 47).

The Śaurasena country seems to be described as extending from Magadha in the south-east upto the territory to the west of the Vindhya. The ancient Śūrasena people lived in the Mathurā region. Greek writers refer to the Sourasenoi and to their cities Methora (Mathurā) and Kleisobora (possibly Kṛṣṇapura = Gokula). This old Śūrasena or Śaurasena country therefore had nothing to do with Magadha, the ancient name of South Bihār. It must, however, be remembered that our author extends Magadha as far as the Varanasi district in the west. He possibly believed that the Śaurasena country extended from the Varanasi region up to the Mathurā district which is to the

1. See *Ind. Ant.*, Vol. XIV. pp. 288 ff. ; Moraes, *Kadamba Kula*, pp. 383-84.

2. See Sircar, *Suc. Sāt.*, pp. 313 ff.

west (actually to the north-west) of the Vindhyan range. It may also be suggested that the text requires emendation and that the intention was to say 'to the north of the Vindhya'. It is interesting to note in this connection that the *Kāmasūtra* (VI. 9. 28) commentary locates Śaurasena *Kauśāmbyāḥ dakṣiṇataḥ* (sic—*paścimataḥ*) *kūle* [*Yamunāyāḥ*].

47. *Kuru* (v. 48).

The Kuru country appears to be described as extending from the Hastināpura region and as lying to the south of the Kuru-kṣetra tract, with Pāñcāla lying in the west. Hastināpura, capital of the Kuru country, has been identified with a locality in the Meerut District, U. P. According to the *Mahābhārata*,[1]

दक्षिणेन सरस्वत्या दृषद्वत्युत्तरेण च ।
ये वसन्ति कुरुक्षेत्रे ते वसन्ति त्रिविष्टपे ।।

that is to say, the land called Kuru-kṣetra lay between the rivers Sarasvatī and Dṛṣadvatī in the Eastern Punjab. This land has been called Brahmāvarta in the *Manusmṛti*.[2] The verse extends the Kuru country to Pāñcāla (i.e. the region of the Pir Panchal range and river) in the west (cf. Nos. 22-23 above).

48. *Siṁhala* (v. 49).

The great country called Siṁhala, the best of all countries, is placed to the east of Maru-deśa and to the south of the Kām-ādri. This Siṁhala cannot be identified with Ceylon. It is evidently in the Punjab-Rajasthan region and reminds us of the kingdom of Siṁhapura mentioned by Hiuen-tsang. The capital of this kingdom has been identified with Khetas or Katas in the Jhelum District, which is, next to Jvālāmukhī, the most frequented place of pilgrimage in the Punjab. Tantric literature locates Sambhala (probably Siṁhala) and Laṅkāpurī in the Swat-Kashmir region.[3] For Kāma, cf. p. 108 above.

1. III. 83. 4 and 205.
2. II. 17.
3. Bagchi, *op. cit.*, pp. 39-40.

49. *Pulinda* (v. 50).

The country of Pulinda is placed to the east of Śilahaṭṭa and possibly to the south of Kāmarūpa. The text reads 'to the north of Kāmarūpa' ; but no locality to the north of Kāmarūpa, the heart of which was the Gauhati region in Assam, could have anything to do with Śilahaṭṭa, i.e. modern Sylhet. The Pulindas were a hill tribe usually connected with the Vindhyan range. Pargiter[1] notices several branches of the Pulinda tribe, viz. (1) a western branch, (2) a Himalayan branch related to the Kirātas and Taṅganas, and (3) a southern branch. The second branch may be intended here, though the plurality of the name may be due either to misreadings in the manuscripts or to the fact that the term was sometimes used to indicate *any* hill tribe. Pulinda of our text may possibly refer to tribes of the hills of Assam, but more probably to the land of the Paliyās or Rājvaṁśīs of North Bengal.

50. *Kaccha* (v. 51).

The Kaccha country is placed to the east of Gaṇeśvara and to the north of the sea, and is possibly also called Sevana. If Gaṇeśvara is the same as Gaṇeśa-giri in the heart of Kāmarūpa, Kaccha may be identified with Kacchapa (possibly Kacchar or Cochbihar) placed by the *Bhaviṣya Purāṇa* in the north-eastern region of Vaṅga.[2] But the expression 'to the north of the sea' seems to indicate Kutch. The identification of Kaccha and Sevana, possibly suggested in the verse, offers another difficulty. Sevana, i.e. Seuna-deśa, was the name of modern Khandesh. It indicated the kingdom of the Yādavas who had their capital first at Sinnar near Nasik and then at Devagiri, modern Daulatābād in the Aurangabad District of the present Maharashtra State. The Seuna country may therefore have been the Nasik-Daulatabad region or even the vast kingdom conquered by the powerful kings of the Yādava dynasty. But it is doubtful if the memory of the Yādava age lingered down to the late medieval period. Kaccha may possibly also indicate the district round Bhṛgu-kaccha, modern Broach near the mouth of the Narmadā.

1. *Mārkaṇḍeya Purāṇa*, trans., pp. 316, 335, 338.
2. See No. 24; cf. No. 37.

The meaning of the word *kaccha* appears to be 'a region on the sea-coast,' and it is possible that different *Kacchas* (including Kutch and Kacchapa), referred to above, have been confused in the verse. As to Kaccha, a medieval tradition refers to the expansion of the Eastern Sea (Bay of Bengal) in the heart of Bengal as far north as Devīkoṭṭa or Bāṇgaḍh in the Dinajpur District, now in East Pakistan.[1] Long ago the sea no doubt touched the central areas of Bengal at least through the mouths of the big rivers falling in it. But the medieval tradition referred to above is apparently due to the existence of big lakes called *bils*, like the Chalan, in the Rajshahi-Bogra-Mymensingh area.

The language of the verse may suggest that Sevana was different from Kaccha, the description of the former having been omitted through oversight. But in that case, the number of countries would be more than fifty-six.[2]

51. *Matsya* (v. 52).

The Matsya country is placed to the north of Pulinda and to the east of Kaccha. This country is described as abounding in fish. If Pulinda of this verse is the same as that in verse 50, and Kaccha is the same as Kacchapa (Kacchar or Cochbihār), this Matsya-deśa may be identified with Virāṭa placed to the north-east of Bengal in a verse of the *Bhaviṣya Purāṇa*.[3] Virāṭa was the celebrated epic king of the Matsya country which has been located in the Alwar-Jaipur-Bharatpur region, and its capital Virāṭa-nagara has been identified with modern Bairāṭ in the former Jaipur State. But this ancient Matsya country seems to be mentioned in our text as Virāṭa (verse 25). It is difficult to say whether the author's intention was to distinguish between Matsya and Virāṭa, both located in the same area or whether he

1. Cf. *IHQ* Vol. XVII, p. 471—यदि पूर्वसमुद्रावधिरार्यावर्तस्तदा गंगायाः पूर्वकूलमपि स्यात्। नैवम्। पूर्वं किल देवीकोट्टसमीपे पश्चिमे पूर्वोदधिरासीत् तदपेक्ष्योक्तम्. This may explain the epic and Puranic tradition regarding the extension of the Prāgjyotiṣa-Kāmarūpa country down to the Bay of Bengal. Al-Bīrūnī speaks of 'the mountains of Kāmrū (Kāmarūpa), which stretch away as far as the sea' (Sachau, *Alberuni's India*, Part I, p. 201).

2. Cf. however p. 71, verse 8 of Group I and verse 5 of Group II.

3. See No. 24 (above, p. 100).

has made a confusion between the actual Matsya country in Eastern Rajasthan and a tract of land in Bengal.

52. *Madra* (v. 53).

The Madra country is located south-eastwise between Vairāṭa and Pāṇḍya. Vairāṭa is evidently the same as Virāṭa (No. 24) and Pāṇḍya the same as Pāṇḍu (No. 25). The ancient Madra people lived about the modern Sialkot District of the Punjab. Their capital Śākala has been identified with Sialkot. Madra however has been used here in a wider sense., as Virāṭa or Matsya lay far to the south of the ancient Madra country.

53. *Sauvīra* (v. 54).

The Sauvīra country seems to be placed to the west of Śaurasena and to the east of Kaṇṭhaka and is called the worst of all countries. Ancient Sauvīra was situated to the east of the Lower Indus and included Multan in the north. Śaurasena, *i.e.* the Mathurā region, was far to the east. Kaṇṭhaka cannot be satisfactorily identified ; but it reminds us of the Kathaioi tribe located by the classical authors between the Jhelam and the Chinab. This Kaṇṭhaka does not appear to be identical with Śrīkaṇṭha in the eastern part of the Punjab. Or, can the intended reading be *Kacchakāt paścime*, 'to the west of Kutch' ?

54. *Lāṭa* (v. 55).

The Lāṭa country seems to be placed to the west of Avanti and to the north-west of Vaidarbha (Berar). The ancient Lāṭa country was the district between the Lower Mahī and the Tāptī ; but sometimes it also extended beyond the Mahī. Bhṛgukachcha (Broach) and Navasārikā (Nausāri) belonged to this country. The *Kāmasūtra* (VI. 5. 26) commentary also locates Lāṭa to the west of Apara-Mālava, *i.e.* Western Malwa.

55. *Varvara* (v. 56).

The great country called Varvara is described as extending from Māyāpura and as lying to the north of Sapta-śṛṅga.

Varvara reminds us of Barbaricum of the *Periplus*, which lay on the sea at the mouth of the Indus and belonged to the country of Scythia (Indo-Scythia of Ptolemy), *i.e.* the Śaka country, about the beginning of the Christian era.[1] It is tempting to identify Māyāpura with the place of that name near Hardwār; but it is far away from the ancient Barbaricum. Sapta-śṛṅga is again a famous place to the north of Nasik. The image of the eighteen-armed goddess of this place is 12 feet high.[2] A verse, discussed in Section III below, mentions both Māyāpura and Saptaśṛṅga; but it is difficult to determine their position with reference to the country of Varvara. Can Māyāpura be the Mayapur Pass in the Jhansi District, U. P.?

56. *Saindhava* (v. 57).

The great country called Saindhava, *i.e.* Sindhu, is described as extending from Laṅkā-pradeśa as far as Makkā. Makkā may indicate here the Muslim lands of Western Asia in general. But Laṅkā is not easy to identify. It is difficult to determine whether the country called Siṁhala = Siṁhapura of the Punjab is intended by the author. Of course, it is possible, in consideration of the travellers' point of view, to identify it with Ceylon. They might have reached Sind from Ceylon which was on the way to Mecca. The ancient Sindhu-deśa lay to the west of the Lower Indus and was not exactly the same as modern Sindh.

III

The Five Prasthas (v. 2).

The *Prasthas* or Tantrik territorial divisions of 'India' are enumerated as five, *viz*, (1) Indra-prastha, (2) Yama-prastha, (3) Varuṇa-prastha, (4) Kūrma-prastha, and (5) Deva-prastha. The word *prastha* means 'a level expanse', 'level plain', 'tableland on the top of a mountain', 'the top of a mountain', etc. The word also indicates 'going on a journey'

1. The editors of the *Lekhapaddhati* (G.O.S.), p. 98, identify Barbara or Varvara with Bābriāvāḍ in Kathiawar. But see below, Chapter XIV, Section ii.

2. *Ind. Cult.*, Vol. VII, p. 450.

and hence possibly, 'pilgrimage.' This fivefold division of India may therefore have been made from a Tantric pilgrim's point of view. The association of *Aśvapati* and *Narapati* with Indra-prastha and Yama-prastha respectively reminds us of the medieval imperial title *Aśvapati-Gajapati-Narapati-rāja-tray-ādhipati*.[1]

1. *Indra-prastha* (vv. 3-6).

Indraprastha, the name used in ancient literature to indicate a city in the suburbs of Delhi (see v. 4), appears to have been connected with Yoginīpura which is another name of Delhi. In the eastern district of our Indra-prastha lay Mathurā and Gokula, the famous pilgrim-spots of North India. Vṛndāvana, which is near Mathurā, is placed at the *kola-deśa*, i. e. side, of Indra-prastha while Hastināpura in the Meerut District lay in its north. Dvārakā is placed to the west and Gadāvarta to the south. Dvārakā is the famous *tīrtha* in Kathiawar. Gadāvarta, quite well-known from the *Harivaṁśa* as the place where the *gadā* or club thrown by Jarāsandha against Kṛṣṇa at Mathurā fell, is also to be placed not very far from Mathurā. This Indra-prastha is therefore bounded by Delhi and Meerut in the north, Gadāvarta in the south, Mathurā in the east, and Dvārakā in the west. Roughly speaking, it included the Meerut region in the north-east and parts of Kathiawar in the south-west with its centre at the city of Indraprastha (Delhi). It is stated to have been the land of the *Aśvapati* or the lord of horses or cavalry.

2. *Yama-prastha* (vv. 7-9).

Yama-prastha seems to be placed in the south (*i.e.* to the south of Indra-prastha) with Someśvara in the west, Sapta-śṛṅga in the east, Māyāpura in the north and Vyaṅkaṭeśa in the south. Someśvara and Vyaṅkaṭeśa are evidently the gods at the celebrated temples at Somanātha in Kathiawar and at Tirupati in the Chittur District of Andhra Pradesh. Māyāpura and Sapta-śṛṅga are no doubt the places of these names mentioned in connection with Varvara (No. 55). There

2. Cf. Sircar, *Indian Epigraphy*, pp. 338-39.

is reference to another locality called Śaṅkhāvarta which is possibly also called Māyā-kṣetra. Roughly speaking, Yama-prastha or the southern *prastha* comprised parts of Kathiawar in the north-west and the district round Tirupati in the south. This was the seat of the *Narapati* or the lord of men or foot-soldiers.

3. *Varuṇa-prastha* (vv. 10-13)

Varuṇa-prastha or Vāruṇa-prastha is said to have Jagannātha [in the Puri District, Orissa] in the east, Makkeśvara [at Mecca] in the west, Hiṅgulā [in Sind] in the north and Trailokyavijayā[1] in the south. The seven *Sāgaras* and *Samudras* (seas and oceans ?) are placed beyond or near this *prastha*. Trailokyavijayā appears to indicate some deity or pilgrim-spot in Ceylon. There is reference also to a place called Rājāvarta. Thus Varuṇa-prastha was evidently connected with the sea and comprised most of the districts on the coast of the Arabian Sea and the Bay of Bengal. This was probably the seat of the *Gajapatī* or the lord of elephants (cf. the Gajapatis of Orissa).

4. *Kūrma-prastha* (vv. 14-16).

Kūrma-prastha is placed to the north of Gokarṇeśa, to the west of Kāmākhyā, to the south of Mānseśa and to the east of Śāradā. Gokarṇa in Nepāl, Kāmākhyā in Assam, the Mān-sarovar in the Northern Himalayas and the Sāradā-maṭha in Kāśmīra are all well known and have already been referred to.[2] This is the Himalayan *prastha* and comprised the northernmost areas of India. This Gokarṇa is not to be confused with the town of this name, having a famous Śiva temple, in the North Kanara District.

5. *Deva-prastha* (vv. 17-19).

Deva-prastha seems to be placed to the west of Vairaja-nātha (or Candranātha), and to the south (east?) of the Amara-kaṇṭaka, and Kāñcīpura and Mohanāvarta are located in this *prastha* which had possibly something to do also with Kāmākhyā. Vairajanātha may be connected with the Virajā-kṣetra or

1. Cf. *The Sakti Cult and Tārā*, ed. Sircar, pp. 147 ff.
2. See above, pp. 93-94 (Nos. 8-9), pp. 103-04 (Nos. 32-33).

Jājpur ; but Candranātha, the celebrated god in the Chittagong District in E.Pakistan, seems to be actually indicated. Kāñcīpura of the verse does not appear to mean the famous Kāñcīpuram near Madras. It is better to identify it with the Kāñcī-pīṭha on the Kopāi river near Bolpur in the Birbhum District of West Bengal. Deva-prastha therefore comprised the south-eastern and eastern regions of India. Bengal and Assam which had the greatest centres of medieval Tantricism were included in this *prastha* which is possibly called *Deva-prastha* significantly.

Chapter VI

GAUḌA

1. *Gauḍa as the Name of a City*

Gauḍa (usually written *Gaur* in English) was the residence of several dynasties of the Muslim rulers of Bengal for many centuries from the establishment of Muhammadan rule in the country about the close of the twelfth century A.D. The ruins of the city lie in lat. 24° 52′, long. 88° 10′ to the north or left bank of the Ganges and south of the modern town of Malda, headquarters of the District of that name in West Bengal. The traces of the ruined city with some buildings of Muhammadan times extend over an immense area now chiefly covered with jungle. As late as the middle of the sixteenth century, a European traveller gives the following account of the city of Gauḍa : "It is situated on the banks of the Ganges and is said to be three of our leagues in length and to contain 200,000 inhabitants. On the one side, it has the river for its defence and on the landward, faces a wall of great height......the streets are thronged with the concourse and traffic of people......that they cannot force their way past......a great part of the houses of this city are stately and well-wrought buildings."[1] In 1683, another European traveller who visited the old city in ruins says, "We spent 3 hours in seeing the ruins especially of the palace which has been......in my judgement considerably bigger and more beautiful than the Grand Seignor's seraglio at Constantinople or any other palace that I have seen in Europe."[1]

In Muslim times, the city of Gauḍa was also known by the name Lakhnautī no doubt a corruption of Sanskrit *Lakṣmaṇāvatī*. This name was apparently coined after that of king Lakṣmaṇasena (*circa* 1189-1206 A.D.) of the Sena dynasty, who was ousted from the western and northern districts of Bengal by the Muhammadans. The Senas thus appear to have had at least one of their secondary capitals at Gauḍa. The city of Rāmāvatī (called Rāmautī by the Muslim writers), named after the Pāla

1. See *Hobson-Jobson*, s. v. *Gour*.

king Rāmapāla (*circa* 1077-1120 A.D.) and the capital or a secondary capital of the Later Pālas of Bengal and Bihar, was probably also situated in the vicinity of the present site of Gauḍa. Thus Gauḍa flourished as an important city of Eastern India for a considerable period of time before the establishment of Muslim rule, and pre-Muslim rulers of the country often built new cities named after them in the vicinity of the site of the old city.

The *Aṣṭādhyāyī* of Pāṇini who flourished in North-Western India about the fifth century B.C. speaks of a city called Gauḍapura. But the rule *pure prācām* (6. 2. 99) which immediately precedes the rule *ariṣṭa-gauḍa-pūrve ca* (6. 2. 100) suggests that both the cities, *viz.*, Ariṣṭapura and Gauḍapura referred to in the *Aṣṭādhyāyī*, were situated outside the eastern part of India[1] in a region which was more or less fully Aryanised before the composition of the grammatical work in question. Ariṣṭapura is actually known to be the capital of the Śivi country[2] in the Jhang District of West Pakistan. Moreover, the early grammarians used the expression 'Oriental' with reference to the south-eastern part of Āryāvarta or Aryan India.[3] There is again no reason to believe that the Aryanisation of Western Bengal (in which the Gauḍa country was situated) made any appreciable progress in the age of Pāṇini.[4] Thus the identification of Pāṇini's Gauḍapura with the city of Gauḍa in Bengal is highly improbable. The name *Gauḍa* is supposed to be derived from the word *guḍa* meaning 'sugar'. The country, of which Gauḍa was the chief city, was also known by the same name possibly because it was famous in ancient times for its production of sugar. But whether the name of the city was applied to the country or that of the country to its chief city cannot be determined in the present state of our knowledge. The fact however that, in the seventh century A.D., the capital of the Gauḍa country was called Karṇasuvarṇa (modern Rāṅgāmāṭi, about eleven miles to the south of Murshidabad, headquarters of the District of that name to the south of the Ganges and Malda) seems to suggest that the city was named

1. Cf. S. Sen, *Vāṅgālā Sāhityer Itihās*, Vol. I, Ist ed., p. 4.
2. Cf. Malalasekera, *Dictionary of Pali Proper Names*, s. v. *Siviraṭṭha.*
3. See Chapter XIX below.
4. Cf. 'Spread of Aryanism in Bengal' in *Journal of the Asiatic Society*, Letters, Vol. XVIII, 1952, pp. 171 ff.; also Sircar, *Stud. Soc. Adm. Anc. Med. Ind.*, Vol. I, pp. 1 ff.

after the country. This fact also indicates that the city of Gauḍa was built in its present site to the south of Malda some time after the seventh century probably in the age of the Pālas. It is interesting to note in this connection that all the *jaya-skandhāvāras* (*i.e.* residences or temporary capitals) of the Pāla kings, including Rāmāvatī-nagara (in the vicinity of the site of Gauḍa) were situated on the Ganges. Urged by their preference for a city on the main course of the river, the Pāla kings may have transferred the headquarters of the Gauḍa country from Karṇasuvarṇa to the present site of Gauḍa in the Malda District. This seems to have been done after the main current of the river had begun to pass through the Padmā and the Bhāgīrathī (the original Ganges), on which Karṇasuvarṇa was situated, had begun to be gradually less important as a water-course. There is moreover reason to believe that originally the Ganges flowed by a route through the middle of the present District of Malda so that the city of Gauḍa lay on its southern or right bank.[1]

About the ancient course of the Ganges, a distinguished writer says, 'The tract between Malda and Murshidabad was the ancient Ganges delta where the river split up into numerous spill channels, the most important of which appear to be the Sarasvatī, the Bhāgīrathī and the Bhairab. Leaving the hills of Rajmahal, the Ganges seems to have passed northwards through the modern Kalindri, and then southwards into the lower course of the Mahānandā, east of the ruins of ancient Gaur. There was also the south-eastern branch of the Ganges (modern Padmā), the bifurcation being pretty old and shown in Ptolemy's map. In the oldest of modern maps, De Barros' (1550) and Gastaldi's (1561), Gaur is shown on the west of the Ganges. Leaving Gaur, the main waters of the Ganges turned southward and flowed through the channel of the Bhairab (as Kṛttivāsa, the reputed author who flourished in the fourteenth century, indicated) and, from at least the twelfth to the sixteenth century, through the Sarasvatī into the Bay, while the ancient eastern branch of the Ganges is traceable in the *jhils* and morasses which extend from Purnea to the sea. The Ganges thus forsook this course in favour of the channel

1. Cf. *Hist. Beng.*, Dacca University, Vol. I, p. 3.

through which the Bhāgīrathī now passes."[1] The situation of the Gauḍa capital Karṇasuvarṇa on its bank seems to suggest that the present Bhāgīrathī carried the main current of the Ganges as late as the seventh century A.D.

2. *Gauḍa as the Name of a Country.*

In modern times, the name Gauḍa is often used in Bengali literature to indicate the whole area inhabited by the Bengali-speaking people. Originally, however, the Gauḍa country seems to have comprised, in a narrow sense, the present District of Murshidabad together with the southernmost areas of the Malda District of Bengal. The Chinese pilgrim Hiuen-tsang who visited India in the first half of the seventh century A.D. speaks of Karṇasuvarṇa as the name of both the capital and the kingdom of king Śaśāṅka who killed king Rājyavardhana of Thānesar about 605 A.D. The king responsible for the death of Rājyavardhana is, however, described in the *Harṣacarita* of Bāṇa, who flourished at the court of Rājyavardhana's younger brother and successor Harṣa, as 'the lord of Gauḍa'. There is thus no doubt that Gauḍa was the usual name of Hiuen-tsang's kingdom of Karṇasuvarṇa apparently so named by the Chinese traveller after the name of the capital of that kingdom. According to the Chinese account, the celebrated Buddhist monastery called Raktamṛttikā-vihāra stood on the suburbs of the city of Karṇasuvarṇa, and the country of that name was about 730 or 750 miles in circuit. As the monastery in question has been located at modern Rāṅgā-māṭi (literally, 'the red earth', the same as Sanskrit *Rakta-mṛttikā*) about eleven miles to the south of Murshidabad,[2] the Karṇasuvarṇa or Gauḍa country has to be located about the present Murshidabad District, although the dominions of the Gauḍa king Śaśāṅka is known to have actually comprised wide regions of Eastern India.

The above location of Gauḍa in a narrow sense is remarkably supported by a late Purāṇic tradition. An interpolated section in some manuscripts of the *Bhaviṣya Purāṇa* locates Gauḍa-deśa, inhabited by the deity Gauḍeśa or Gauḍeśī, in the land

1. Radhakamal Mookerjee, *The Changing Face of Bengal*, 1938, pp. 141-42.

2. For epigraphic evidence, see *Ep. Ind.*, Vol. XXXVII, pp. 25ff.

between the river Padmā and Vardhamāna. Thus the Gauḍa country is placed exactly about the Murshidabad District bounded in the north by the Padmā and in the south by the Burdwan or Vardhamāna District. The same authority regards Gauḍa as only one of the seven *deśas* forming Puṇḍra-deśa, viz. (1) Gauḍa, (2) Varendra (Malda-Rajshahi-Bogra region), (3) Nīviti, (4) Suhma-deśa (*i.e.*, Rāḍha), (5) Jhārīkhaṇḍa (Santal Parganas District) called *jāṅgala* apparently meaning 'jungly', (6) Varāhabhumi (Barābhūm in the Purulia District), and (7) Vardhamāna (Burdwan).[1] The localities forming the Gauḍa country according to the same tradition are : (1) Navadvīpa (in the Nadia District), (2) Śāntipura (in the Nadia District), (3) Maulapattana (Mollāi in the Hooghly District), and (4) Kaṇṭakapattana (Katwa in the Burdwan District). The tradition would thus comprise in the Gauḍa country the present Murshidabad District together with parts of the Nadia, Burdwan and Hooghly Districts of West Bengal. It may be pointed out in this connection that the Puṇḍra country in this tradition comprised Western and Northern Bengal together with some eastern districts of Bihar. Nīviti is said to have included Bardhankoṭ in the Rangpur District. It is sometimes believed that the same name is found in the *Trikāṇḍaśeṣa*. But the *Trikāṇḍaśeṣa* passage *Puṇḍrāḥ syur=Varendrī Gauḍa-nīvṛti* seems to mean that the Puṇḍra country was the same as Varendrī lying in the realm (*nīvṛt*) of Gauḍa. The name Gauḍa is here used in a wider sense. It is however not improbable that the *Bhaviṣya Purāṇa* tradition is based on a misunderstanding of the *Trikāṇḍaśeṣa* passage.

The late medieval work entitled *Ṣaṭpañcāśaddeśavibhāga* which is incorporated in the *Śaktisaṅgama Tantra* describes the Gauḍa country as lying between the country of Vaṅga and Bhuvaneśa, *i.e.* Bhubaneswar in the Puri District of Orissa.[1] The same work describes Vaṅga as the land extending from the sea to the Brahmaputra.[2] The sea is apparently the Bay of Bengal to the south of Vaṅga; the Brahmaputra placed on the northern boundary of Vaṅga seems to indicate that portion of the river which bifurcates from the Jamuna. Thus the eastern half of Bengal has been called Vaṅga and its western half to-

1. See above, pp. 105-06.
2. Cf. above, p. 90.

gether with parts of Orissa has been designated Gauḍa in the work in question. This broad division of the Bengali-speaking area into two halves, viz. Eastern Bengal called Vaṅga and Western Bengal called Gauḍa, is echoed by some Muslim writers of the Mughul period, who speak of the country as Gaur-Bangāla or Gaur-wa-Bangāla, i.e. Gauḍa-Vaṅga.[1] Vaṅgāla (Bangāl) was originally the name of the Buckergunje region of the Vaṅga country ; but later the name came to be applied to the whole of Vaṅga or East Bengal and still later to the whole of the Bengali-speaking area. At present, Gauḍa, Vaṅga and Vaṅgāla are indiscriminately used to indicate the wide area of East India where the Bengali language is spoken. The major part of the country lies in West Bengal in the Indian Union and East Bengal in Pakistan ; but parts of it belong to the other neighbouring states like Bihar.

We have seen that the Chinese pilgrim Hiuen-tsang describes Karṇasuvarṇa or Gauḍa as a small tract. It is to be noticed that he distinguishes the above country from Puṇya-vardhana or Puṇḍravardhana in North Bengal, Samataṭa in South-East Bengal and Tāmralipti in South-West Bengal. In Indian literature also, Gauḍa is likewise separated from other tracts of Bengal. The Kauṭilīya *Arthaśāstra* (Chs. 32-33) mentions the textile products of Vaṅga and Puṇḍra and the silver of Gauḍa. Varāhamihira's *Bṛhatsaṁhitā* (XIV) mentions side by side Suhma, Samataṭa, Lauhitya (the Brahmaputra valley), Gauḍaka (*i.e.* Gauḍa), Pauṇḍra (*i.e.* Puṇḍravar-dhana), Tāmraliptika (*i.e.* Tamralipta) and Vardhamāna. There is reason to believe that the northern and south-eastern parts of Bengal were Aryanised considerably earlier than the western part of the country including Gauḍa. The mention of Gauḍa in the *Arthaśāstra*, which in its present form has to be assigned to about the third century A. D., points to the growing importance of the country in the economic life of Eastern India. In the fourth century A.D., the Gauḍa region became an integral part of the Gupta empire ; but with the decline of the imperial power of the Guptas, the Gauḍas established an independent monarchy in the sixth century. The

1. See *Numismatic Supplement*, No. XXXIV, referring to the *Humāyūn Nāma* of Gulbadan Begam. See also Firishta's history, Nawal Kishore Press ed., Vol. I, p. 21; Vol. II, p. 293. Firishta also speaks of Gaur as the capital of Bangāla. I am indebted to Dr. Z.A. Desai for the references.

importance of Gauḍa became established in the social and cultural life of India during the days of these independent monarchs of Gauḍa. We have already referred to king Śaśāṅka of Gauḍa, who flourished about the first quarter of the seventh century A.D. and ruled over extensive territories in Bengal, Bihar and Orissa. An earlier king of Karṇasuvarṇa was Jayanāga. Three other kings of Gauḍa appear to have been Dharmāditya, Gopacandra and Samācāradeva who ruled in the sixth century A.D. Under the vigorous rule of these monarchs, Gauḍa appears to have extended its power over the neighbouring territories. One of the above kings of Gauḍa fought with the Maukhari king Īśānavarman about the middle of the sixth century and another conquered and annexed Kāmarūpa to his empire about the close of that century.[1]

The Maukhari court-poet, while referring to the struggle between the Maukharis and the Gauḍas, speaks of the latter as having been compelled to be *samudr-āśraya*, *i.e.* a people whose shelter is the sea, owing to their defeat at the hands of the Maukhari king.[2] This shows that in the sixth century A.D. the Gauḍas had already become famous as a sea-faring people. That the Gauḍa country about this time produced many seafaring merchants is suggested by the discovery of an inscription[3] of *Mahānāvika* (*i.e.* Captain) Buddhagupta of Raktamṛttikā (near the Gauḍa capital) in the Wellesley District of the Malay Peninsula. The Dubi inscription of Bhāskaravarman of Kāmarūpa seems to refer to the Gauḍa army as specially strong in naval warfare.[4]

It may be mentioned in this connection that, besides Gauḍa in Bengal, some other tracts in different parts of India were also known by this name. One such Gauḍa was the modern Gonda District of Uttar Pradesh, in which the celebrated city of Śrāvasti was situated.[5] It is, however, possible

1. Cf. *Ep. Ind.*, Vol. XXX, pp. 293 ff.; also *Sel. Ins.*, 1965, pp. 530-31 for the annexation of parts of Orissa.

2. Cf. Sircar, *Stud Soc. Adm. Anc. Med. Ind.*, Vol. I, pp. 47-48.

3. Chatterjee and Chakravarti, *India and Java*, Part II, p. 7.

4. *Ep. Ind.*, Vol. XXX, pp. 293 ff.

5. Cf. *Kūrma Purāṇa*, I. 20. 19. For a controversy on this point, see *History of Bengal*, Dacca University, Vol. I, p. 579, note.

to think that Gauḍa as the name of Goṇḍā was a later modification of the older name of the area in imitation of the famous land of East India. The name of the Goṇḍ people of Central India was often Sanskritized as Gauḍa and the land inhabited by the Goṇḍs came also to be known by this name. Kannaḍa *Gauḍa* or *Gavuḍa* meaning 'a village headman' has nothing to do with Bengal. The same is the case with Oriya *Gaüḍa* meaning 'a milkman' and written as *Gokuṭa* in some inscriptions.

3. *Gauḍa as the Collective Name of the Eastern Countries of India*

About a century after the establishment of the powerful kingdom of Gauḍa, its name began to be used in a general sense to indicate the countries of Eastern India. In Daṇḍin's *Kāvyādarśa* (seventh century) one of the two principal styles of Sanskrit composition is described as Gauḍa or *prācya*, *i. e.* eastern.

Daṇḍin draws a distinction between the Gauḍa or eastern and the Vaidarbha or southern styles. Among the characteristics of the former is the love of long compounds not only in prose, in which they were also used by the southerners, but also in poetry. The easterners also favoured alliteration and harsh sound effects, recondite etymologizing phraseology and strength often resulting in bombast and affectation. Jacobi believed that Sanskrit poetry in the east had developed the evil effects of old age before the art became current in the western and southern parts of India. But it should be noticed that in Bharata's *Nāṭyaśāstra*, which is a few centuries earlier than the *Kāvyādarśa*, the qualities which Daṇḍin ascribes to the Vaidarbha style is assigned to the Kāvya style in general. Keith is no doubt right when he says, "This is a strong suggestion that at the time of the *Nāṭyaśāstra* there had not developed the characteristics of the Gauḍa style and that they emerged gradually with the development of poetry at the courts of princes of Bengal."[1] These 'princes of Bengal' appear to have been no other than the kings of Gauḍa who flourished in the sixth century and the earlier part of the seventh, to whom reference has been made above. But the fact that the Gauḍa style came to be ascribed to the whole of Eastern India may

1. *A History of Sanskrit Literature*, p. 60.

suggest that poets at other East Indian courts were imitating those at the Gauḍa court. This led ultimately to the application of the name Gauḍa to East India generally.

Like the literary style of Eastern India named after Gauḍa, the East Indian alphabet was named after the same country. Bühler says, "Towards the end of the 11th century, the Nāgarī inscriptions of Eastern India shew such distinct traces of changes leading up to the modern Bengali writing and these changes become so numerous in the 12th century that it is possible to class their alphabets as Proto-Bengali. Only a few among the Proto-Bengali letters are new local formations. The great majority occur already in other older scripts, be it in exactly the same or in similar shape."[1] Bendall and Bühler also noticed the influence of this script on the alphabet used in the Nepalese manuscripts of the period between the 12th and 15th centuries.[2] Now, in coining the name Proto-Bengali for the script in question, Bühler seems to have ignored several factors. In the first place, 'the eastern variety of the Nāgarī alphabet' found in 'the Nāgarī inscriptions of Eastern India', out of which Proto-Bengali developed in the twelfth century, should better have been distinguished from the 'Nāgarī' found in the inscriptions discovered in other parts of India by applying a more specific name to it. Secondly, when, even earlier than the eleventh century A.D.,[3] the same script was used in Bengal, Bihar, Assam and Orissa and when even now Bengal and Assam use practically the same alphabet while the Maithili and Oriya alphabets are nearly the same, the name Proto-Bengali can hardly be regarded as quite appropriate, in spite of the fact that, along with its literature, Bengal's alphabet has come to occupy a more important position. Thirdly, as early as the first half of the eleventh century, the same East Indian script was named after Gauḍa. Al-Bīrūnī, who wrote his work on India about 1030 A.D., speaks of the following alphabets: (1) Siddha-

1. *Ind. Ant.*, Vol. XXXIII, App., p. 58.

2. *Ibid.*, p. 60.

3. Ojha (*Bhāratīya Prācīn Lipimālā*, p. 77) has traced letters like *e* and *kh* of the Bengali type in the inscriptions of the time of Nārāyaṇapāla (*circa* 854-908 A. D.), that is to say, in records belonging to the ninth and tenth centuries. Bühler himself does not deny such facts.

mātṛkā used in the Kashmir, Banāras and Kanauj regions, (2) Nāgara used in Mālava, (3-5) Ardhanāgarī, Malwārī and Saindhava, used in Sindh, (6-9) Karṇāṭa, Āndhrī, Drāviḍī and Lārī used respectively in the Kannaḍa, Andhra, Drāviḍa and Lāṭa countries, (10) Gauḍī used in Pūrva-deśa, *i.e.* the eastern country, and (11) Bhaikṣukī which was the writing of the Buddha (*i.e.* the Buddhists) used in Udunpur (possibly Uddaṇḍapura, *i.e.* modern Biharsharif in the Patna District) in Pūrva-deśa.[1] This shows that the East Indian script, called Proto-Bengali by Bühler, was named after Gauḍa at least by the beginning of the eleventh century. Many of the names found in the list of sixty-four alphabets in the *Lalitavistara* (translated into Chinese in 308 A.D.) are apparently imaginary and doubtful; but the separate mention of the Aṅgalipi, Vaṅga-lipi, Magadha-lipi, Drāviḍa-lipi, Kanāri-lipi, Dakṣiṇa-lipi, Apara-Gauḍ-ādi-lipi, etc.,[2] seems to suggest that the tendency towards the growth of special characteristics in the alphabets of Southern and Eastern India was noticed even in an earlier age.

What has been said above would show that the name Gauḍa was specially applied to the literary style and script of Eastern India and that Gauḍa in these cases indicates the East Indian countries generally. In this connection, attention may be invited to the language of the *caryā* songs assigned by linguists to the period between the tenth and twelfth centuries A.D. They no doubt offer the earliest stage of the East Indian dialect just emerged from the Apabhraṁśa stage. But there is a great controversy among scholars as regards the closeness of the language of the *caryās* with the different dialects now spoken in Eastern India such as Bengali, Assamese, Oriya, Maithili and Eastern Hindi. Linguisticians whose mother-tongue is any of the above languages have tried to prove the special affinity of the *caryā* dialect with their mother-tongue. Thus H. P. Sastri, S. K. Chatterji, P. C. Bagchi, S. Sen and other Bengali scholars believe that the *caryās* are written in old Bengali. K. L. Barua and other Assamese writers regard the language of the *caryās* as old Assamese which they call Kāmarūpī. J. K. Mishra

1. Sachau, *Alberuni's India*, Vol. I, p. 173.
2. G. H. Ojha, *op. cit.*, p. 17, note 3.

and others from Mithila take the *caryā* language to be old Maithili. There are also similar claims on behalf of Oriya and Eastern Hindi.[1] We are however inclined to believe that the characteristics of the *caryā* language may be easily traced in all the languages now spoken in Eastern India because it was the language of Eastern India generally. The authors of the songs may be attributed to particular areas and their compositions may exhibit in some cases certain regional peculiarities; but it will hardly be correct for the matter of that to regard the language of the songs to be any one of those that later developed in the areas in question especially in view of the fact that the modern languages of Eastern India have a good deal in common amongst them. It is very probable that, in the tenth, eleventh and twelfth centuries A.D., the difference among these languages was not as remarkable as in modern times and that the literary language of the *caryā* was easily understood in different parts of Eastern India. The proper name for the *caryā* langnage should therefore be Gauḍī or East Indian and not old Bengali, Assamese, Maithili, Oriya or Eastern Hindi. We may then have a Gauḍa language besides a Gauḍa style of Sanskrit composition and a Gauḍa alphabet. East Indian linguists of particular areas may successfully trace affinities of their mother-tongue with the *caryā* language ; but they should better not regard it as the origin exclusively of their mother-tongue.[2]

4. *Gauḍa as the Name of Āryāvarta or Northern India.*

The South Indian Brāhmaṇas are usually divided into five sections, viz., Drāviḍa (Tamil), Karṇāṭa, Gurjara, Mahārāṣṭra and Tailaṅga. These are collectively called the Pañca-Drāviḍa.[3]

1. See S. Sen, *op. cit.*, pp. 45 ff.; K. L. Barua, *Early History of Kāmarūpa*, pp. 318 ff. ; J. K. Mishra, *History of Maithila Literature*, p. x ; etc.

2. The Gauḍī dialect (*vibhāṣā*) is recognised in Puruṣottama's *Prākṛtānuśāsana* (12th century). See Sircar, *Gram. Prak. Lang.*, p. 118.

3. See Apte, *Sanskrit-English Dictionary*, s. v. *Drāviḍa*. The *Śabdakalpadruma*, s. v., quotes a faulty stanza from the *Skanda Purāṇa* in support of the list of the five classes of Drāviḍa Brāhmaṇas :

कणाटाश्चेंव तेंलंगा (द्राविडा) गुजंरा राष्ट्रवासिनः।
आंघ्राश्च द्राविडः पंच विंघ्यदक्षिणवासिनः ॥

An inscription of 1425 A.D. speaks of four out of the five classes of Drāviḍa Brāhmaṇas as Kannaḍiga, Tamiḷa, Teluṅga and Iāḷa (*i.e.* Lāṭa=Gurjara).[1] Now these social groups of the Brāhmaṇas were developed out of the principal linguistic sub-divisions of the people of South India. On the analogy of the above divisions, the North Indian Brāhmaṇas were also divided into five groups under the general name of Gauḍa. The *Śabdakalpadruma*, s. v. *Gauḍa*, quotes the following verse from the *Skanda Purāṇa* :

सारस्वताः कान्यकुब्जा गौडमैथिलिकोत्कलाः ।
पंचगौडा इति ख्याता विंध्यस्योत्तरवासिनः ॥

The five classes of the Gauḍa or North Indian Brāhmaṇas were thus the Sārasvata (associated with the Sarasvatī valley in the Eastern Punjab), Kānyakubja, Gauḍa, Maithila and Utkala.[2] Although this seems to be a rather arbitrary classification, there is no doubt that the name Gauḍa has been applied in this case to North India generally. That the said classification is not very late is suggested by the mention of the Pañca-Gauḍīya community in an inscription of 926 A.D.[3]

The application of the name Gauḍa in the general sense of Āryāvarta or North India can also be traced elsewhere in literature. There is a tradition regarding king Bhoja (*circa* 1000-55 A.D.) of the Paramāra dynasty of Malwa recorded in the following verse :

1. *South Indian Inscriptions*, Vol. I, pp. 82-84.

2. Wilson's *Glossary of Judicial and Revenue Terms* gives an account of the Gauḍa Brāhmaṇas under the entries *Brāhmaṇ* and *Gauḍ*. Under *Gaur-Brāhmaṇ*, Wilson says, "The Brāhmaṇ of the Gaur tribe or caste ; one of the five Gauṛs, but located in the upper provinces throughout the Subah of Delhi to the hills. There are many sub-divisions of these Gauṛ-Brāhmaṇs of Hindustān, who are apparently unknown in Bengal, as the Ādh-Gauṛ, Kaithal-Gauṛ, Gūjar-Gauṛ, Sidh-Gauṛ, etc., amounting in all to forty-two." He also speaks of the Gauṛ Kāyath said to be settled from Bengal in the upper provinces by Nāṣir-ud-dīn (son of Balban) in the thirteenth century, Gauṛ-Rājpūt numerous in the North-Western Province (U. P.), Gauṛ-Tagā a tribe of Brāhmaṇical descent in the north-west of India, and Gauṛ-Ṭhākur a tribe of Rājpūts settled in the Farrukhabad District. The Gauṛ-Tagās claim that they were originally invited from Bengal by Rājā Janamejaya, the Kaurava king of Hastināpura, for the purpose of exterminating the Takṣakas or snakes.

3. Cf. *Ep. Ind.*, Vol. XXXII, p. 48. See also the *Rājataraṅgiṇī* (*circa* 1150 A. D.), IV. 468.

पंचाशत्पंचवर्षाणि सप्तमासदिनत्रयम् ।
भोजराजेन भोक्तव्यः सगौडो दक्षिणापथः ॥[1]

According to this tradition, Bhoja ruled over both Gauḍa and Dakṣiṇāpatha for a little over 55 years. The real implication of the verse has so long been misunderstood by scholars. There is no doubt that it speaks of Bhoja's lordship over Gauḍa in the sense of North India and over Dakṣiṇāpatha or South India; that is to say, Bhoja claimed to have ruled over the whole of India. including both Northern and Southern India. This was merely the conventional way of saying that Bhoja was a *cakravartin*, *i.e.* an imperial ruler. The Indian *cakravartins* were conventionally represented as the ruler of the 'whole earth' which only meant the *cakravarti-kṣetra* bounded by the Himalayas and the sea. Sometimes Āryāvarta or North India was conceived as an independent *cakravarti-kṣetra* for North Indian rulers and South India bounded by the three seas (the Indian Ocean, the Bay of Bengal and the Arabian Sea) for the rulers of the Deccan.[2] In the verse quoted above, Bhoja is represented as the lord of both the *cakravarti-kṣetras* of the north and the south of India. Like similar claims of other monarchs, the claim on Bhoja's behalf does not mean that the Paramāra king actually ruled over any part of South India or over Gauḍa in Bengal in the narrow sense of the term. Here Bhoja merely claims to have been a *cakravartin* which means nothing more than an imperial ruler of any part of India.

1. *Bhojaprabandha*, Calcutta ed., p. 3; cf. Ray, *DHNI*, Vol. II, p. 858.

2. For the conventional *cakravarti-kṣetras*, see *JRASB*, Letters, Vol. V, 1939, pp. 497 ff. See also above, pp. 15-16.

Chapter VII

VAṄGA AND VAṄGĀLA

I

European (mainly Portuguese) writers of the 16th and 17th centuries place a certain City of Bengala near the Meghnā estuary, *i.e.* the joint mouth of the Padmā and the Meghnā. This estuary extends over the wide area between the Districts of Buckergunge and Chittagong in East Pakistan. As Bengala (like the modern name *Bengal*) is a foreign corruption of *Vaṅgāla*, a celebrated historian has suggested[1] that this late medieval City of Bengala (which he locates near modern Chittagong) was the capital of the ancient Vaṅgāla-deśa and 'gave its name to the kingdom, or *vice versa*, and, in either case, the old kingdom of Vaṅgāla must be located in the region round the city'.[2] He also suggests that the celebrated Pāla and Candra dynasties of Bengal originally ruled in the Chittagong region.[3]

The above theories appear to be unwarranted. The City of Bengala, mentioned by foreign travellers in the late medieval period,[4] seems to have nothing to do with the early medieval kingdom of Vaṅgāla, which originally denoted a smaller area, but whose geographical sense gradually expanded so as to include ultimately the whole of the land of the Bengali-speaking people.

In this connection, we have to trace the gradual expansion of the geographical connotation of the name Vaṅgāla. The

1. See R. C. Majumdar in *IHQ*, Vol. XVI, pp. 227-35.

2. *Ibid.*, p. 229 ; cf. p. 232, : "......this original kingdom must be in this region [of Chittagong and Dianga]."

3. *Ibid.*, p. 235.

4. On the strength of Ibn Batūta's reference to 'Sudkawan (Sātgāon) and Bengala' and to 'Lakhnautī and Bengala', it has been suggested that in these two early cases Bengala refers to the city and not to the country. There can be no doubt however that the names refer to provinces. According to Muslim authors both Rāl (Rāḍha) and Barind (Varendra) formed parts of Lakhnautī (Raychaudhuri, *Stud. Ind. Ant.* p. 191). Similarly Bengala no doubt means the province of South-Eastern Bengal, sometimes also called Sonārgāon after its chief city. Orissa was usually called Jājnagar by Muslim authors. Such reference to Bihār, Lakhnautī, Bang and Jājnagar (*IHQ*, *op. cit.*, p. 235) would, if we follow the said line of argument, suggest the existence also of a city called Bang. But this was never the case. Some later Muslim writers refer to East Bengal as *Bang* and to the whole of Bengal as *Bangāl*.

earliest reference to Vaṅgāla has been traced in the Nesarī plates (805 A.D.) of Rāṣṭrakūṭa Govinda III, which speak of Dharma (*i.e.*, Dharmapāla) as the king of Vaṅgāla.[1] This shows that the name Vaṅgāla was not unknown in earlier times and that the Pālas originally ruled in the tract called by that name. It is interesting to note that Dharmapāla is apparently called 'lord of Vaṅga' in the Sāgar Tāl inscription of Bhoja and 'king of Gauḍa' in the Sanjan plates of Amoghavarṣa. With the expansion of their territories, the Pālas preferred to style themselves as 'lords of Gauḍa'. This seems to be the reason why Vaṅgāla did not become popular in the sense of the Pāla empire.

The name is often found in records and works of the period later than the 10th century A.D. Reference to Vaṅgāla is found in the Tirumalai inscription (1025 A.D.) of Rājendracoḷa,[2] which speaks of king Govindacandra as the lord of Vaṅgāla-deśa. The Ablur inscription[3] of Kalacurya Vijjala (1157-67 A.D.) mentions Vaṅga and Vaṅgāla separately. This differentiation seems to be supported by a work called *Ḍākārṇava* which makes separate mention of Vaṅgāla and Harikela (Vaṅga, according to the *Abhidhānacintāmaṇi* of the 12th century lexicographer Hemacandra),[4] and by the *Hammīramahākāvya* of Nayacandra Sūri (15th century), which mentions Vaṅga and Vaṅgāla side by side.[5] In the 16th century, however, Abul Fazl says, "The original name of Bangāl (Vaṅgāla) was Bang (Vaṅga). Its former rulers raised mounds measuring ten yards in height and twenty in breadth, throughout the province, which were called *āl*. From this suffix the name (*i.e.* Bangāl) took its rise and currency."[6] This identification of Vaṅga and Vaṅgāla has to be reconciled with the evidence referring to them as two different countries.

The *Raghuvaṁśa* reference to the defeat of the Vaṅga people in the land watered by the lower streams of the Ganges[7] and

1. *JAS*, Letters, Vol. XXII, 1956, pp. 133-34.
2. *Ep. Ind.*, Vol. IX. p. 229 ff.
3. *Ibid.*, Vol. V, p. 257.
4. Raychaudhuri, *op. cit.*, p. 189.
5. *IHQ*, *op. cit.*, p. 237.
6. Jarrett, '*Ani-i-Akbari*, trans., Vol. II, p. 120.
7. वंगानुत्खाय तरसा नेता नौसाधनोद्यतान् ।
निचखान जयस्तंभान्गंगास्रोतोन्तरेषु सः ॥ (*Raghuvaṁśa*, IV. 36)

epigraphic references to Vaṅga comprising the Vikrampur region of Dacca and Faridpur[1] and to the *Nāvya* region of Vaṅga very probably in the Faridpur and Buckergunge Districts,[2] leave hardly any doubt that Vaṅga certainly included at least parts of the present Dacca, Faridpur and Buckergunge Districts. In the lower part of this region very high roads or earthen embankments are constructed even today in order to prevent the tide of floods and to facilitate communications during the rainy season. It may therefore be suggested that the southern part of old Vaṅga thus at first came to be known as Vaṅgāla. It is interesting to note that this view is supported by other evidences.

As said above, the Tirumalai inscription refers to Govindacandra as the lord of Vaṅgāla-deśa. The inscriptions of the Candra dynasty of East Bengal, to which Govindacandra belonged, however, say that Trailokyacandra, the first king of the family, became lord of Candra-dvīpa and flourished as the mainstay of the fortune (*i.e.* as an ally) of the king of Harikela (Vaṅga according to Hemacandra's *Abhidhānacintāmaṇi*).[3] Thus the Candra kingdom is called Candra-dvīpa according to one evidence and Vaṅgāla-deśa according to another. Candra-dvīpa and Vaṅgāla-deśa thus appear to be more or less identical. As Candra-dvīpa is no other than the celebrated Bāklā-Candra-dvīp (*i.e.* parts of the Buckergunge District and the adjoining region), the Buckergunge area was apparently included in Vaṅgāla-deśa.

Mallinātha explains *Gaṅgā-sroto-'ntareṣu* as *Gaṅgāyāḥ srotasāṁ pravāhānām=antareṣu dvīpeṣu*; cf. *gīt-āntareṣu*, 'in the intervals of singing', *vāṣpa-salil-āntareṣu*, 'in the interval of weeping'; *na mṛṇāla-sūtraṁ racitaṁ stan-āntare*, '......in the space between the breasts',etc. See below, Chs. X and XI.

1. Cf. *Vaṅge Vikramapura-bhāge*, etc. (*Ins. Beng.*, Vol. III, p. 125).

2. Cf. *Vaṅge Nāvye Rāmasiddhi-pāṭake*, etc., *Nāvye Vinayatilaka-grāme pūrve samudraḥ sīmā*, etc. (*ibid.*, p. 146). *Nāvya* means a region accessible by boats. That this region was then not far from the sea (*samudra*), *i.e.* the estuary, is also clear. The word means apparently the same thing as Bengali *bhāṭi*, 'with the tide, the region towards the sea' (from *bhāṭā*, floods of the tide running towards the sea). In the Muslim period, the district called Bhāṭi sometimes indicated the coastal strip between the estuaries of the Hooghly and the Meghnā or, according to some, the valleys of the Brahmaputra and Meghna and their tributaries. The late work called *Śaktisaṅgama Tantra* places Vaṅga-deśa between the *ratnākara* (sea) and the Brahmaputra (above, p. 91). Cf. *N.K. Bhatt Com. Vol.*, 322.

3. *Ind. Cult.*, Vol. VII, p. 411. According to Keśava's *Kalpadrukoṣa*, Harikela or Harikeli was the name of the Sylhet (Śrīhaṭṭa) region. Cf.

The suggestion is further supported by the identification of Vaṅgālavaḍā-bhū in the Rāmasiddhi-pāṭaka of the *Nāvya* region of Vaṅga, mentioned in the Sāhitya Pariṣad plate[1] of Viśvarūpasena, with modern Bāṅgroḍā in the region of Rāmsiddhi under the Gaurnadī Police Station of the Buckergunge District. Vaṅgāla thus appears to have been popularly mentioned as a separate political unit, after the Candras had established a kingdom in Vaṅgāla, *i.e.* the southern part of Vaṅga. Śrīcandra, son of Trailokyacandra, ruled over the Dacca, Faridpur and Sylhet Districts, and apparently also over the Tipperah District. Thus Vaṅga proper was now included in the new kingdom of Vaṅgāla. As a result of this, the name Vaṅgāla could be optionally used in an expanded sense to indicate a large area in East Bengal that formed part of the kingdom of Śrīcandra and his successors. By this time, Vaṅga and Vaṅgāla apparently signified more or less the same territory.

Early Muslim authors refer to the Muhammadan kingdom of Bengal first as Lakhnautī or Gaur, as their first settlement did not include East Bengal. The kingdom of East Bengal was sometimes referred to as Bang, Bangāl or Sonārgāon ; and later sometimes the whole of Bengal was mentioned as Gaur-Bangāl.[2] Sātgaon was often recognised as a separate political unit. But, as early as the reign of Ghiyāsuddīn Tughluq Shāh (1320-25 A.D.), some Muslim authors are known to speak of Bangāl in a wide sense so as to include Lakhnautī, Sonārgāon and Sātgāon.[3] In the 16th century, the name Bangāl loosely indicated a wider area. It may or may not have geographically included the Chittagong region which was included in the Mughul Subah of Bangāl only during the reign of Aurangzīb in the 17th century. There is no proof at all

श्रीहट्टो हरिकेलिः स्याच्छ्रीहटोपि क्वचिद् भवेत् ।

This name came to be applied to Vaṅga apparently due to the temporary annexation of Vaṅga to the Harikela kingdom. Cf. *Ep. Ind.*, Vol. XXVIII, pp. 338-39.

1. N. G. Majumdar, *Ins. Beng.*, Vol. III, pp. 840 ff.

2. Cf. *Num. Suppl.*, No. XXXIV, pp. 200-01. The *Śaktisaṅgama Tantra* divides Bengal into two halves, viz. Vaṅga and Gauḍa (lying between Vaṅga-deśa and Bhuvaneśa or Bhubaneswar). See above, pp. 91, 105, 123 note. The *Mānasollāsa* (I.2.175) speaks of Gauḍa-Vaṅgāla.

3. Raychaudhuri, *op. cit.*, p. 191 ; Raverty, *Tabaqāt-i-Nāsiri*, p. 590 n.

that the Chittagong area formed part of the old kingdom of the Candras and the Pālas. It must be noted that not a single inscription of those dynasties has so far been discovered in the Chittagong District.[1]

We now come to the City of Bengala mentioned by the European (mainly Portuguese) writers of the 16th and 17th centuries A.D. Duarte de Barbosa, Barthema (1510 A. D.), Ovington (1639 A.D.) and several other writers refer to the city. Some authorities identify the city with modern Chittagong or locate it near about that place. It is shown in a map drawn by Gastaldi in 1651 and in another in the *Travels of Cornelius le Bruyan* (published in 1701). In these maps and works as well as in those of Blaev, Sausson, Purchas and others, the City of Bengala is indicated in the Chittagong region.[2] But Barbosa, one of the earliest Portuguese writers on Indian geography,

Muslim authors popularised the use of *Bangāl* in preference to other names to denote the whole of Bengal. A Nepal inscription of 1346 A.D. refers to the army of Sulṭān Shamsuddīn Ilyas as *Vaṅgāla-bahula bala* (*JBORS*, Vol. XXII, p. 81; *IHQ*, *op. cit.*, p. 227).

1. The Arab merchant Sulaiman (851 A. D.) refers to the powerful kingdom of Rahma which has been identified with the kingdom of the Pālas. See Section II below. But the identification of Rahma, Rahmī and Ruhmī of the Arabs with Rāmu, a few miles east of Cox's Bazar (*IHQ*, *op. cit.*, pp. 232-34) is wrong. Apart from the fact that there is absolutely no proof of Pāla occupation of Ramu, it seems unconvincing that the Bihar-Bengal empire of the Pālas (even if it included Rāmu), with the emperors often residing at *jayaskandhāvāras* in Bihar, would be named after Rāmu which then probably formed a part of Arakan. Marco Polo who placed Bangala 'tolerably close to India' and under the rule of the king of Mien (Burma), and Fakir Muhammad who placed Bangāl to the east of Bhāṭi appear to have made some confusion. The Bengali ballad called *Māṇikcandra-rājār Gān*, which says *Bhāṭi haite āila Vāṅgāl lambā lambā dāḍi*, 'the Vāṅgālas (people of Vaṅgāla proper) with their long beards came from Bhāṭi (the southern country)'. proves that the people of East Bengal, who were responsible for the original composition of the ballad, knew Vaṅgāla to be identical with Bhāṭi (see C.C. Banerji, *Caṇḍīmaṅgala-bodhinī*, Vol. II, p. 765). Mukundarāma (16th century) in his *Caṇḍīmaṅgala* (C.U. ed., p. 655) speaks of the Vāṅgāls as good sailors. As the Chittagong people are reputed sailors, it may be suggested that Mukundarāma identifies the land of the Vāṅgāls with Chittagong. But Mukundarāma was a man of West Bengal and to the people of his part of the country, the inhabitants of any District of East Bengal are Vāṅgāls even today. Moreover, the passages in question are certainly interpolated in the *Caṇḍīmaṅgala*. The *Māṇikcandra-rājār Gān*, an East Bengal production, is therefore more important in this connection. Therein we see that the people of Bhāṭi were known even to the people of East Bengal as Vāṅgāls, i. e. inhabitants of Vaṅgāla.

2. Raychaudhuri, *op. cit.*, p. 189; *IHQ*, *op. cit.*, p. 229n.

says that the Bay of Bengal is 'a gulf which enters towards the north and at its inner extremity there is a great city inhabited by Moors (Muhammadans), which is called Bengala, with a very good harbour.'[1] The expression 'at the inner extremity' may point to a locality far to the north of Chittagong, and Moreland suggests that the Portguese meant Sonārgāon by the City of Bengala.[2] But the city apparently belonged to what the Portuguese called the Porto Grande=Portus Mangnus=Great Port, and we have to determine the location of the latter.

Portuguese writers, when they speak of a voyage to Bengal, usually refer to the Great Port as well as the Porto Pequeno= Portus Parvus=Little Port. Portguese *porto* is usually understood in the sense of English *port*, and the Little Port is identified with Hooghly or Sātgāon while the Great Port is located near Chittagong. Moreland has however shown that 'the word *porto* in the mouth of the sea-faring Portuguese at the period referred primarily to a gap in the coast line and not, as landsmen are apt to assume, a town on the sea-shore; in other words, *porto* might signify a gulf or estuary which might contain several sea ports.'[3] He has pointed out that sometimes the Little Port indicated the Hooghly estuary, and the Great Port meant the Meghnā estuary and included both Chittagong and Śrīpur (in the Dacca District). The location of the Great Port therefore does not help us materially in locating the City of Bengala.

1. *IHQ, op. cit.*, pp. 229-30.

2. *India at the Death of Akbar*, p. 309.

3. *Ibid.*, 307-08. Father F. Fernandus, the Jesuit missionary who was sent to Bengal in 1598, speaks of the danger when his ship ran aground in the Portus Parvus; but it was afloat and, after sailing for eight days within the Portus, it reached the Portuguese station of Hooghly. By *Potrus Parvus* therefore Fernandus meant the Hooghly river and not any port. He next went to the Portus Magnus and first reached Śrīpur which he describes as a station in the Portus Magnus. Śrīpur was an important port on the Padmā about 18 miles from Sonāragāon. the eastern capital of Bengal. The Father then arrived at Chittagong which was also a station belonging to the Portus Magnus. According to Fernandus therefore the Great Port extended from the mouth of the Karṇaphuli river to the immediate neighbourhood of Dacca. By the word *porto* some writers however actually meant a port; Fitch, *e. g.*, identifies Porto Pequeno with Sātgāon. According to Moreland, the change in the meaning may be attributed to the fact that 'the Portuguese did not to any great extent trade directly with Śrīpur ; their communications were with either Hooghly or Chittagong, that is, with only one *Station* in each *Porto*, and, in these circumstances, the transfer of the name from the *Porto* to the *Station* might easily take place, just as the *Mersey* has become a synonym for Liverpool' (*op. cit.*, p. 309).

The chief points of interest in this problem are two. Firstly, there is not a single reference to this City of Bengala, so famous to the Portuguese, in the whole range of medieval Bengali literature, not even in the Bengali history of the Tripurā royal family which often mentions Chittagong. The Bengalis therefore did not know any place of that name, or knew it by a different name, *i.e.* a name of their own. Secondly, while some European writers like Rennell could not trace the city or its site, others denied the very existence of the city called Bengala. In 1689, Ovington remarked, "A late French geographer (Baudrand) has put Bengala into his Catalogue of imaginary cities, and such as have no real existence in the world."[1] It therefore seems quite clear that even Europeans other than the Portuguese sometimes considered the name of the City of Bengala as a misnomer and had no memory of it after the collapse of the Portuguese power in Bengal. The City of Bengala thus appears to be a name given by the Portuguese to a city which was usually known to others by its Bengali name. As however there is absolutely no proof (in the Bengali literature and Bengal records of any period) of the existence of any city called Vaṅgāla in the south-eastern part of Bengal, the Portuguese name 'City of Bengala' seems to have originally indicated the city *par excellence* (*i.e.* the chief city) of the country of Vaṅgāla, *i.e.* East Bengal. Names like Andhrapura, Magadhapura or Māgadhapura, etc., are known to have been alternative names of the chief cities derived from the names of countries, and we have seen that the south-eastern, north-western and south-western provinces of Bengal were also known to the Muslim writers as Sonārgāon (Vaṅgāla), Lakhnautī (Gauḍa) and Sātgāon (Rāḍha) respectively after the chief cities of those provinces. As the chief city of East Bengal lay not far from the Meghnā estuary,[2] it might have been called the City of Bengala by the Portuguese. But appa-

1. *Bengal Past and Present*, Vol. XIII, p. 262 ; *IHQ*, *op. cit.*, p. 230n.

2. Sonārgāon was the capital of East Bengal during early Muslim days. Dacca was made the capital of the Bengal Subah early in the 17th century. The remark of Purchas that 'Gauro (Gaur or Lakhnautī), the seat royal, and Bengala are fair cities' (*IHQ op. cit.* p. 230) apparently refers to the fact that Bengala was the chief city of East Bengal and was looked upon as a secondary capital of Bengal.

rently that name was used by foreigners (chiefly Portuguese), and the Bengalis continued to use the local or Bengali name. That is why the name could not be traced after the Portuguese collapse in Bengal.[1]

It must however be admitted that some authorities locate the City of Bengala near about Chittagong.[2] It seems that originally the chief city of East Bengal was known to the Portuguese as the City of Bengala ; but after the Portuguese port in the Chittagong region had become the most flourishing centre of trade in Bengal, they began to call it the City of Bengala. This again seems to have been done when Sonārgāon passed out of the picture due to the transference of the provincial capital first to Rājmahal and then to Dacca. The fact that Portuguese sailors came by sea and first reached this flourishing port after days of hardship in the sea may also have emphasized its importance to them as the city *par excellence* of the country of Bengal.

Whatever, however, the location of the City of Bengala may be, there is no doubt that it had nothing to do with the old country of Vaṅgāla known to have been the kingdom of the Candra kings of East Bengal.[3]

There has been some discussion on the derivation of the

1. Quite different is the case with Meghnā (pronounced *Magnā* in East Bengal, with the first vowel sounding like *a* in *man*) apparently derived from the Latin word meaning 'great', which has supplanted the older name of the river.

2. It has been pointed out that Ortelius places Bengala in the same place where Hommanus places Chatigam or Chittagong. Yule refers to a chart of 1743 and concludes that Chittagong is probably identical with the City of Bengala. Campos also identifies Chittagong and Bengala. Ovington places Bengala to the south of Chittagong. In Blaev's map and Sausson's chart, the city is located on the southern bank of the Karṇaphuli near about the place where Broucke places Dianga. This location is supported by Vignorla's map of 1683. But in an old map in Thevenot, the city is placed above Xatigan or Chittagong. Hosten and Majumdar think that Dianga, opposite Chittagong, represents the site of Bengala. It is also suggested that originally both Dianga and Chittagong were included in the city. Hosten identifies Dianga with a place now called Bandar on the left bank and near the mouth of the Karṇaphuli river. He also takes Dianga to have been the Porto Grande. Blochman identifies Dianga with Dakhin-ḍāṅgā or Brāhmaṇ-ḍāṇgā both on the Sāṅgu river to the south of Chittagong (*IHQ*, *op. cit.*, pp. 231-32 and notes). Blochman may be right, as the foreign name seems to correspond to the Bengali word *dāṅgā* which is pronounced in East Bengal as *ḍaṅgā*, with the first vowel sounding like the English *a* in *man*.

3. See also S. H. Hodivala in *Num. Supl.*, No. XXXIV (*JASB*), pp. 199-211. Cf. *Ind. Cult.*, Vol VII, pp. 407-08.

name *Bengal.*[1] But the conclusions are not satisfactory. Since the emergence of modern Hindi from its Apabhraṁśa stage, the speakers of Hindi and other languages have been calling the Beṅgali-speaking area by the name *Vaṅgāl* (without the final *a* in the last consonant), which has been transliterated by the English in their script as *Bengal.* The Muslims first came to India when the final *a* in the last consonant of Sanskritic words in North Indian dialects was pronounced. The earlier pronunciation of *Vaṅgāl,* as is well known, was *Vaṅgāla* which was naturally transliterated by the Muslims in their script as *Baṅgālah* (pronouncing *Baṅgāla*). This Muslim pronunciation of the name is directly responsible for the name *Bāṅgālā* now applied by the Bengalis to their country. The first *ā* in the name is comparable to the first *ā* in Bengali words like *hājār* (thousand) derived from Persian *hazār.* The change of *a* into *ā* is due to the fact that *a* pronounced outside East India resembles *ā* as pronounced by the Bengalis. The last *ā* in the name is again comparable to Bengali words like *khānā* derived from *khānah.*[2] The people of Bengal is known elsewhere in India as *Vaṅgālī* which the English transliterated as *Bengali* or *Bengalee* and which the Bengalis have made *Bāṅgālī* in their language.. This word is also influenced by such Muslim words as *Kābulī.*[3] It is interesting to note in this connection that, according to a general ethnological principle, the specific name of a tribe often originates among neighbouring tribes and is eventually adopted by the tribe to which it is applied.

1. See N. N. Das Gupta in *IHQ*, Vol. XXII, pp. 277-85.

2. Cf. also *bandah* ((slave) = Bengali *bāndā*, *khajānah* (revenue) = Bengali *khājānā*, etc. Just as the word orginally spelt and pronounced in Bengali as *khājānā* has now been modified to *khājnā* (with the elision of the central *ā*), the earlier Bengali spelling and pronounciation of the name *Bāṅgālā* have now given way to the modified form *Bāṅglā* (sometimes softened into *Bāṅlā*). That, during Muslim rule, the Bengalis adopted the Muslim pronunciation of the name of their country is probably indicated by the fact that today Bengalis of any education speak, in their common speech, usually of the English *India*, *Bengal* and *Calcutta* in preference respectively to the names *Bhāratavarṣa*, *Bāṅgālā* (or *Bāṅglā* or *Bāṅlā*) and *Kalikātā*. Cf. also the popularity of the Anglicised surnames Banerji, Mukherji, etc. The introduction of a large number of Persian and Arabic words in the vocabulary during Muslim rule has further to be compared with the later adoption of numerous English words during the British period.

3. Cf. the *yā-yi nisbati* in Persian and the corresponding Arabic suffix to form relatives (D.C. Phillott, *Higher Persian Grammar*, pp. 400 and 714-17).

To sum up the discussion, the Candra kings of South-East Bengal are sometimes represented as lords of Candra-dvīpa (Bāklā-Candradvīp in the Buckergunge region) and sometimes as lords of Vaṅgāla-deśa. This fact points to the location of Vaṅgāla in its original geographical sense in the coastal region of Southern Bengal. The connotation of the name began to expand with the expansion of the Candra kingdom of Vaṅgāla over wide areas of South-East Bengal. The Nesarī plates of 805 A.D. suggest that the name Vaṅgāla (apparently indicating the original territory of the Pālas in the Buckergunge region) was known even in earlier times, though it does not appear to have been popular in the sense of the eastern part or the whole of Bengal before the rise of the Candras. The name was popularised in the sense first of East Bengal, and then in that of the whole of Bengal, by the Muslims. The location of Vaṅgāla proper in the Buckergunge region near the Bay of Bengal, that is to say, in the southern part of the ancient Vaṅga country, seems to be supported by Abul Fazl's identification of Vaṅga and Vaṅgāla and his derivation of the latter name from *Vaṅga+āl* (Sanskrit *āli*) on the grounds that 'its former rulers raised mounds measuring ten yards in height and twenty in breadth throughout the province, which were called *āl*.' Although the derivation may be wrong and *Vaṅgāla* may have sprung from *Vaṅga*+ the Prakrit suffix *āla* in the sense of a notable district belonging to Vaṅga,[1] we have seen that the interesting mention of the earthern mounds, primarily meant for keeping off the encroachment of sea-water from the corn-fields, refers to a condition prevailing in the Buckergunge region of the coastal area of Bengal even to-day. The name Vaṅgāla is very rarely found in records earlier than the rise of the Candras in the tenth century A.D. The mention of the name in the Goharwa inscription of Karṇa (1041-71 A.D.) in the description of the exploits of his predecessor Lakṣmaṇarāja proves its importance in the eleventh century. Of course Lakṣmaṇarāja, who is represented in the Goharwa inscription as having defeated the Vaṅgāla country or people, ruled about the middle of the tenth century when the Candra power had already been established in Vaṅgāla.

1. Cf. Sircar, *Grammar of the Prakrit Language*, p. 33.

As we have seen, the most important point in regard to the supposed existence of a 'city of Bengala' is that no such name of a great city is found in any of the numerous works of medieval Bengali literature. This question, usually ignored, has to be answered. Our answer to this question fully tallies with that of Hodivala who reached the same conclusion after throughly ransacking the available Muslim sources. In an attempt to locate the Mughul mint-town of Bangāla in Akbar's empire, he says, 'Briefly, there would appear to be *fairly good* grounds for thinking that *Bangāla* was not the real or fixed name of any town or city, but an alternative or honorific designation by which the *capital of the province at the time being* was known. Thus the *Bangāla* of Mun'im Khān's time *might have been* Gaur, and it is *not impossible* that during the subsequent twenty years the name was sometimes applied to Tānda. The *Bangāla* of the coins of the 39th and following years of Akbar's reign would, by parity of reasoning, be Akbarnagar."[1]

II

Dharmapāla (c. 770-810 A.D.), the second emperor of the Pāla dynasty, was one of the mightiest Indian monarchs of his time. It is well known that the two other great Indian powers in Dharmapāla's age were the Rāṣṭrakūṭas of the Deccan and the Gurjara-Pratīhāras who originally ruled over the Rājasthān region but transferred their capital to Kanauj before 833 A.D. and that Dharmapāla had to fight with both of them. Dharmapāla consolidated Pāla hold on Bengal and Bihar and extended his political influence over Uttar Pradesh for a short time. His relations with Orissa are unknown, although his son and successor, Devapāla (c. 810-50 A.D.), is vaguely credited with having made the *Utkala-kula* (i. e. the people

1. See Hodivala, *op. cit.*, p. 211. It will be seen that the mint-town of Bangāla, whence Akbar's coins were issued, could not have been situated in the Chittagong region (where the 'city of Bengala' is usually located), as that area lay outside the Mughul empire during Akbar's rule. See also A. Cortesao in *JRASB*, Letters, Vol. XI, pp. 10-14; cf. A.B.M. Habibullah, *ibid.*, pp. 33 ff. See Sircar, *Stud. Ind. Coins*, pp. 315 ff.

or royal family of Utkala or Orissa) *utkīlita*.[1] The meaning of the word *utkīlita* is 'unfastened' and it may be that Devapāla claimed to have saved the people or royal family of Orissa from foreign aggression. There is another statement in a Pāla record that the king of Utkala fled from his capital at the approach of Jayapāla, the general and cousin of Devapāla.[2]

The account of India written by the Arab merchant Sulaimān in 851 A.D. (completed by Abū Zayd in 916 A.D), well known from the translation of parts of Sulaimān's work in the *History of India as told by its own Historians*, Vol. I, by Elliot and Dowson, speaks of three important and rival Indian powers, viz. Juzr, Balharā and Ruhmī.[3] As will be seen below, information about these three important powers is also found in the works of some other early Arabic writers as well as in the Persian work *Hudūd-al-'Alam* written in 982 A.D., although each of the three names have variants. The identification of Juzr and Balharā respectively with Sanskrit Gurjara (i.e. the Gurjara-Pratīhāra king) and Vallabharāja (i.e. the Rāṣṭrakūṭa king) is quite obvious, while most scholars believe that the third name refers to the Pāla king of Eastern India.[4] But the significance of Sulaimān's *Ruhmī* could not be determined.

Many writers have located the kingdom of Ruhmī in Bengal ; but a few of them placed it outside. Thus Yule had once identified Ruhmī with Rahmaniya or Pegu, though he later preferred Rāmu near Cox's Bazar in the Chittagong District, now in East Pakistan.[5] Some other authors associated the place with Mrohaung or Myohaung in the Akyab District of Burma, which was one of the old capitals of Arakan.[6] But Hodivala has shown that these suggestions are unconvincing.[7] Moreover, as will also be shown by us below, the said kingdom bordered on the one hand the empire of the Balharā or the

1. Cf. *Gauḍalekhamālā*, p. 74, verse 13.
2. *Ibid.*, p. 58, verse, 6.
3. Elliot and Dowson, *op. cit.*, p. 5.
4. See *IHQ*, Vol. XVI, p. 232.
5. *Cathay and the Way Thither*, ed. Cordier, Vol. I, p. 243; *Travels of Marco Polo*, trans., Vol. II, p. 100.
6. Cf. Hodivala, *Studies in Indo-Muslim History*, p. 4.
7. *Loc. cit.*

Rāṣṭrakūṭa ruler of the Deccan and, on the other, it abutted on the kingdom of Kāmarūpa or Assam. In the age in question, this kingdom could have only indicated the empire of the Pālas of Eastern India.

Hodivala draws out attention to Yule's estimate of Sulaimān's account of the Indian kingdoms as 'a medley of disjointed notes put together at random', the information being 'extremely vague'. According to Yule, Sulaimān's knowledge of India was 'slight and inaccurate and he had no distinct conception of the magnitude of the country.'[1] On Sulaimān, Hodivala himself says, "He does not appear to have personally known anything of Bengal and he is repeating only what he had heard or read in some previous author. This is clear from the expressions, 'It is said' and 'It is stated', which he prefixes to his averments."[2]

On the strength of Mas'ūdī's statement that 'Rahma (i.e. Ruhmī) is the title for their kings and generally, at the same time, their name', Hodivala comes to the following conclusion: 'Rahma which is said by Mas'ūdī to have been the title or name of the king as well as of his kingdom, is to be explained by the fact that the kingdom was described in the original writing, to which Sulaimān and Mas'ūdī were indebted for their knowledge, as *mulk al Darhmī*. This phrase is equivocal and may mean 'the kingdom of Dharma' and also 'the king Dharma'. The *dāl* was subsequently supposed to be a *re* and the *re* a *waw*. The phrase was thus misread as *mulk al-Ruhmī* or *al-Ruhmī*, 'kingdom of Ruhmī'......It seems almost certain that he (Sulaimān) found the name of the kingdom or the king only in some manuscript and read it wrongly as *Al-Ruhmī* instead of *al-Dharmī* or *al-Dharma*."[3] This Dharma is of course the shortened form of the name of the Pāla king Dharmapāla as sometimes found in inscriptions.[4]

There can be no doubt that Hodivala's suggestion on the subject is the most ingenious and constructive.[5] Unfortunately,

1. *Cathay*, etc., *op. cit.*, intro., p. ciii.
2. *Op. cit.*, p. 5.
3. *Loc. cit.*
4. See *Ep. Ind.*, Vol. XVIII, p. 245, verse 23; Vol. XXXIV, p. 131, verse 23.
5. There are many such brilliant suggestions in Hodīvala's works. Unfortunately, some of them have been appropriated by others without

he does not appear to have known that there are variants of the name *Ruhmī* in the works of the Arabic writers, such as *Dahmā* etc., which go unquestionably in favour of his suggestion. R. C. Majumdar's comment that Hodivala's opinion regarding *Rahma* being a mistake for *Dharma* 'is improbable as the term continued in use long after Dharmapāla's death'[1] ignores the use of *dāl* instead of *re* as the first letter of the name in many manuscripts, and the nature of Sulaimān's work as indicated above. Indeed, the same is the case with the works of later Arabic authors who gathered most of their information from older sources.

V. Minorsky has shown that the name of the rival of both the Gurjara-Pratīhāra and Rāṣṭrakūṭa kings, whose empire comprised wide areas on the east coast of India, is written variously in the manuscripts as *Rhmī* (i.e. *Rahmī*, *Rahmā*, *Rahmay*, *Ruhmī*, etc.), *Rūhmī*, *Dhm* (i.e. *Daham*, *Dahum*, *Dham*, etc.), *Dhmī* (i.e. *Dahmī*, *Dahmā*, *Dahmay*, *Dahmā*, *Dhamā*, etc.), *Ūhm* and *Ẓhm*.[2] S. M. H. Nainar points to the variants as *Dahmā*, *Damnī*, *Dumī* and *Rahmā*.[3] There are also similar variants of the names *Gurjara* and *Vallabharāja* in the manuscripts.

This East Indian king 'is at war with the Balharā (Vallabharāja) as he is with the king of Jurz (Gurjara)' (Sulaimān).[4] 'On one side, the country of the Balharā called Kamkar (*sic*—Kuntala or Karṇāṭa) is exposed to the inroads of the king of Jurz; on another side, it is exposed to the attacks of Rahmā' (Idrisī).[5] The East Indian kingdom, which was

proper acknowledgment. Thus his suggestion that the names of Yādava Rāmacandra's sons were not Śaṅkaradeva and Bhīmadeva but Siṅghaṇadeva and Bhillamadeva (*op. cit.*, pp. 372-73) are adopted in *The Delhi Sultanate*, ed. Majumdar, pp. 48-49, without giving him the credit due to him.

1. *IHQ*, Vol. XVI, p. 232, note 38.

2. *Hudud al-'Alam, The Regions of the World, A Persian Geography*, 372 *A.H.*=982 *A.D.*, pp. 236-37.

3. *Arab Geographers' Knowledge of Southern India*, pp. 156-57, 160 and 167.

4. Elliot and Dowson, *op. cit.*, p. 5; Hodivala, *op. cit.*, p. 5.

5. Nainar, *op. cit.*, p. 160. Mas'ūdī once speaks of Ba'urah, the king of Qannawj (i. e. Kanauj), who is no doubt the Gurjara-Pratīhāra king usually mentioned by the Arab writers as Gurjara (Juzr, etc.). Maś'ūdī's description of the ruler proves the identification beyond doubt. "He is an enemy of the Balharā, the king of Hind. The king of Qannawj has four armies corresponding with the four cardinal winds and each army is composed of

famous for its gold, silver, aloes, yak-tails, elephants, extremely fine muslins, spotted rhinoceroses and cowrie-coins, lay in the immediate neighbourhood of the dominions of Vallabha rāja and Gurjara. It comprised both land and sea (Mas'ūdī, 943 A.D.). Its king, though not of a noble extraction, was a powerful monarch, his army containing 50,000 elephants and 10,000 to 15,000 fullers and washermen. Ya'qūbī (891 A.D.), who locates the kingdom of Balharā after his, considers him the most important king ruling over a very extensive gold-producing country lying on the sea. Ibn Khurdādbīh, the first draft of whose work was prepared in 846 A. D. and the second draft in 885 A. D., says, "After [the Ballahrā] [comes] the king Jurz......after him Ghāba ; after him Rhmī or Dhmī, between whose [possessions] and those of the other [kings] is a distance of a year (*sic*), and he is said to possess 50,000 elephants as well as cotton and velvet garments and Indian aloes. Then after him [comes] the king of Qāmrūn (i.e. Kāmarūpa or Assam) whose kingdom adjoins China."[1]

Idrisī (1154 A.D.) says, "Among the kings of India, there are the Balharā (Vallabharāja), Jāba, Tāfar, al-Hazr (i. e. Gurjara), Ghāna (or Ghāba) and Qāmrūn (i.e. Kāmarūpa). These names are only taken by the prince who reigns over the province or the country ; no other has the right to assume them; but whoever reigns takes the name."[2] This statement exhibits considerable confusion in the author's mind as well as in his sources. *Vallabharāja* was the hereditary title of the Rāṣṭrakūṭa monarchs who inherited it from their predecessors, the Cālukyas of Bādāmi.[3] The Arabs were familiar

seven hundred thousand men, also said as nine hundred thousand men. The army of the north has to oppose the king of Multan and his allies ; the army of the south has to defend the country against the Balharā, the king of Mānkīr ; and in the same manner are the other armies engaged against the other neighbouring powers." Cf. Nainar, *op. cit.*, pp. 159-60. The name *Ba'urah* is written also as *Barūrah* and *Brūzah*. According to Hodivala, the correct reading of the name is Bozah, Bozoh or Bodzah (*op. cit.*, p. 25). There is thus little doubt that the name is really *Bhoja*, the Gurjara-Pratīhāra monarch of Kanauj, who ruled in c. 836-885 A.D.

1. For these details, see Minorsky, *op. cit.*, pp. 237-38; cf. Hodivala, *op. cit.*, pp. 4 ff.; Nainar, *op. cit.*, p. 157.

2. Nainar, *op. cit.*, p. 156.

3. The original title was *Śrīpṛthivīvallabha* which was shortened as *Śrīvallabha* or *Pṛthivīvallabha* and also merely as *Vallabha*. Both the Cālukya kings of Bādāmi and the Rāṣṭrakūṭa monarchs were popularly known as Vallabha or 'the Vallabha king.'

with the Rāṣṭrakūṭas and wrongly supposed that the other names also indicated hereditary titles as in the case of the Rāṣṭrakūṭa kings. *Gurjara* (*Gurjara-Pratihāra*) was a dynastic or clan name and not a hereditary title like *Vallabharāja*, so that it was never a title assumed by a ruler. Still at least the Gurjara-Pratīhāra king could have been called 'the Gurjara', even though the Imperial Gurjara-Pratīhāra dynasty ceased to rule more than a century before Idrisī's time. But the case of Qāmrūn, i. e. Kāmarūpa or Assam, is far worse, since it was the name of a country and was neither a hereditary royal title nor a dynastic designation. It is apparently this sort of confusion in the minds of the Arab writers, which led them to mention *Dahmā*, etc., sometimes as a royal designation and sometimes as the name of a kingdom. But there can be no doubt that *Dahmā*, etc., stand for *Dharma*, a shortened form of the name of the Pāla emperor Dharmapāla. His mention by the Arabs long after his death is explained by similar references like Idrisī speaking of the Gurjara-Pratīhāra emperor in 1154 A.D. But it seems that Devapāla (c. 810-50 A. D.), the son and successor of Dharmapāla and the contemporary of the Gurjara-Pratīhāra king Bhoja (c. 836-85 A. D.) and the Rāṣṭrakūṭa king Amoghavarṣa (814-78 A. D.) who transferred his capital to Mānyakheṭa (Mānkīr of the Arabs), is mentioned by the Arab writers by the name of his father as Dharma apparently through the confusion referred to above.

Minorsky identifies the East Indian king, whom he calls Dahum, with the Candella king Dhaṅga (c. 950-1002 A. D.) while K. C. Panigrahi takes him to be a ruler of the Bhauma (Bhauma-Kara) dynasty of Orrisa.[1] Both these suggesstions are utterly unconvincing, since neither the Candella nor the Bhauma-Kara kingdom adjoined Assam. Moreover, it is impossible to explain how the less important ruler of Bundelkhand and Orissa would be mentioned as the greatest king of Eastern India and the mighty Pāla emperor, who was one of the most powerful monarchs of his age, should be passed over in silence by the Arab writers while describing the political condition of India in the age of the Rāṣṭrakūṭas. As regards the Candella king, it has to be noticed that, even after

1. Minorsky, *op. cit.*, pp. 236-38; Panigrahi, *Chronology of the Bhauma-Karas and Somavaṃśīs of Orissa*, pp. 64ff.

the composition of Sulaimān's work in 851 A. D., in the year 854 A. D. when the Khajurāho inscription was engraved, Dhaṅga was a feudatory of the Gurjara-Pratīhāra emperor Vināyakapāla, and claimed to have been the ruler of the territory bounded by Kālañjara in the Banda District, Bhāsvat on the Mālava-nadī (Bhīlsā on the Vetravatī or Betwa in the old Gwalior State), the Gopa hill or Gwalior, the Kālindī (Yamunā) and the Cedi country in the Jabalpur region.[1] This territory, again, did not lie on the sea as required by the Arabs' description of the East Indian kingdom.

There are two interesting passages in the *Hudūd al-'Alam* in its description of the towns of Hindustān, which may throw some light on the relations of the said East Indian king (i. e. Dharmapāla or Devapāla) with the contemporary ruler of Orissa. They run as follows in Minorsky's translation:[2]

I. "Nmyās, Harkand, Ūrshīn, Smndr, Andrās—these five large towns (lands ?) are situated on the sea-coast, and the royal power in them belongs to Dahum. Dahum does not consider any one superior to himself and is said to have an army of 300,000 men. In no place of Hindustān are fresh aloes found but in [the possessions] of the king of Qāmrūn and of Dahum. These countries produce in large quantities good cotton which [grows] on trees yielding their produce during many years. The product of this country is the white conch which is blown like a trumpet and is called *sanbak* (*shankh*). In this country there are numerous elephants."

II. "Urshfin, a town with a district protruding into the sea like an island. Its air is bad. That sea is called there the Sea of Gulfs. The royal power belongs to a woman who is called Rāyinā. Extremely large elephants are found there, such as in no other place in India. From it come large quantities of pepper and rotang."

Minorsky thinks that Ūrshīn mentioned in the first passage is different from Ūrshfīn mentioned in the second.[3] But V. V. Barthold,[4] Nainar[5] and Panigrahi[6] take the two names,

1. *Ep Ind.*, Vol. I, pp. 124 ff, text lines 26 and 28.
2. *Op. cit.*, pp. 87-88.
3. *Op. cit.*, p. 243.
4. *Ibid.*, p. 27.
5. Cf. *op. cit.*, p. 85, note 188.
6. *Op. cit.*, pp. 64ff.

also written as *Ūrnshīn*, to be identical. If the identification is accepted, it would appear that Orissa, ruled by a queen, formed a part of the empire of the EastIndian monarch. In that case, the ruling queen of Orissa must have acknowledged the suzerainty, or at best was a subordinate ally, of the Pāla emperors, Dharmapāla and Devapāla.

Several queens are known to have sat on the throne of the Bhauma-Kāras of Orissa, whose era started from 831 A.D., the earliest of the queens being Tribhuvana-mahādevī I, who issued a charter in the year 120 =951 A. D.[1] She therefore flourished about a century after Devapāla. But an earlier queen named Gosvāminī is mentioned in the charter of Tribhuvanamahādevī I,[2] though it is not possible to determine whether this Gosvāminī or any other ruling queen of Orissa is referred to by the Arab writers as a contemporary of Dharmapāla or Devapāla. The name *Rāyinā*, also written as *Rābiyā*, may be a wrong Arabic transliteration of Sanskrit *rājñī* meaning 'the queen'.

Before closing the subject, it may be pointed out that the Arab writers sometimes confused the Pāla empire (Rahma) with the Orissan principality ruled by a queen. Thus Ibn-al-Fakih (902 A. D.) says, "In India lies a realm called Rahma bordering on the sea. Its ruler is a woman. It is ravaged by the plague and any man who comes from elsewhere in India and enters the country (Rahma) diest here. Yet many come by reason of the great profits to be made."[3] About 902 A. D., the Pāla empire was ruled by king Nārāyaṇapāla (c. 854-908 A. D.) and not by a woman.

1. Cf. *Ep. Ind.*, Vol. XXIX, p. 80, note 6; p. 191, note 2.
2. B. Misra, *Orissa under the Bhauma Kings*, p. 25.
3. Harvey, *History of Burma*, p. 10,

CHAPTER VIII

SAMATAṬA

I

The Ashrafpur (Dacca District) copper-plate grants[1] of king Devakhaḍga (middle of the seventh century A. D.) of the Khaḍga dynasty of East Bengal were issued from the royal residence at Karmānta which may have been the capital of the Khaḍgas. Because a gold-covered Śarvāṇī image, caused to be made by Devakhaḍga's queen Prabhāvatī, was discovered at Deulbāḍī about 14 miles to the south of Comilla (Tippera District) and the official designation *Karmāntapāla* occurring in the Narteśvara image inscription found at Bhārellā in the Baḍkāmtā Police Station of the Tippera District was understood as 'the lord of Karmānta' (though it really means 'the superintendent of the royal barns'), N. K. Bhattasali was inclined to believe that Karmānta was the capital of the Samataṭa country (which, in his opinion, covered the Tippera and Noakhali Districts together with the eastern half of the Mymensingh and Dacca Districts and the greater part of the Sylhet District) and that the city stood at the site of modern Baḍ-Kāmtā (literally 'the bigger Kāmtā) about three miles from Bhārellā.[2] It is difficult to accept all these views. In the first place, 'the eastern half of the Dacca and Mymensingh Districts and the greater part of the Sylhet District' do not appear to have formed any part of Samataṭa. Secondly, the recent discovery of the Kailāṇ copper-plate inscription[3] of king Śrī-dhāraṇarāta of Samataṭa, shows that, about the middle of the seventh century A.D., when the Khaḍgas were ruling over the Dacca region in Vaṅga, the Rātas were holding sway over Samataṭa, although very soon Devakhaḍga extirpated Rāta rule and annexed Samataṭa to his dominions. It, therefore, seems that Devakhaḍga's capital was somewhere in the Dacca region. Thirdly, the said Kailāṇ inscription and the

1. Cf. Bhandarkar's List of Inscriptions, Nos. 1394 and 1588.
2. *Ep. Ind.*, Vol. XVII, pp. 349 ff.
3. *IHQ*, Vol. XXIII, pp. 221 ff.

recently discovered Tippera copper-plate grant[1] of Bhavadeva suggest that the capital of Samataṭa in the seventh and the following centuries was not at Karmānta but at the city of Devaparvata on the river Kṣīrodā. The Kṣīrodā is the modern Khīrā or Khīrnai which is a dried-up river course traceable as branching off from the Gomatī just west of the town of Comilla. It flows by the eastern side of the Maināmatī hills and skirts their southern end near the Caṇḍīmuḍā peak where another branch of the river meets it flowing by the western side of the hills. The river thus surrounds the southern end of the Maināmatī hills where the ancient hill-fort of Devaparvata seems to have been situated.

Further light has now been thrown on Devaparvata, the capital of the Samataṭa country during the early medieval period by the Paścimbhāg copper-plate grant of king Śrīcandra (c. 925-75 A. D.) of the Candra dynasty (which originally ruled over Candradvīpa or Vaṅgāladeśa in the Buckergunge region, but later extended its power over Vaṅga, Samataṭa and Śrīhaṭṭa) recently edited by Sri Kamalakanta Gupta Chaudhury in the *Nalini Kanta Bhattasali Commemoration Volume* published by the Dacca Museum in 1966.[2] Unfortunately, the illustration of the writing on both sides of the plate as published by Gupta Chaudhury is not satisfactory while his reading and interpretation of the new stanzas occurring only in the Paścimbhāg inscription contain errors.

Verse 7 of the Paścimbhāg plate, which mentions the city of Devaparvata on the Kṣīrodā river in the Samataṭa country while describing the achievements of Trailokyacandra (c. 905-25 A.D.) the father of Srīcandra, runs as follows :

> *Kṣīrodām=anu Devaparvata iti śrīmat=tad=etat=puraṁ*
> *yatr=āgantu-janasya vismaya-rasaḥ Kamboja-vārt-ādbhutaiḥ* |
> *Lālambī-vanam=atra nāvika=śatair=anviṣya siddh-auṣadhi-*
> *vyāhārā iti ha śrutāś=Samataṭan=nirjitya yat-sainikaiḥ* ||

The stanza may be interpreted as follows: After having conquered Samataṭa, Trailokyacandra's soldiers exclaimed,

1. *Journ. As. Soc.*, Letters, Vol. XVII, pp. 83 ff.
2. See pp. 166 ff., Plates XXXVI-XXXVII.

"That prosperous Devaparvata lying on the Kṣīrodā is this city where the visitor has the feeling of astonishment at the wonderful reports about the Kambojas," and, having searched the Lālambī forest in the area through hundreds of boatmen, they heard the tales about superbly efficacious medicinal herbs. Unfortunately, Gupta Chaudhury commits two errors in reading the verse : (1) he reads *Kṣīrodām=anu* as *Kṣīrodāmavu* (which is meaningless) and then makes it *Kṣīrod-āmbu* (which violates the metre), and (2) the word *nāvika* is read by him as *vātika*. His translation of the stanza consequently runs as follows : "In consequence of the strange news of Kamboja, the new-comers to this illustrious capital, like the venerable mountain (*i.e.*, the Mandara Mountain) in the waters of the Kṣiroda (sea) were struck with feelings of wonder, whose soldiers conquered Samataṭa where was situated the forest of Lālamvī, traditionally said to have been filled with sure medicinal herbs sought for by hundreds of persons suffering from the morbid affection of the nervous system."

The verse offers much valuable information. In the first place, it speaks of the conquest of the Samataṭa country, *i.e.*, the present Tippera-Noakhali region of South-East Bengal within East Pakistan, by the Candra king Trailokyacandra, Verse 8 of the Paścimbhāg, plate refers to the invasion also of the Vaṅga country by Trailokyacandra when it says that his forces enjoyed, out of curiosity, the famous curds of the Vaṅga country at the village of Kṛṣṇaśikharin and its hamlets. Unfortunately Gupta Chaudhury reads *caṅga* for *Vaṅga* in the passage *bhuktvā Vaṅga-dadhīni Kṛṣṇaśikhari-grāmeṣu* and translates it as "drinking nice coagulated milk out of curiosity in villages ensconced amidst black hills." It has to be remembered in this connection that Trailokyacandra is sometimes described in the Candra records as 'the pearl of the Vaṅga country' and also as 'the mainstay of the royal fortunes of the kings of Harikela (*i. e.*, Sylhet)' even though he is represented as the king essentially of Candra-dvīpa. The claims of Trailokyacandra have now to be read along with his son Śrīcandra's rule over Vaṅga, Samataṭa and Śrīhaṭṭa with his capital at Vikramapura in Vaṅga.

In the second place, the reference to Devaparvata in the Paścimbhāg plate supports the evidence of the Kailāṇ and

Tippera plates and suggests that the said city lying on the bank of the Kṣīrodā river was the capital of the Samataṭa country. The reference to hundreds of boatmen in verse 7 of the Paścimbhāg plate reminds us of the description of the Kṣīrodā river in the Kailāṇ plate as *naubhir=aparimitābhir=uparacita-kūlā*.

Thirdly, shortly before the Candra invasion of Samataṭa, the city of Devaparvata seems to have been devastated by the Kambojas. The name Kamboja is probably the Sanskritised form of the name of the Koch people of North Bengal, a few kings of this clan being known from their inscriptions to have been ruling in the northern and south-western regions of Bengal in the tenth century A. D. They must have carved out their kingdoms at the expense of the Pāla emperors. It may be remembered in this connection that an ethnically allied tribe called Mec (Sanskritised as *Mleccha*) ruled Assam from the middle of the seventh to the beginning of the eleventh century A. D.

Lastly, Lālambī-vana, which is no doubt the present Lālmāi hills near Comilla, seems to have been famous for its medicinal herbs. Its mention in the present context throws welcome light on the controversy regarding the identification of Rohitāgiri mentioned in Candra records as the original home of the Candras. Most scholars identified Rohitāgiri with modern Rohtāsgaḍh in the Shahabad District of Bihar. But N. K. Bhattasali suggested that *Rohitāgiri* might be a Sanskritised form of *Lālmāi* (*Lālmāṭi* or red earth). Of course, the modification of *Lāl-māṭi* to *Lāl-māi* seems to be philologically unsound. In any case, the Paścimbhāg plate mentioning *Lālmāi* as *Lālambī* shows that Bhattasali's suggestion is wrong and that it offers no challenge to the identification of Rohitāgiri with Rohtāsgaḍh.

II

There is a confusion regarding the identification of the geographical name variously given as Saknāt, Sankāt and Sanknāt in the manuscripts of the *Tabaqāt-i-Nāṣirī* by Minhājuddīn of Sirāj in his account of Muḥammad-i-Bakhtiyār's conquest of Nūdīa (Nava-dvīpa, i.e. Nadīā on the Bhāgīrathī in South-West Bengal) in the dominions of the Sena king Lakṣmaṇasena of Bengal. The text of Minhājudīn's work was published in Calcutta in 1864 and was translated by Raverty in 1881. A translation of parts of the *Tabaqāt-i-Nāṣirī* was also incorporated in Elliot's and Dowson's *History of India as told by its own Historians*, Vol. II, pp. 259 ff.

Minhājuddin says that, on the eve of Bakhtiyār's invasion, many people of Nūdīa 'went away to the country of Saknāt-Sankāt-Sanknāt, and to the cities of Bang and Kāmrūp; but Rāi Lakhmanīa did not like to leave his territory.'[1] In Raverty's translation, the passage in question runs as follows: 'Most of the Brāhmaṇas and inhabitants of that place retired into the province of Sankanāt-Saknāt, the cities and towns of Bang, and towards Kāmrūd ; but to begin to abandon his country was not agreeable to Lakhmanīa.'[2] On the fall of Nūdīa, 'Rāi Lakhmanīa went towards Saknāt-Sankāt and Bang, where he died. His sons are to this day rulers of the territory of Bang.'[3] In Raverty's translation we have, 'Lakhmanīa got away towards Sankanāt-Saknāt and Bang, and there the period of his reign shortly came to a termination. His descendants up to this time are rulers in the country of Bang.'[4]

Stewart in his *History of Bengal* took Saknāt-Sankāt-Sanknāt to be identical with Jagannāth (Jagannātha-pūrī), i.e. modern Purī in Orissa, and Dowson drew the attention of scholars to this view. But Minhājuddīn makes it clear that Saknāt-Sankāt-Sanknāt was the name of a country or province and not a city like Purī and that it was adjacent to Bang (Vaṅga having in this age its headquarters at Vikramapura in the Dacca District of East Bengal) or at least lay in the direction of Bang from the Nadīā side. It appears moreover that, with the loss of the Nadīā-Lakhnautī (Gaur) region, i. e. the western part of the Sena kingdom, Lakṣmaṇasena took shelter in the eastern part of his dominions. It is impossible to infer from Minhājuddīn's evidence that the Sena king left his dominions and took shelter at the court of the contemporary Gaṅga king of Orissa, who was one of his enemies.

1. Elliot and Dowson, *op. cit.*, p. 308.
2. *Op. cit.*, p. 558.
3. Elliot and Dowson, *op. cit.*, p. 309.
4. Cf. N. G. Majumdar, *Ins. Beng.*, Vol. III, p. 107.

Commenting on the passages quoted from Elliot's and Dowson's work in his *Studies in Indo-Muslim History*, Hodivala says, "Dowson notes that Sanknāt is also written Sankāt and Saknāt and he asks if it is not Jagannāth. But Minhāj states a few lines lower down that Lakhmaṇīa himself fled to *Sanknāt and Bang* and that 'his sons are to this day rulers in the territory of Bang.' Now Vaṅga or Baṅga is the specific name of *Eastern* Bengal and we possess epigraphic evidence of Lakṣmaṇasena's descendants having ruled for at least three generations (sic) at Vikrampur near Sonārgāon in Dacca. Sanknāt may be a mistake for Sonārgāon or Songāon. A still nearer phonetic approach (sic) would be Satgāon (better—Sātgāon) and it is possible that Minhāj who knew little or nothing of Bengal geography has confused the two names. Hooghly district in which Satgāon lay was under Hindu rule for long after the Muhammadan conquest of Lakhnautī."[1] Unfortunately these suggestions about the location of Saknāt-Sankāt of the *Tabaqāt-i-Nāṣirī* are equally untenable, although Hodivala is right in his criticism of the Jagannāth theory. Minhājuddīn stayed at Lakhnautī for over two years in 1242-45 A.D.[2] and could hardly have been as ignorant of the geography of Bengal as Hodivala takes him to have been. Sonārgāon was not a country and, since it was a city in Bang, it is certainly absurd to speak of one's flight towards Sonārgāon *and* Bang. The arguments against the identification of Saknāt-Sankāt-Sanknāt with Jagannāth or Purī, offered by Hodivala himself, are equally applicable against its identification with Sātgāon which lay to the south of Nadīā unlike the Vikramapura region lying to its east.

The *History of Bengal*, published by the Dacca University, offers conflicting suggestions regarding the identification of Minhājuddīn's Saknāt-Sankāt-Sanknāt. In Volume I of this work, R. C. Majumdar refers to Saṅkaṭa-grāma, mentioned in Sandhyākaranandin's *Rāmacarita* as the territory of one of Rāmapāla's feudatories named Caṇḍārjuna, and observes, "*Ain-i-Akbarī* refers to *pargana* Sakot in *Sarkār* Satgāon. The name *Sakot* resembles *Saṅkaṭa*...........Saṅkaṭa-

1. *Op. cit.*, 207.
2. Cf. Elliot and Dowson, *op. cit.*, p. 260.

grāma is probably the same as Saṅka-koṭa referred to in *Vallālacarita* and Sankanāt referred to in *Tabaqāt-i-Nāṣirī*."[1] It will be seen that the identification of Saknāt-Sankāt-Sanknāt with a locality near Sātgāon is as untenable as its location at Sātgāon itself.

In Volume II of the above work, K. R. Qanungo says, "The region to which Rāi Lakhmanīa fled from Nadīā is named in the *T. N.* as *Bang wa S-n-k-n-āt*. Bang means East Bengal and the second place-name is a copyist's error for Sil-hāṭ, i.e. Sylhet. The invariable practice of Persian writers is to designate a district by joining together two well-known places in it, which might be even 50 miles apart; *e. g.* Sultanpur-Nandurbar, Vani-Dindori, Trimbak-Nasik, Dholpur-Bari, Kora-Jahanabad, Kara-Manikpur."[2] Of all the identifications of Saknāt-Sankāt-Sanknāt of the *Tabaqāt-i-Nāṣirī* proposed by scholars and discussed above, Qanungo's suggestion appears to be the least objectionable, as Sylhet may be regarded as a district adjacent to Bang. But the statement that Bang and Saknāt-Sankāt-Sanknāt have been mentioned jointly to indicate a district like Sultanpur-Nandurbar is clearly wrong. The passage 'to the country of Saknāt-Sankāt-Sanknāt and to the cities of Bang and Kāmrūp' quoted above from Elliot's and Dowson's translation of the *Tabaqāt-i-Nāṣirī* shows beyond doubt that Saknāt-Sankāt-Sanknāt did not form an integral part of Bang. The other passage quoted from the same work, however, suggests that this territory lay adjacent to Bang, and the Sylhet District no doubt suits this position. As regards the possibility of scribal errors in the transliteration of Sil-hāṭ (actually *Silhaṭ* from *Śrīhaṭṭa* or *Śrīhaṭa*) as Saknāt, Sankāt or Sanknāt in the Perso-Arabic script, we may cite many other instances of the type.[3]

It should, be pointed out that, though the Śrīhaṭṭa or Sylhet

1. *Op. cit.*, p. 158.
2. *Op. cit.*, p. 4.
3. Cf. Ray, *DHNI*, Vol. II, pp. 599ff., 698, etc. Cf. '*Kanauj* and *futuh*, when spelt without diacritical points, assume the same form : a good illustration of the difficulty of reading accurately oriental names:—here two words of the same form have not a letter in common' (Elliot and Dowson, *op. cit.*, p. 45).

territory is reffered to in the Paścimbhāg plate of Śrīcandra mentioned above, the more popular name of the district as mentioned in the Chittagong plate (about the ninth century A.D.) of king Kāntideva as the Harikela (sometimes called Harikeli or Harikelā) *maṇḍala* (province). Moreover, both Bang and Saknāt Sankāt-Sanknāt appear to have formed a part of the dominions of Lakṣmaṇasena while there is hardly any evidence in favour of the inclusion of Sylhet in the Sena kingdom. Viewed from this angle, the identification of Saknāt-Sankāt-Sanknāt with Sylhet does not appear to be entirely beyond doubt. On the other hand, there was another territory near Bang (Vaṅga), which was not only well known in Indian literature between the fourth and the thirteenth century A.D. but also probably formed a part of Lakṣmaṇasena's kingdom. Its name again may be regarded as the origin of Saknāt-Sankāt-Sanknāt of the Perso-Arabic script with equal plausibility as Silhaṭ (Sylhet) if not with more justification. This is the country of Samataṭa which was the name of the Tippera-Noakhali region of South-East Bengal as late as the thirteenth century when the *Tabaqāt-i-Nāṣirī* was composed. It has to be remembered that Samataṭa as a territory near Vaṅga was considerably more important than Śrīhaṭṭa or Sylhet The intended reading for Minhājuddīn's *Saknāt-Sankāt-Sanknāt* thus appears to have been *Samtaṭ* or *Santaṭ*

The Allahabad pillar inscription of Samudragupta (*circa* 340-76 A.D.) mentions the Samataṭa country while Varāhamihira's *Bṛhatsaṁhitā* (sixth century A.D.) distinguishes it from Vaṅga. The country is mentioned in the seventh century in the accounts of the celebrated Chinese pilgrims Hiuen-tsang and I-tsing as well as in the Kailāṇ plate[1] of king Śrīdhāraṇa of the Rāta dynasty, who is described as the lord of Samataṭa. The same country is further mentioned in the Bāghāurā inscription[2] of Mahīpāla I (*circa* 988-1038 A. D.) and in the Mehār plate[3] (1234 A.D.) of Dāmodara. The Mehār plate mentions

1. *IHQ*, Vol. XXIII, pp. 221 ff.
2. *Ep. Ind.*, Vol. XVII, p. 355.
3. *Ibid.*, Vol. XXVII, pp. 182 ff.

Samataṭa as a *maṇḍala* or province. The name Samataṭa disappeared from Indian geography after the thirteenth century A.D.

King Lakṣmaṇasena ruled from his headquarters at Vikramapura in the Dacca District in Vaṅga at least up to 1205-06 A.D. when Śrīdharadāsa compiled his *Saduktikarṇāmṛta*. King Viśvarūpasena, son and successor of Lakṣmaṇasena, ruled after his father at least for fourteen years as known from the Madanpāḍā plate.[1] The *Tabaqāt-i-Nāṣirī* speaks of Sena rule in Bang as late at least as 1242-45 A. D. when Minhājuddin was staying at Lakhnautī. But about this time the erstwhile feudatories of the Senas in the Samataṭa or Tippera-Noakhali region asserted their independence. A king named Harikāladeva Raṇavaṅkamalla was ruling over the kingdom of Paṭṭikerā in the Tippera District in 1221 A.D. while another king named Dāmodara is known to have established his suzerainty in the Tippera-Noakhali-Chittagong region and to have been ruling from 1231 A.D. at least up to 1243 A.D. Dāmodara belonged to the Deva family and was preceded on the throne by his father Vāsudeva, grandfather Madhusūdana and great-grandfather Puruṣottama. It seems that the earlier members of this family acknowledged the supremacy of the Senas of Vikramapura. Some of Dāmodara's epithets as found in his inscriptions are clearly imitated from those of the Later Senas, used in the copper-plate grants of king Viśvarūpasena. This shows that, even if the Senas were continuing their precarious existence at Vikramapura as late as the fifth decade of the thirteenth century (as indicated by the *Tabaqāt-i-Nāṣirī*), they were no better than subordinate allies of the Deva king. Dāmodara's son, Daśaratha, issued his charters from Vikramapura which had been previously the capital of the Senas in Vaṅga. This points to the complete overthrow of the Senas. Daśaratha's title *Arirājadanujamādhava* suggests that he is identical with Danuj Rāī, Rājā of Sonārgāon (near Vikramapura) close to Dacca, who, according to Ziāuddīn Baranī, entered into an agreement with Sulṭān Ghiyāsuddīn Balban of Delhi about 1281 A.D. that he should guard against the escape of the rebellious Sulṭān Mughisuddīn Tughril

1. See *JAS*, Letters, Vol. XX, pp. 209 ff.; cf. pp. 201 ff. There was no king named Keśavasena in the Sena dynasty. See also *Ep. Ind.*, Vol. XXXIII, pp. 315ff.

Khān of Lakhnautī by water.[1] Thus the Devas of Samataṭa were the successors of the Senas at Vikramapura in Vaṅga.

Earlier expeditions of the Muslim rulers of Lakhnautī against the country of Bang do not appear to have been successful. But Sulṭān Tughril Khān, who began to rule in 1268 A.D., is doubtfully said to have built a fort at Loricol about 25 miles due south of Dacca and to have invaded the kingdom of Hill Tippera.[2] In course of his Bengal compaign, Sulṭān Ghiyāsuddīn Balban claimed to have subdued Iqlim-i-Lakhnautī and 'Arsah-i-Bangāla while he advised his son Nāṣiruddīn Bughra Khān, whom he left as the governor of Muslim Bengal, to exert for the conquest of Diyār-i-Bangāla.[3] It is probable that the dominions of Daśaratha or Danuj Rāī of Sonārgāon in Bang were included in this Dīyār-i-Bangāla.[4] When exactly the Deva dynasty of Sonārgāon was overthrown by the Muslims is not known. Jalāluddīn, a son of Sulṭān Shamsuddīn Firūz-Shāh (1301-22 A. D.) is known to have issued coins from the Lakhnautī mint 'from the spoils of Bang' while a coin of Shamsuddīn Fīrūz Shāh issued from his mint at Sonārgāon itself bears the date 710 A. H. =1310-11 A.D.[5] Thus the Deva dynasty of Sonārgāon seems to have been extirpated by the Muhammadans shortly after the reign of Daśaratha. A copper-plate grant issued by a ruler named Vīradharadeva in his 15th regnal year has been recently discovered at Maināmatī in the Tippera District of East Pakistan.[6] He seems to have been the successor of Daśarathadeva.

1. *Hist. Beng.*, Vol. II, p. 65.
2. *Ibid.*, p. 59.
3. *Ibid.*, p. 67. In Minhājuddīn's *Tabaqāt-i-Nāṣirī*, the land of Vaṅga is called Bilād-i-Bang, Wilāyat-i-Bang and Mamālik-i-Bang, while Lakṣmaṇāvatī (Gauḍa) is called, besides Shahr-i-Lakhnautī, also Mamalakat-i-Lakhnautī, Mamālik-i-Lakhnautī, Diyār-i- Lakhnautī, Bilād-i-Lakhnautī, Mulk-i-Lakhnautī and Khiṭṭa-i-Lakhnautī. See *op. cit.*, pp. 43, 55, 59, 63-65, 71, 73-75, 83, 85, 107, 131, 143, 161, 165.
4. About this time, the lower portion of South-West Bengal seems to have formed a part of the dominions of the Gaṅga kings of Orissa. From the twelfth to the sixteenth century, the Bhāgīrathī was claimed to have been the eastern boundary of the kingdom of Orissa. The upper part of that area, called Rāl (Rāḍha), formed one wing of the Muslim territory of Lakhnautī, its other wing being Barind (Varendra), according to Minhājuddīn. The city of Devkoṭ was situated in Barind.
5. *Hist. Beng. op. cit.*, pp. 80-81.
6. F. A. Khan, *Further Excavations in East Pakistan : Mainamati*, p. 26.

Chapter IX

PRĀGJYOTIṢA

F. E. Pargiter published a paper entitled 'Ancient Countries in Eastern India' in the *Journal of the Asiatic Society of Bengal*, 1897. Many of the results of this study were later incorporated in his English translation of the *Mārkaṇḍeya Purāṇa* (1904), in which he made a laudable attempt to locate the peoples or countries mentioned in the geographical section of the said Purāṇa.

Unfortunately, some of Pargiter's identifications of localities are clearly wrong. Thus he locates the Puṇḍras in the northern areas of Chota Nagpur in the southern region of Bihar and the Utkalas in the land extending from the southern part of Chota Nagpur to the Balasore District of Orissa. As regards the latter, Kālidāsa's *Raghuvaṁśa* (Canto IV) makes it clear that they lived in the land to the west of the Kapiśā (modern Kaśāi) running through the Midnapore District in South-West Bengal between the deltaic region of South Bengal inhabited by the Vaṅgas and the country of the Kaliṅgas whose king is called *Mahendranātha*, i.e., the lord of the Mahendragiri peak in the Ganjam District of Orissa. The Utkalas therefore lived in the Balasore District and the adjoining coastal regions. There is really no evidence suggesting the inclusion of Southern Chota Nagpur in the Utkala country. Likewise, in locating the Puṇḍras, Pargiter ignored the evidence of the *Divyāvadāna* and the records of the age of the Pālas and their contemporaries, and we have now inscriptions of the Maurya and Gupta epochs —all clearly indicating that this people lived in North Bengal. According to the *Divyāvadāna*,[1] the city of Puṇḍravardhana lay beyond Kajaṅgala the eastern border of the land which was the early sphere of Buddhism (i.e. Bihar and Eastern U. P.), while epigraphic records of the early mediaeval period apply the name Puṇḍravardhana-bhukti to North Bengal, its chief city being

1. Ed. P. L. Vaidya, p. 13. *Puṇḍravardhana* is explained as 'the city of the Puṇḍras', *vardhana* being the same as Old Persian *vardana*. See Monier-Williams, *Sans.-Eng. Dict.*, s. v. *vardhana*.

Puṇḍravardhana, identified with modern Mahāsthān in the Bogra District.[1] The Damodarpur plates of the Gupta age show that Koṭivarṣa (the area around modern Bāṇgaḍh in the Dinajpur District) was a *viṣaya* or district of the *bhukti* or province of Puṇḍravardhana.[2] The Mahasthan inscription of the third century B.C. not only proves that Puṇḍravardhana was also called Puṇḍranagara, 'the city of the Puṇḍras', but also supports its identification with Mahasthan.[3] Hiuen-tsang, the Chinese pilgrim of the seventh century A.D., travelled from Kajaṅgala (modern Rajmahal in East Bihar) to the east to Puṇḍravardhana and thence to the east to Kāmarūpa (Prāgjyotiṣa).[4] Thus Pargiter's location of the Puṇḍras in Northern Chota Nagpur is palpably untenable.

It is a matter of regret that the wrong location of the Puṇḍras influenced Pargiter's identification of the land of the Prāgjyotiṣa people and this has also affected writers on the early history of Assam.

According to Pargiter, the early kingdom of Prāgjyotiṣa comprised the major part of modern Assam together with the Jalpaigurī, Cochbihar, Rangpur, Bogra, Mymensingh, Dacca and Tippera Districts and parts of the Pabna District in Bengal and probably also the eastern areas of Nepal.[5] He points out that the Prāgjyotiṣa country bordered on the lands of the Kirātas and Cīnas forming the retinue of Bhagadatta of the *Mahābhārata*, etc., who also drew his troops from the dwellers of Sāgarānūpa (marshy region near the sea) and is even represented as dwelling in the Eastern Sea. It is suggested that 'these marshy regions can only be the alluvial tracts and islands near the mouths of the Ganges and the Brahmaputra

1. See Maitreya, *Gauḍalekhamālā*, pp. 15, etc.; Majumdar, *Ins. Beng.*, Vol. III, pp. 2, etc.

2. *Select Inscriptions*, pp. 283-84, etc.

3. *Ibid.*, pp. 82-83.

4. Watters, *On Yuan Chwang's Travels in India*, Vol. II, pp. 184ff. The pilgrim noticed a *stūpa* built by the Maurya emperor Aśoka (third century B.C.) at the city of Puṇḍravardhana though no Aśokan *stūpa* was found by him in Kāmarūpa. Considering the tradition that Aśoka built 84,000 *stūpas* throughout his empire, it seems that North Bengal was included in his dominions, but Assam was not.

5. *JASB*, 1897, p. 106.

as they existed anciently', so that 'Prāgjyotiṣa comprised the whole of North Bengal proper'.[1]

K. L. Barua, in his *Early History of Kāmarūpa*,[2] refers to Pargiter's views and suggests that the western boundary of Prāgjyotiṣa was the Kosī river in Bihar and that the Purnea District was included in it. This, however, clearly goes against the epic and Puranic tradition regarding Pauṇḍraka Vāsudeva (*i.e.* Vāsudeva, king of the Puṇḍras), the epigraphic and literary evidence about the location of the land of the Puṇḍras in North Bengal and of Prāgjyotiṣa or Kāmarūpa in the Brahmaputra Valley, and the Chinese *Tang-shu* and the *Kālikā Purāṇa* and *Yoginī Tantra* stating clearly that the western boundary of Prāgjyotiṣa-Kāmarūpa was the river Karatoyā. Some scholars believe that the Karatoyā was originally a very big river uniting in its bed the streams of the present Tīstā, Kosī and Mahānandā.[3]

The Nidhanpur (Sylhet District) charter, originally issued by Mahābhūtavarman of Prāgjyotiṣa in the sixth century and renewed by his descendant Bhāskaravarman about the middle of the following century, records the grant of land near the Kauśikā (sometimes called *śuṣka* or dried up), and Barua identifies it with the Kosī in Bihar, though others prefer its identification with the Kuśiyārā in the Sylhet region near Assam.[4]

1. *The Mārkaṇḍeya Purāṇa*, p. 328, note. Pargiter even goes so far as to say, "The *Raghuvaṁśa* places it seemingly beyond the Brahmaputra (IV. 81); but Kālidāsa was a little uncertain in distant geography." Since Kālidāsa flourished at the court of the Gupta emperors whose dominions included the province of Puṇḍravardhana (North Bengal) and bordered on the *pratyanta* kingdom of Kāmarūpa, another na me of Prāgjyotiṣa, it is impossible to think that the poet was ignorant of the location of this country. But Kālidāsa says that the king of Prāgjyotiṣa-Kāmarūpa trembled when Raghu crossed the Lauhitya (Brahmaputra) which is thus not represented as a boundary of the country. The Aphsad inscription (*CII*, Vol. III, pp. 203, 206) also places it in the valley of the Lohitya (Lauhitya) while describing the victory of Mahāsenagupta over Susthitavarman. It may however be noted that, while Mahāsenagupta invaded Prāgjyotiṣa-Kāmarūpa from the east or south-east, Raghu entered the country from the Himalayan region in the north.

2. Cf. pp. 2-3.

3. Gait, *History of Assam*, 2nd ed., pp. 10-11.

4. Barua, *op. cit.*, pp. 50-51; *IHQ*, Vol. VII, p. 743 (823).

But Barua does not notice that, if the Purnea District of Bihar formed a part of Prāgjyotiṣa-Kāmarūpa in the days of Bhāskaravarman, contemporary Chinese writings would not have mentioned the Karatoyā as the boundary between Puṇḍravardhana and Prāgjyotiṣa-Kāmarūpa.[1] Barua's belief that Puṇḍravardhana was a part of Prāgjyotiṣa-Kāmarūpa in the first quarter of the sixth century or about 525 A.D. when Mahābhūtavarman is supposed to have been ruling totally ignores the evidence of the Damodarpur inscription of 543 A.D., which includes the province of that name in the Gupta empire.[2] If Māḍhāśālmalī in the Puṇḍravardhana-bhukti, mentioned in the Khalimpur plate of Dharmapāla, is regarded as identical with or situated near Mayūraśālmalī in the Candrapurī-viṣaya as known from the Nidhanpur plates[3] of Mahābhūtavarman and Bhāskaravarman, the inclusion of Puṇḍravardhana in Prāgjyotiṣa-Kāmarūpa is scarcely proved since certain border areas of Prāgjyotiṣa-Kāmarūpa may have been annexed to the Puṇḍravardhana-bhukti by the Pāla emperor. It is well known that, during the age of the Pālas and their contemporaries, wide areas of East and South Bengal were comprised in the *bhukti* called Puṇḍravardhana or Pauṇḍra.

Accarding to the *Kālikā Purāṇa* (chapter 38) compiled in the Assam region during the early mediaeval period, Naraka, son of the god Viṣṇu and the goddess Earth, was taken to Prāgjyotiṣapura near the temple of Kāmākhyā in the heart of Kāmarūpa. He drove out the Kirāta inhabitants of the country from the area between the Karatoyā in the west and the Dikkaravāsinī and Lalitakāntā in the east for settling the twiceborn, while the Kirātas were rehabilitated in the sea coast extending from Lalitakāntā in the east. Similar information is supplied by the late mediaeval *Yoginī Tantra* according to which Kāmarūpa was bounded in the north by the Kañjagiri or Kāñcan-ādri (probably the Kanchenjunga) in Nepāla, in the west by the Karatoyā, in the east by the Dikṣu river or Dikkaravāsinī, and in the south by the Brahmaputra-

1. Watters, *On Yuan Chwang's Travels in India*, Vol. II, pp. 184ff. (cf. p. 186 and note 2).

2. *Select Ins.*, pp. 337ff.

3. Cf. P. N. Bhattacharya, *Kāmarūpaśāsanāvalī*, p. 6.

saṅgama or the confluence of the Brahmaputra and the Lākṣā.[1] The Dikṣu is no other than the modern Dikhu falling in the Brahmaputra near Sibsagar in Assam, while the Lākṣā is the modern Lakhyā which joins the Brahmaputra in the Mymensingh District of East Pakistan. The temple of Dikkaravāsinī is often located at Dikrang near Sadiya, and Lalitakāntā is likewise associated with the hill streams of Sandhyā, Lalitā and Kāntā not far from Gauhati, the chief city of Assam, lying a few miles from the Kāmākhyā temple[2] and identified with the ancient city of Prāgjyotiṣapura.

It will be seen from the above discussion that the Karatoyā was regarded as the western boundary of Prāgjyotiṣa-Kāmarūpa from the seventh century A.D. while the same appears to have been the case when North Bengal was included in the Magadha empire under the Guptas from the fourth to the sixth century A.D. and the Nandas and Mauryas between the fourth and second century B.C. There is again no evidence to prove that the position was otherwise between the fall of the Mauryas and the rise of the Guptas. It may be mentioned in this connection that North Bengal was also called Varendra or Varendrī from the age of the Pālas and that Sandhyākaranandin's *Rāmacarita* mentions Pauṇḍravardhanapura as the chief city of Varendrī which it locates between the Ganges and the Karatoyā.[3]

Prāgjyotiṣa-Kāmarūpa is represented as a land sometimes of the north and sometimes of the east.[4] This is because the country extended from the hilly region regarded as a part of the Himalayas in the north to near about the junction of the Brahmaputra and the Lakhyā in the south. The *Mahābhārata*[5] represents Bhagadatta as *Pūrvasāgaravāsin* (dwelling in the Eastern Sea) while the *Rāmāyaṇa*[6] states that Prāgjyotiṣapura, the city of Bhagadatta's father Naraka, was situated on the Varāha mountain in the sea. The *Mahābhārata*[7] also represents Bhagadatta as the leader of the Cīnas,

1. See *Ep. Ind.*, Vol. XII, p. 68; Sircar, *The Śākta Pīṭhas*, p. 13 and note; cf, *JAIH*, Vol. I, pp. 19-20, locating the Dikkaravāsinī temple near Paya in the Lohit District.

2. *The Śākta Pīṭhas*, p.. 17, note 3.

3. See V.R.S. ed., pp. 84 (III. 10), 153 (V. 1).

4. *The Mārkaṇḍeya Purāṇa*, p. 328, note; Sircar, *Cosm*, etc., pp. 65.77.

5. V 4. 11; XI. 23. 10.

6. IV. 42. 30-31.

7. II. 26. 9; 33. 9-10.

Kirātas, Mlecchas and Sāgarānūpavāsins. It is therefore interesting to note that, in the course of Arjuna's *digvijaya* in the north, he is said to have defeated Bhagadatta leading the Cīnas, Kirātas and Sāgarānūpavāsins[1] while Bhīma in his expedition in the east is said to have reached the Lauhitya valley and defeated the Mleccha rulers and the Sāgarānūpavāsins.[2]

This sea or Eastern Sea not far from Prāgjyotiṣa-Kāmarūpa requires some explanation. As we have seen, Pargiter supposed that it was below the mouths of the Ganges and the Brahmaputra. But the deltaic region of South Bengal watered by the mouths of the Ganges was inhabited by the ancient people, called the Vaṅgas by the Indians and Gangaridae by the Greeks, from very early times.[3] Since the *Kālikā Purāṇa* seems to locate the sea to the east (or at least the south-east) of Prāgjyotiṣa-Kāmarūpa, P. N. Bhattacharya considers it possible that Bhagadatta's empire extended upto the South China Sea.[4] But it is extremely doubtful if we can take the *Kālikā Purāṇa* evidence so literally and conclude that such a large area of South-East Asia was actually included in Prāgjyotiṣa-Kāmarūpa. There was really no sea to the east or south-east of the country. K. L. Barua points to the application of the name *hāor* (Sanskrit *sāgara*) to the marshy parts of Sylhet and Mymensingh and refers to the Bhatera copper-plate inscription of Govinda Keśavadeva of Śrīhaṭṭa (Sylhet) mentioning the sea as the boundary of a plot of the gift land.[5] He therefore concludes that the sea referred to was the low-lying and water-logged land to the south of the Assam range and suggests that the said area may have been connected in olden times with the Bay of Bengal (Eastern Sea) through the estuaries of the Brahmaputra.

Considering the fact that the traditional southern boundary

1. *Ibid.*, II. 26. 9.
2. *Ibid.*, II. 30. 26-27.
3. See below, Ch. XIII.
4. *Kāmarūpaśāsanāvali*, Intro., pp. 4, note 2; 11, note 2. For the country to the east of this land, cf. Watters, *op. cit.*, p. 186; Beal, *Life of Hiuen Tsiang*, pp. 132-33.
5. *Early History of Kāmarūpa*, p. 2; *Ep. Ind.*, Vol. XIX, p. 282, text line 32. The word used here is *sāgara*. An inscription of the Senas speaks of the *samudra* as the boundary of one of the plots of land granted (Majumdar, *Ins. Beng.*, Vol. III, p. 146, text line 47).

of Prāgjyotiṣa-Kāmarūpa was the confluence of the Brahmaputra and the Lakhyā in the Mymensingh District, Barua's suggestion seems to be justifiable. In this connection, it may also be noted that the epigraphic records of the kings of ancient Assam speak of the Lauhitya or Brahmaputra as 'the sea', while there is a tradition that, in ancient times, the Eastern Sea extended upto Devīkoṭṭa in the Dinajpur District of North Bengal.[1] This also explains the mention of the Lauhitya in the *Mahābhārata* side by side with the Sāgarānūpavāsins.

Reference may be made here to another point not entirely unconnected with our topic. In an inscription of king Jayadeva Paracakrakāma of Nepal, probably dated in 737 A.D., his queen Rājyamatī is described as born in Bhagadatta's family (to which all kings of ancient Assam claimed to have belonged) and as the daughter of Harṣadeva, called *Gauḍ-Oḍr-ādi-Kaliṅga-Kosala-pati*, 'lord of Gauḍa, Oḍra and other lands as well as of Kaliṅga and Kosala (apparently South Kosala).'[2] This Harṣadeva, father-in-law of the Nepalese king Jayadeva, is generally identified with king Harṣavarman, a descendant of Sālastambha who occupied the throne of Prāgjyotiṣa-Kāmarūpa in the third quarter of the seventh century A.D.[3] But it is not noticed that, although Harṣa was the king of Prāgjyotiṣa-Kāmarūpa, neither Prāgjyotiṣa nor Kāmarūpa occurs in the above list of his territories and that this fact clearly points to the degree of historical accuracy to be expected in the description. We have similar other cases of poetical excesses. Thus the Aihole inscription (634 A.D.) of Pulakeśin II of Badami describes his father Kīrtivarman I as the vanquisher of the Nalas, Mauryas and Kadambas, and nobody ever doubted the genuineness of the claim. But the same Kīrtivarman I is described in the Mahakuta pillar inscription (602 A.D.) of his younger brother Maṅgalīśa as the subduer of Vaṅga, Aṅga, Kaliṅga, Vaṭṭūra, Magadha, Madaraka, Kerala, Gaṅga, Mūṣaka, Pāṇḍya, Dramila, Coliya, Āluka and Vaijayantī. It is interesting that, in this list, Kīrtivarman's genuine victory over the Nalas and Mauryas is conspicuously omitted, while his imaginary success against

1. Cf. *Ep. Ind.*, Vol. XXIX, p. 151; above, p. 112, note 1.
2. Gnoli, *Nepalese Inscriptions in Gupta Characters*, p. 117, verse 15.
3. See Barua, *op. cit.*, pp. 112-13; but cf. Gait., *op. cit.*, p. 30.

the Vaṅgas, Aṅgas, Madrakas and others has been mentioned. The conquest of Vaijayantī, the capital of the Kadambas, is the only genuine claim of victory in the fictitious description of the Mahakuta pillar inscription.[1]

The statement in the Nepalese inscription regarding the lordship of Harṣavarman of Prāgjyotiṣa-Kāmarūpa over a number of countries in Eastern India should therefore be taken with a grain of salt.

1. *Bomb. Gaz.*, Vol. I, Part ii, pp. 345-46; *The Classical Age*, ed. Majumdar, pp. 232-33.

Chapter X

ODIŚĀ

I

Orissa is spelt *Oḍiśā* in the language of the Oriyas. This name seems to be derived from an earlier form like *Auḍrīya-viṣaya* through forms like *Oḍḍī-visaa* and *Oḍī-visā*. The Tibetan author Tāranātha mentions the name as *Oḍi-viśa*.[1] The tribal name Uḍra or Oḍra lies at the root of all these forms. But Uḍra, Oḍra or Auḍra cannot be regarded as the earliest name of Orissa.

In ancient times a powerful people called the Kaliṅgas lived in the present Orissa region. In the third century B.C., the Maurya emperor Aśoka (*circa* 269-232 B.C.) conquered the Kaliṅgas and annexed the Kaliṅga country to his empire. The province of Kaliṅga in the Maurya empire was divided into two administrative units. The north-eastern part of the country had its headquarters at Tosalī (although the land around the city was later often called Tosala) which is the modern Dhauli near Bhubaneswar in the Puri District of Orissa. In a later age, kings of the Ārya-Mahāmeghavāhana family of the Cedi clan, which is represented by Khāravela, described as 'the supreme lord of Kaliṅga', appear to have had their capital at the same place. In Maurya times, South-Western Kaliṅga had its headquarters at the city of Samāpā near modern Jaugaḍa in the Ganjam District. Indian literature intimately associates Kaliṅga with the Mahendra-giri now standing on the borders of the Ganjam District of Orissa near those of the Srikakulam District of Andhra Pradesh. But there is no doubt that the Godāvarī or the Kṛṣṇā was often regarded as the south-western boundary of the Kaliṅga country. This is indicated by the fact that, about the fifth century A.D., some rulers, enjoying the title *Kaliṅg-ādhipati*, not only had their capital at Piṣṭapura, modern Pithapuram in the East Godavari District, but sometimes even claimed lordship over

1. *Ind. Ant.*, Vol. IV, pp. 365-66.

the whole coast land between the Kṛṣṇā and the Mahānadī.[1] Actually however, the north-eastern boundary of the ancient Kaliṅga country lay even beyond the Mahānadī, although Kaliṅga did not include the whole of modern Orissa. The north-eastern part of coastal Orissa was included in early times in the land of the Utkalas while the Patna-Sonpur region in the upper valley of the Mahānadī formed the eastern part of the country named Dakṣiṇa-Kosala till the early medieval period. But the Utkalas and some of their neighbours may have been ethnically related to the Kaliṅgas. Indeed, an ancient tradition recorded in the *Mahābhārata* and some of the Purāṇas[2] regards the progenitors of the Aṅgas of East Bihar, the Vaṅgas of South Bengal, the Kaliṅgas, the Puṇḍras of North Bengal and the Suhmas of West Bengal as full brothers. Early European writers sometimes represent the people of South Bengal (Gangaridae) as a branch of the Kaliṅga people.[3]

We have seen how the river Kṛṣṇā was sometimes regarded as the south-western boundary of the ancient Kaliṅga country. But in the fourth and fifth centuries A.D., the Śālaṅkāyanas ruling over the coast land between the Kṛṣṇā and the Godāvarī with their capital at the city of Veṅgī (modern Peddavegi near Elūru in the West Godavari District, Andhra Pradesh) did not claim to be lords of Kaliṅga. In the sixth and seventh centuries the Viṣṇukuṇḍins were ruling over the same area. They also did not claim to be rulers of the Kaliṅga country. If the identification of the *Andhr-ādhipati*, mentioned in the Haraha inscription[4] of the time of Maukhari Īśānavarman, dated Vikrama 611 = 553-54 A.D., with a Viṣṇukuṇḍin monarch is accepted, the kingdom of the Viṣṇukuṇḍins was probably known as Andhra. During the second quarter of the seventh century, the Eastern Cālukyas established themselves at Piṣṭapura (Pithapuram in the East Godavari District) and ruled over the entire coast country from the Visakhapatnam District in the north-east to the Guntur District

1. Cf. *Ep. Ind.*, Vol. XXX, p. 144.
2. *Mahābhārata*, I. 104. 53; *Bhāgavata*, IX. 23. 5; *Vāyu*, 99, For the separate mention of the Uḍras and Utkalas in the *Bṛhatsaṁhitā*, see below.
3. Pliny, *Hist. Nat.*, VI. 21 ff.
4. *Ep. Ind.*, Vol. XIV, pp. 115ff.

in the south-west for many centuries. But their kingdom became famous under the name of Veṅgī. During this period, the major portion of the Visakhapatnam District formed a part of Veṅgī, although the Yelamanchili Taluk of that District was called Elamañci-Kaliṅgadeśa. The Early Eastern Gaṅgas were ruling over the area about the Srikakulam District (Andhra Pradesh) with their capital at Kaliṅganagara, which has been identified with modern Mukhaliṅgam near Śrīkākuḷam, from the close of the fifth century A.D. They also enjoyed the title *Kaliṅg-ādhipati* or lord of Kaliṅga. In the early medieval period, it is only the kingdom of these Early Eastern Gaṅgas that was exclusively known as Kaliṅga, because, as will be seen below, the kingdoms in the Ganjam-Puri-Cuttack region assumed different names such as Koṅgoda, Tosalī, Uḍra, etc., since the latter part of the sixth century A.D. With the rise of the Imperial branch of the Eastern Gaṅgas, and especially with the conquest of the coast land between the Godāvarī and the Ganges (Bhāgīrathī) by the Imperial Gaṅga monarch Anantavarman Coḍagaṅga (1078-1147 A.D.) about the beginning of the twelfth century, the major part of ancient Kaliṅga came under the Kaliṅga king. But the old name did not get time enough to become popular again as the successors of Anantavarman Coḍagaṅga soon transferred their capital to the Cuttack District (Orissa) far away from the Śrīkākuḷam region that had become famous under the name Kaliṅga during the many centuries' rule of the Early Eastern Gaṅgas.[1] The Eastern Gaṅgas originally lived in the present Kannaḍa-speaking area of Mysore, whence they migrated to and settled in the present Telugu-speaking area of Śrīkākuḷam. The Kannaḍa origin of the Eastern Gaṅgas is not only supported by the copper-plate grants of Anantavarman Coḍagaṅga and his successors,[2] but also by the following statement in the description of Coḍagaṅga in the *Mādalā Pāñji*[3] :

1. The tradition recorded in Yaśodhara's commentary on the *Kāmasūtra* (VI.6), composed about the middle of the thirteenth century (cf. Kieth, *Hist. Sans. Lit.*, p. 469), that Kaliṅga lay to the south of Gauḍa seems to be based on the Imperial Gaṅga occupation of parts of South-West Bengal.

2. Cf. *Ep. Ind.*, Vol. XXVIII, p. 239.

3. Ed. Mahanti, p. 23. The statement is however anachronical.

Cuḍagaṅga Karṇāṭa-desaru āsi, etc. The mother tongue of the Imperial Gaṅgas was originally Telugu. This is clearly indicated not only by the great part Telugu plays even in their later documents, but also by the popularity of names like *Aniyaṅkabhīma* which was later Sanskritised as *Anaṅgabhīma* through the intermediate form *Anaṅkabhīma*. After the transference of their capital to the heart of Orissa, the Imperial Gaṅgas began to become Oriyas; but most of their matrimonial alliances were still contracted with South Indian royal families and a large number of their officials and proteges were South Indians.

With reference to the geography of Kaliṅga, mention has to be made of the views of certain scholars based on the repeated reference to the country in the Ceylonese chronicles. Chapter 59 of the *Cūlavaṁsa* says how king Vijayabāhu I (1055-1110 A.D.) of Laṅkā married the Kaliṅga princess Trilokasundarī and offered befitting maintenances to her relations Madhukārṇava (Madhukāmārṇava ?), Bhīmarāja and Balātkāra of Siṁhapura, capital of Kaliṅga, for settling them in his kingdom. The younger sister of the Kaliṅga princess, Sundarī by name, was given in marriage to Vijayabāhu's son Vikramabāhu. In this connection, Geiger observes, "Sīhapura (Siṁhapura) is the town which according to the legend (cf. *Mhvs*, 6.35) was founded in Lāḷa (Rāḍha) by Vijaya's father Sīhabāhu. Lāḷa borders in the north of the Kaliṅga kingdom, the home of Tilokasundarī, as must be inferred from *Mhv.*, 6.1-5. The south-eastern district of Chutea Nagpur to the west of Bengal is still called Singbhum."[1] It has however to be noticed that, in the age of Vijayabāhu (actually from about the end of the sixth to at least about the beginning of the twelfth century), the name Kaliṅga was exclusively applied to the kingdom of the Eastern Gaṅgas of Kaliṅganagara near Śrīkākuḷam, who styled themselves as *Kaliṅg-ādhipati*. Siṁhapura (modern Singupuram in the same neighbourhood) was, however, the capital of the *Kaliṅga-ādhipatis* in the fourth and fifth centuries A.D. and was no longer the capital of Kaliṅga although it may have been the residence of some scions of the Gaṅga family. Rāḍha and Kaliṅga do not appear to have had contiguous boundaries in any period of history. Siṁhapura in Rāḍha (probably modern Singur in the Hooghly District of

1. *Cūlavaṁsa*, trans., Part I, p. 211.

West Bengal) cannot be regarded as the same as the Kaliṅga capital of that name, identified with modern Siṅgupuram near Śrīkākuḷam. The representation of Siṁhapura as the capital of Kaliṅga in the *Mahāvaṁsa* tradition seems to be due to the fact that the chronicle was composed about the fifth century while the *Cūlavaṁsa* appears merely to have continued the same tradition even though the later capital of the country was at Kaliṅga-nagara (modern Mukhaliṅgam near Śrīkākuḷam) and not at Siṁhapura.[1]

As regards the north-eastern limit of ancient Kaliṅga, a well-known passage in the Tīrtha-yātrā section of the Vana-parvan (114.3) of the *Mahābhārata* has—*eṣa Kaliṅgaḥ Kaunteya yatra Vaitaraṇī nadī*. This shows that the river Vaitaraṇī forming the eastern border of the Cuttack District of Orissa was regarded in ancient times as the boundary of the Kaliṅga country. Kālidāsa's *Raghuvaṁśa* (IV. 38) speaks of the Utkala country lying between the land of the Kaliṅgas and that of the Vaṅgas. The eastern boundary of Utkala can be determined only when we know the exact area inhabited by the Vaṅga people.

Early Greco-Roman writers represent the emperors of the Nanda dynasty of Magadha as the rulers of the Prasii and the Gangaridae and speak of Palibothra (*i.e.* Pāṭaliputra, near modern Pāṭnā, Bihar) as the capital of the country of the Prasii. There is little doubt that the Prasii represented the *Prācyas* (*i.e.* the people of the Eastern Division of ancient Bhāratavarṣa) of Indian literature. But there is some confusion about the Gangaridae, their name being often Indianised as *Gaṅgā-rāṣṭra*, *Gaṅgā-rāḍha* and *Gaṅgā-hṛdaya*. Greek Gangaridae is however the plural form of *Gangarid* from a base like *Gange* or *Ganges*; cf. *Sassan—Sassanid—Sassanidae*; *Akhamenes—Akhamenid—Akhamenidae*. The word *Gangaridae* therefore means 'the Gangetic people'. The land inhabited by this people is clearly indicated by the author of the *Periplus of the Erythraean Sea* (*circa* 80 A.D.) and the *Geography* of Ptolemy (*circa* 145 A.D.). Ptolemy[2] mentions the five estuaries of the river Ganges and says, "All the region about the mouths of the Ganges is occupied by the Gangaridae with the following city—Gange, the royal city (*i.e.* the capital of the country)." It is clear from this that the

1. Kaliṅga of the Ceylonese chronicles is sometimes identified with Śrīvijaya (Nicholas and Paranavitana, *A Concise History of Ceylon*, p. 198).
2. *Geog.*, VII. I. 18 and 81. See Chapter XIII below.

Gangaridae or Gangetic people received their name from their chief city called Gange, apparently named after the river Ganges. The *Periplus* (para. 63) however applies the name Ganges not only to the river and a city standing on the bank of its principal mouth but also to the country, of which the city was apparently the capital. This book says about the country called Ganges that 'there is a river near it called the Ganges' and that 'on its bank is a market-town which has the same name as the river Ganges.' The principal product of this country is stated to have been the Gangetic muslin which reminds us not only of the world-wide celebrity of the Dacca muslin in the medieval age, but also of the *dukūla*, *kauśika*, *patrorṇa* and *prāvara* mentioned in the Sabhā-parvan of the *Mahābhārata* as presents made to Yudhiṣṭhira by the Vaṅgas, Kaliṅgas, Tāmaliptas and Puṇḍras of Eastern India.[1] But a people called Gāṅga or Gāṅgeya inhabiting Lower Bengal and having their capital at a city called Gaṅgā (Greek *Gange* or *Ganges*) is not known from ancient Indian literature. This powerful people, known to foreign writers from the fourth century B.C. to the second century A.D., were apparently known to the Indians by a different name. Curiously enough Kālidāsa, who flourished in the fourth and fifth centuries A.D., locates the Vaṅga people, well-known in ancient Indian literature, exactly in the same region where the Gangaridae or Gangetic people are placed by the early European writers. Canto IV (verses 36-37) of his *Raghuvaṁśa* describes how Raghu defeated the Vaṅgas in a naval battle and raised pillars of victory in what is called *Gaṅgā-sroto-'ntara* no doubt in the land of the defeated people. The expression *Gaṅgā-sroto-'ntareṣu* has been explained by the celebrated commentator Mallinātha as *Gaṅgāyāḥ srotasāṁ pravāhānām = antareṣu dvīpeṣu.*[2] Thus the country of the Vaṅgas is located by Kālidāsa in the deltaic region of Southern Bengal, which is intersected by the mouths of the river Ganges. This further proves that the Vaṅga people were identical with the Gangaridae who, according to the Greco-Roman writers, lived in the region about the mouths of the Ganges and had their chief city about the confluence of the Ganges and the Bay of Bengal. The modern representative

1. See Motichandra, *Geog. Ec. Stud. Mahābh.*, pp. 112-13.
2. See above, p. 132, note 7; below, Ch. XIII (pp. 213 ff.)

of this ancient city seems to be the holy place at the junction of the Gaṅgā and the Sāgara, called Sāgara or Gaṅgā-sāgara. The name *Gaṅgā*, suggested by the early Greco-Roman writers, may be regarded as an *eka-deśa* of the name *Gaṅgā-sāgara*.[1] The name Vaṅga, originally applied to wide areas of Southern Bengal, came in the medieval age to be confined to the eastern parts of that region together with the adjoining areas. But later the name came to be used to indicate the whole of Bengal.

The above discussion will show that in the early centuries of the Christian era, the Vaṅgas lived in the deltaic region of Southern Bengal watered by the mouths of the Ganges and had their capital at the city of Gaṅgā near the junction of the Bhāgīrathī and the Bay of Bengal and that Gaṅga-sāgar is the modern representative of the ancient capital of the Vaṅgas. After the name of the capital, the country was also often called Gaṅgā. Early European writers mention the Vaṅgas as the Gangaridae, *i.e.* the Gāṅga or Gāṅgeya people. The Greek name of the Vaṅgas seems to be the result of a confusion the foreigners made between the sounds of the two names *Vaṅgāḥ* and *Gaṅgā*. The identification of the Vaṅgas and the Gangaridae and the location of their habitat are clearly indicated by the evidence supplied by Kālidāsa's *Raghuvaṁśa*, the *Geography* of Ptolemy and the *Periplus of the Erythraean Sea*. Ptolemy and Kālidāsa place the Cambyson or Kapiśā river, identified with the present Kaśāi running through the Midnapur District (West Bengal), about the western border of the country of the Gangaridae or Vaṅga people. This is supported by the Jain *Prajñāpanā*, according to which Tāmralipti in the present Tamluk region of Midnapur once formed a part of the Vaṅga country.[2] It appears therefore that the Cambyson or Kapiśa, *i.e.*, the modern Kaśāi, formed the boundary between the land of the Vaṅgas and that of the Utkalas. Thus it may be said that the Utkala country lay between the Kaśāi and the Vaitaraṇī rivers. Roughly speaking therefore, Utkala comprised the present Balasore District of Orissa together with parts of the Cuttack District of that State and of the Midnapur District of

1. See *Proc. IHC*, Bombay, 1947, pp. 91 ff., and below, Ch. XIII. (pp. 213 ff.).

2. Raychaudhuri, *Stud. Ind. Ant.*, p. 186

West Bengal. The Puri, Ganjam and Cuttack Districts of Orissa then formed parts of the north-eastern area of the Kaliṅga country.

A copper-plate inscription[1] was discovered at Sumandala near Khallikot in the Ganjam District. It records the grant of a piece of land in the Khallikot area by a chief named Dharmarāja in 569 A.D. The said chief acknowledged the suzerainty of a king named Pṛthivīvigraha who is stated to have been ruling the Kaliṅga *rāṣṭra* as a nominal feudatory of the Gupta emperors.[2] This inscription shows beyond doubt that the south-western part of coastal Orissa enjoyed the ancient name of Kaliṅga as late as the second half of the sixth century A.D. Soon however the name of the kingdom of the Vigraha dynasty, to which Pṛthivīvigraha belonged, had to be changed.

Another copper-plate inscription[3] of the Vigraha dynasty has been discovered in a locality in the Puri District. The inscription records the gift of a village situated in Dakṣiṇa-Tosalī in 599 A.D. by an independent monarch named Lokavigraha who was apparently one of the successors of Prthivīvigraha of the Sumandala plates. By this time, all vestiges of Gupta suzerainty in Orissa were a thing of the past. But what is more important is that Lokavigraha's kingdom has been mentioned in the epigraph as Tosalī and not as Kaliṅga while he is represented as granting a village in South Tosalī. This shows that Lokavigraha was not only holding sway over Dakṣiṇa-Tosalī, but that he also claimed the lordship of Uttara-Tosalī, *i. e.* North Tosalī. The inscriptions of the Bhauma-Karas of a later age make it clear that the ancient Utkala country in the Balasore region was roughly known as Uttara-Tosalī while Dakṣiṇa-Tosalī comprised the Ganjam-Puri-Cuttack area. We have to determine the reason leading to the introduction of these names about the second half of the sixth century. This however does not appear to be entirely unintelligible from what we know of the history of Orissa in the period in question.

1. *Ep. Ind.*, Vol. XXVIII, pp. 79-85.

2. Jain traditions seem to suggest that the Guptas ended their rule in U.P. and Bihar in Gupta 231 (550 A.D.) and in Bengal and Orissa in Gupta 255 (574 A.D.). See *J.N. Sarkar Com. Vol.*, II, pp. 343 ff.

3. *Ep. Ind.*, *op. cit.*, pp. 328 ff.

As already indicated above, shortly before 500 A.D., the Gaṅgas founded a kingdom about the present Srikakulam District of Andhra Pradesh. They assumed the title of *Kaliṅg-ādhipati*, 'lord of Kaliṅga', and had their capital at the city of Kaliṅganagara situated in the vicinity of modern Śrīkākuḷam. The Vigrahas were holding sway over the Ganjam-Puri-Cuttack area lying immediately to the north-east of the Gaṅga kingdom. Naturally they wanted a new name for their own kingdom to avoid confusion. The new name of their kingdom, *viz.*, Tosalī, seems to have been coined after the name of their capital city. That is to say, the Vigrahas probably had their capital at the ancient city of Tosalī *i.e.* the modern Dhauli in the Puri District. The name Tosalī was also extended to the ancient Utkala country probably due to the expansion of Vigraha power over that area. These facts appear to supply the reason underlying the renaming of the south-western part of coastal Orissa as Dakṣiṇa-Tosalī and the application of the alternative name Uttara-Tosalī to the country of the Utkalas. The names Uttara-Tosalī and Dakṣiṇa-Tosalī were popular in the age of the Bhauma-Karas who flourished between the ninth and eleventh centuries. Later however the name Tosalī lost its popularity and the names Utkala and Uḍra, Oḍra or Auḍra gradually came to be applied to the whole of coastal Orissa and ultimately to the entire Oriya-speaking area.

We have seen how the Utkalas lived in the present Balasore District and its neighbourhood. But the original habitat of the Uḍras cannot be determined. They are not mentioned in very early works. Some manuscripts of the *Manusmṛti* (about the third century A.D.; cf. X. 44) no doubt mention the Uḍras; but many manuscripts of the work read in its place the name of the Aṅgas or Cholas,[1] and either of these two may have been the original reading. The *Nāṭyaśāstra* ascribed to Bharata-muni mentions the people or land called Uḍra; but the work in its present form does not appear to be much earlier than the sixth century A.D.[2] Whatever the antiquity of the name Uḍra may be,

1. Jha, *Manusmṛiti—Notes*, Part I, p. 465.

2. Varāhamihira's *Bṛihatsaṁhitā* (XIV), composed about the first quarter of the 6th century, mentions the Uḍra people separately from the Utkalas and Kaliṅgas (cf. verses 6-8). The *Mahābhārata* mentions the Uḍra, Oḍra or Auḍra people. See Sörensen's *Index*.

it is sometimes used to indicate the whole of coastal Orissa from the sixth and seventh centuries A.D. Probably the Uḍras originally lived in the land lying to the north of the ancient Utkala country. The conquest of Utkala by a king of the Uḍra country may have led to the use of *Uḍra* as a synonym of *Utkala* and the extension of the power of a king of this Uḍra-Utkala region over the south-western part of coastal Orissa at a later date may have been at the root of the use of both the names to indicate the whole of coastal Orissa. The known facts of early Orissan history appear to support this conjecture.

According to a copper-plate inscription[1] discovered at Soro in the Balasore District, in 579 A.D., Uttara-Tosalī, *i.e.* ancient Utkala, was under the rule of a *Mahārāja* named Śaṁbhuyaśas who belonged to the Mudgala or Maudgalya *gotra*. Another inscription[2] of the same monarch, viz. *Paramabhaṭṭāraka* Śaṁbhuyaśas, which was found at Patiakella in the Cuttack District, says that in 602 A.D. his feudatory, *Mahārāja* Śivarāja, was ruling in Dakṣiṇa-Tosalī. This record suggests further that the king Śambhuyaśas, ruler of both Uttara° and Dakṣiṇa-Tosalī, *i.e.* the entire coastal region of Orissa, was born in the Māna family. The name of the present Mānbhūm or Mānabhūmi seems to have been derived from the rulers of this Māna family, also known from a few other records. The Bhauma-Kara monarch Śāntikara II (middle of the tenth century) is known to have married Hīrāmahādevī who was the daughter of king Siṁhamāna probably belonging to the same family as Śaṁbhuyaśas. The Mānas appear to have been ruling over the present Mānbhūm-Singbhūm region together with the adjacent areas of Orissa.[3]

1. *Ep. Ind.*, Vol. XXIII, pp. 201 f.
2. *Ibid.*, Vol. IX, pp. 287 f.
3. It is not impossible that they had their capital at the city of Khijjiṅga, modern Khiching in the northern part of the Mayurbhanj District of Orissa (cf. *Indo-Asian Culture*, Vol. VIII, 1958, pp. 429-30). R.P. Chanda was inclined to assign the earliest antiquities discovered at Khiching to the age of the Ādi-Bhañjas who began to rule from about the beginning of the eleventh century A.D J.N. Banerjea assigns some of these sculptures to the tenth century and others vaguely to the early medieval period. There is little doubt that some Khiching sculptures are earlier than the eleventh century and this fact shows that Khiching was the seat of some pre-Ādi-Bhañja rulers, since the excellent art of Khiching could have scarcely flourished without royal patronage. We have now inscriptions of the pre-Ādi-Bhañja rulers, Dhruvarāja and

We have seen how in 569 A.D. Pṛthivīvigraha was ruling over Kaliṅga, *i.e.* the north-eastern areas of Kaliṅga roughly identical with the later Dakṣiṇa-Tosalī, and how in 599 A.D. Lokavigraha, another monarch of the same family, not only ruled over South Tosalī, but also claimed lordship over North Tosalī. It has also been shown how *Paramabhaṭṭāraka* Śambhuyaśas belonging to the Mudgala *gotra* and the Māna family ruled over Uttara-Tosalī, in 579 A. D. and over Dakṣiṇa-Tosalī in 602 A.D. It is clear from these facts that in the latter half of the sixth century there was a struggle between the Vigrahas and the Mānas for the sovereignty of coastal Orissa and that the Vigrahas, who were at first ruling over the whole of Tosalī, both Uttara and Dakṣiṇa, were gradually ousted by the Mānas, first from Uttara-Tosalī and then from Dakṣiṇa-Tosalī. In this way, the whole of coastal Orissa came under the suzerainty of the Mānas about the beginning of the seventh century. If these Mānas may be regarded as belonging to the Uḍra clan, we can explain the popularity of the name Uḍra in the sense of the whole of coastal Orissa from the sixth or seventh century. The fact that they conquered Utkala or Uttara-Tosalī sometime before the exapansion of their power over Dakṣiṇa-Tosalī may be the reason underlying the use of *Uḍra* as a synonym of *Utkala* first in the sense of the Balasore region and then to indicate also the Cuttack-Puri-Ganjam area. If the Māna family belonged to the Uḍra clan, it may be supposed that the Uḍras originally lived in the Mānbhūm-Singhbhūm region and the adjoining parts of Orissa.

In the first quarter of the seventh century, the greatest monarch in Eastern India was Śaśāṅka, king of Gauḍa, who had his capital at Karṇasuvarṇa near modern Murshidabad in West Bengal. The Gauḍas ousted Māna rule from Orissa and extended their suzerainty as far as Koṅgoda about the borders between the Districts of Puri and Ganjam. In the second quarter of the seventh century, the king of Gauḍa, probably a successor of Śaśāṅka, was disastrously defeated by Harṣavardhana of Kanauj and his ally, Bhāskaravarman of Kāmarūpa. As a result of this humiliation of the Gauḍa king, his Śailodbhava

Kumāravarman, who ruled in the area in the tenth century A.D. Cf. *Ep. Ind.*, Vol. XXXIII, pp. 82 ff.

feudatories in Koṅgoda declared their independence while his Datta feudatories in the Balasore-Cuttack region began to rule semi-independently. Some inscriptions of Somadatta and Bhānudatta of this Datta dynasty have been discovered. It seems that the dominions of these Dattas have been described as Uḍra by the Chinese pilgrim Hiuen-tsang who visited Orissa about 638 A.D. His accounts appear to suggest that the pilgrim applied the name Uḍra to the Balasore-Cuttack-Puri region. King Harṣavardhana, who now probably regarded the Gauḍa king as his subordinate ally, invaded Koṅgoda in 643 A.D. in order to punish the Śailodbhavas on behalf of the Gauḍa monarch.

During the eighth century A.D., the Śailodbhavas continued their independent rule in Koṅgoda in the Puri-Ganjam region. But the political condition of the Balasore-Cuttack area in this age is not quite clear. According to Chinese evidence, a Buddhist king having a name like Śubhaṅkarasiṁha ruled over the Wu-ch'a or Uḍra country in 795 A.D.[1] He seems to have been a contemporary of the Later Śailodbhavas of the Ganjam-Puri area and ruled over the Cuttack-Balasore region. In 831 A.D., the Bhauma-Karas established their capital at the ancient city of Virajā which is the modern Jājpur (Yāyapura or Yājapura). They founded a new city called Guheśvarapāṭaka or Guhadevapāṭaka in the suburbs of Jājpur and this remained the Bhauma-Kara capital till the end of Bhauma-Kara rule in the eleventh century. It was apparently the Bhauma-Karas who overthrew Śailodbhava rule from Koṅgoda which now formed a part of Dakṣiṇa-Tosalī.

The Somavaṁśīs established their power in the Patna-Sonpur region in the Upper Mahānadī valley in the tenth century. The Somavaṁśī king Mahābhavagupta I Janamejaya (*circa* 935-70 A.D.) issued most of his charters from Suvarṇapura (Sonpur). His son and successor, Mahāśivagupta Yayāti I (*circa* 970-1000 A.D.) transferred his capital to the new city of Yayātinagara (modern Binka) built by and named after himself. In the second quarter of the eleventh century, the Somavaṁśī king Mahāśivagupta Yayāti III Caṇḍīhara (*circa* 1025-60 A.D.)

1. Cf. *Ep. Ind.*, Vol. XV, pp. 363-64; Vol. XXIX, p. 84. A king of Orissa named Śubhakarasiṁha, who became a Buddhist monk, is said to have left India in 715 A.D. and reached China the following year when he was 80 years of age (A. Getty, *Gaṇeśa*, pp. 73-74).

extended his power over the coastal regions of Orissa. The story of the transference of the lordship of that region from the Bhauma-Karas to the Somavaṁśīs is not clearly known.[1] But there is little doubt that Yayāti III built a city, named Yayātinagara after himself, in the erstwhile Bhauma-Kara kingdom. This city is mentioned in the *Mādalā Pāñji* as Abhinava-Yayātinagara (*i.e.* the *new* Yayātinagara) in its description of the Gaṅga kings who conquered coastal Orissa from the Somavaṁśīs, but had originally little to do with the upper valley of the Mahānadī. The Gaṅga king Anaṅgabhīma III seems to be described in this work as *Abhinava-Yayātinagara-Viṣṇu.*[2] The Muslim authors of the thirteenth and fourteenth centuries mentioned the Gaṅga kingdom as Jājnagar (no doubt a corruption of the name *Yayātinagara*) apparently after this Abhinava-Yayātinagara of the *Mādalā Pāñjī*, which seems to have continued to remain the capital of coastal Orissa under the Gaṅgas for some time.

About the beginning of the twelfth century A.D., the Gaṅga king Anantavarman Coḍagaṅga (1078-1147 A.D.) of Kaliṅganagara conquered the Puri-Cuttack-Balasore region from the Somavaṁśīs. There is evidence to show that his descendants later transferred the Gaṅga capital to Cuttack. According to the Nagari plates[3] of Anaṅgabhīma III (*circa* 1211-38 A.D.), great-grandson of Anantavarman Coḍagaṅga, that king had his headquarters at Abhinava-Vārāṇasī-kaṭaka, *i.e.* the present Barabati area of Cuttack. The *Mādalā Pāñjī* seems to suggest that Anaṅgabhīma III at first had been residing at Caudvārakaṭaka and later transferred his capital to Vārāṇasī-kaṭaka built by himself on the site of the village of Barabati on the opposite bank of the Mahānadī. As the *Mādalā Pāñjī* uses the name Abhinava-Yayātinagara in the description of this king, it may be supposed that it was the earlier name of Cuttack, which was therefore the source of the name Jājnagar used by the Muslim writers of the early medieval period to indicate the Gaṅga kingdom covering coastal Orissa. But it seems to us that Abhinava-Yayātinagara or Jājnagar should better be identified with modern

1. See *The Struggle for Empire*, pp. 209-10.
2. Ed. Mahanti, p. 28. *Yayāti* is pronounced *Jajāti*.
3. *Ep. Ind.*, Vol. XXVIII, pp. 235 ff.

Jājpur. This is because Jājpur seems to be a corruption of the name Yayātipura which is again practically the same as Yayātinagara, both meaning 'the city of Yayāti'. That Jājpur was once the headquarters of the Imperial Gaṅgas seems to be supported by the following tradition in the description of Anaṅgabhīma in some manuscripts of the *Mādalā Pāñjī* : *rājā Yājapura-nagara caudvāre Kaṭake vije kari thānti.*[1] In the description of the occupation of the Kesarī (*i.e.* Somavaṁśī) kingdom by Anantavarman Coḍagaṅga also we have the passage : *Yājapura Kaṭake praveśa hoilā.*[2]

We have seen that the Bhauma-Kara emperors had their capital at Virajā, *i.e.* modern Jājpur, in the suburbs of which they built their new capital named Guheśvarapāṭaka or Guhadevapāṭaka. In the rule of coastal Orissa, these Bhauma-Karas were succeeded by the Somavaṁśī king Mahāśivagupta Yayāti III Caṇḍīhara. It is possible to think that Yayāti III retained the headquarters of this newly acquired territory at its old capital and that it was he who renamed Guheśvarapāṭaka or Guhadevapāṭaka after himself as Yayātinagara or Yayātipura. The very name of modern Jājpur (*Yayātipura*>*Yayāipura*>*Yāyaipura*>*Yāyipura*>*Jājpur*) appears to support this conjecture. It seems that the capital of coastal Orissa was retained at the same city for some time even after the overthrow of Somavaṁśī rule from that area by the Gaṅgas of Kaliṅganagara. If the Gaṅga conquerors of coastal Orissa ruled the country from Yayātinagara or Yayātipura, *i.e.* modern Jājpur, before the transference of its headquarters to the Cuttack region, we can easily explain why the Muslim writers of the thirteenth and fourteenth centuries mentioned the kingdom of the Gaṅgas in Orissa as Jājnagar (Yayātinagara).

About 1360 A.D., during the reign of the Gaṅga king Bhānu III (*circa* 1352-78 A.D.), Sulṭān Fīrūz Shāh of the Tughluk dynasty of Delhi led an expedition against the Gaṅga capital. An account of this expedition is found in the *Tārīkh-i-Fīrūz Shāhī* by Shams-i-Sirāj 'Afif. In this work, the kingdom of the Gaṅga monarch has been called Jājnagar, but his capital has been mentioned as Vārāṇasī, *i.e.* Vārāṇasī-kaṭaka which is the present

1. Mahanti, *op. cit.*, p. 34.
2. *Ibid.*, p. 22.

Barabati area of Cuttack.[1] It seems that, if the earlier name of Cuttack was Yayātinagara and if that was the reason for the early Muslim writers applying the name Jājnagar to the Gaṅga kingdom, Shams-i-Sirāj would not have used two different names to indicate the capital and kingdom of the Gaṅga king without any comment. Indeed it becomes rather difficult in that case to understand why he uses the new name for the city, but its old name to indicate the kingdom of which it was the capital. The description of the Gaṅga capital and kingdom in the *Tārīkh-i-Fīrūz Shāhī* seems to suggest that Jājnagar as the name of the Gaṅga kingdom had nothing to do with Vārāṇasī or Cuttack which was then its capital.[2]

Muslim rule was established in the western and northern parts of Bengal about the beginning of the thirteenth century. From that time, the Muslim rulers of Bengal often led expeditions against the Gaṅga kingdom. This may have been the cause underlying the transference of the Gaṅga capital from Jājnagar or Jājpur, which was nearer the borders of the Muslim territories of Bengal, to Cuttack which lay further away. But the Muslim writers appear to have continued the use of the name Jājnagar to indicate the Gaṅga kingdom for some time even after the transference of the Gaṅga headquarters from the city of that name. The name Kaṭaka or Cuttack suggests that it was originally a camping ground of the Gaṅga king's forces. It is not impossible that the story of the Gaṅga king's attempt to check Muslim aggression from the east is hidden under this name of the new capital of his kingdom.

II

In the seventh century A.D., the Chinese pilgrim Hiuen-tsang visited, among numerous territories, in the course of his travels in the Indian subcontinent, the two countries called (1) Wu-chang-na in the valley of the river Su-p'o-fa-su-tu in the north-west and (2) Wu-t'u on the shore of the ocean in the

1. See *Ep. Ind.*, Vol. XXVIII, pp. 247-48.

2. It has to be noticed in this connection that the transfer of the capital from Jājnagar to Cuttack seems to have led to the separate mention of Jājnagar and Orissa among the conquests of 'Alāuddīn Husain Shāh (1493-1519 A.D.) in 1493-94 A.D. Cf. H. N. Wright, *Cat. C Ind. Mus.*, Vol II, pp. 144, 173.

east.[1] There has been no doubt that Wu-chang-na lay in the Swat (Suvāstu) valley in the present West Pakistan while Wu-t'u is the same as Uḍra or Oḍra, *i.e.* modern coastal Orissa. The two geographical names are found in Tibetan sources as (1) O-rgyan, U-rgyan or O-ḍi-ya-na, and (2) O-ḍi or O-ḍi-vi-śa, while the first name is found as Yue-ti-yen in the T'ang annals which locate the territory to the south-east of Chitral and to the north of the Indus.[2] It has been rightly identified with the land called Oḍiyāna, Uḍḍiyāna, etc., in Sanskrit literature, a Buddhist monk of the place being mentioned in a Mathurā inscription of the year 77 of the Kaṇiṣka era, which probably corresponds to 155 A.D.[3]

There can really be no confusion between Uḍḍiyāna and Oḍra (Orissa) although a few writers even now regard the two names as identical and as referring to Orissa.[4] This is because these writers consider the important evidence referred to above less convincing than certain doubtful traditions.[5] It is however not usually noticed that the confusion between Uḍḍiyāna and Orissa is a legacy of the early medieval period. Uḍḍiyāna does not appear to have been so familiar to some of the writers of that age as Oḍra, etc., so that the less known was replaced by the well-known exactly as in the case of the substitution of *Vaṁkṣu* (the Oxus) by *Sindhu* (the Indus) in Kālidāsa's Raghuvaṁśa, IV. 67.[6]

The *Kālikā Purāṇa*, probably incorporated originally in the *Rudrayāmala Tantra*, is believed by scholars to have been compiled in Assam sometime before 1000 A.D.[7] This work recognises the following four *Pīṭha-sthānas* : Oḍra or Uḍḍiyāna in the

1. Watters, *On Yuan Chwang's Travels in India*, Vol. I, pp. 225ff.; Vol. II, pp. 193 ff.; Beal, *Buddhist Records of the Western World*, Calcutta reprint, Vol. II, pp. 166ff.; Vol. IV, pp. 410 ff.

2. Lévi, *Journal Asiatique*, 1915, pp. 105 ff ; Bagchi, *Studies in the Tantras*, pp. 37 ff.

3. Lüder's List, No. 62.

4. Cf. Bagchi, *loc. cit.*; Sircar, *The Śākta Pīthas*, p. 12, note 3.

5. Bagchi, *loc. cit.*

6. The name of the Vaṁkṣu is found in Vallabha's commentary while Mallinātha and others have *Sindhu* instead.

7. Cf. *Journal of Oriental Research*, Vol. X, pp. 289 ff.; *Journal of the Orissa Academy*, Vol. II, p. 60; Hazra, *Studies in the Puranic Records on Hindu Rites and Customs*, p. 53; Sircar, *The Śākta Pīṭhas*, p. 12, note 5; p. 17, note 4.

west, Jālaśaila or Jālandhara in the north, Pūrṇagiri in the south and Kāmarūpa-Kāmagiri in the east. Of the two sections on this topic, the first speaks of the western *Pīṭha* as Oḍra the place of Jagannātha, the *Oḍreśa*, and Kātyāyanī-Śivā, the *Oḍreśvarī*,[1] while the second mentions Uḍḍiyāna as the seat of Kātyāyanī in place of Oḍra.[2] Since this confused Oḍra = Uḍḍiyāna has been located in the western part of the Indian subcontinent and not in the east or south, the intended country cannot really be Orissa.

There is moreover clear evidence in Indian literature to show that Orissa was different from Uḍḍiyāna. The geographical name written as *Orissa* in English is spelt as *Oḍiśā* in the language of the Oriyas. This form seems to have been derived from *Auḍrīya-viṣaya* through intermediate forms like *Oḍḍī-visaa* (cf. Tibetan *Oḍi-viśa* referred to above).[3] In medieval Sanskrit literature, Orissa is sometimes called *Uḍḍīśa*[4] which looks like re-Sanskritised from *Oḍiśa*.

The *Jñānārṇava Tantra* which was composed considerably earlier than the sixteenth century A.D., contains two lists of *Pīṭha-sthānas*, one enumerating only eight and the other no less than fifty names.[5] The second of these two lists is also found in the *Tantracūḍāmaṇi* as well as in Brahmānanda's *Śāktānanda-taraṅgiṇī*.[6] It is again closely followed in a section of Kṛṣṇānanda Āgamavāgīśa's *Tantrasāra*.[7]

1. See Vaṅgavāsī ed., p. 410;
   Oḍi-ākhyam prathamam pīṭham dvitīyaṁ Jālaśailakam |
   tṛtīyaṁ Pūrṇapīṭham tu Kāmarūpam caturthakam ||
   Oḍra-pīṭham paścime tu tath=aiv=Oḍreśvarīm Śivām |
   Kātyāyanīṁ Jaganāthan=Oḍreśañ=ca prapūjayet ||
2. *Ibid.*, pp. 79-80 :
   Devīkūṭe pāda-yugmaṁ prathamaṁ nyapatat=kṣitau |
   Uḍḍiyāne c=oru-yugmam hitāya jagatām tataḥ ||
   Kātyāyanī c=Oḍḍiyāne Kāmākhyā Kāmarūpiṇī [illegible] |
   Pūrneśvarī Pūrṇagirau Caṇḍī Jālandhara-girau ||
3. See above, p. 167.
4. Sircar, *The Śākta Pīṭhas*, p. 97, *s.v.*
5. *Ibid.*, pp. 18, 20-21; see also *Jñānārṇava Tantra*, Ānandāśrama ed. Paṭalas V (verses 66-67) and XIV.
6. Sircar, *The Śākta Pīthas*, p. 21, note 1.
7. *Ibid.*, p. 23, note 3.

This list has the following two stanzas:

(1) *Pīṭham=Ujjayinīñ=c=aiva vicitraṁ Kṣīrik-ābhidham* ।
Hastināpura-pīṭhañ=ca Uḍḍīśañ=ca Prayāgakam ॥

(2) *Mahendraṁ Vāmanañ=c=aiva Hiraṇyapuram=eva ca* ।
Mahālakṣmīpura-pīṭham=Uḍḍīyānam=ataḥ param ॥

The separate mention of Uḍḍīśa (Orissa) and Uḍḍīyāṇa (Uḍḍiyāna) in the same list of *Pīṭha-sthānas* contained in a well-known popular text of medieval India shows beyond doubt that the two cannot be identical even if the Chinese and Tibetan evidence regarding the location of Uḍḍiyāna in the Swat valley is considered unsatisfactory by some in spite of the fact that it has completely satisfied a large number of scholars.

Chapter XI

DAŚĀRṆA, AŚMAKA AND KUNTALA

I

Our attention has been drawn to a verse contained in Lakṣmīdhara's *Kṛtyakalpataru*, which refers to a pious lady named Vasundharā who was the daughter of a certain Vikrama and was an inhabitant of Daśārṇa.[1] In this connection, several suggestions have been offered; but none of them is supported by any evidence or argument of any kind. Vikrama has been identified with Candragupta II Vikramāditya and Vasundharā with Prabhāvatīguptā, daughter of Chandragupta II and queen of Vākāṭaka Rudrasena II. We wonder why the daughter of any person of the name of Vikrama has to be identified with the only known daughter of only one of the numerous Vikramas known to history.

The suggestion that the Vākāṭaka territory was known as Daśārṇa which is identified with the Chattisgarh Division of M. P. as well as the reference to '*Vijaya*-Daśanapura in the Andhra country mentioned in some inscriptions as the capital of a Vākāṭaka principality' in Jayaswal's *History of India*, p. 136, is also untenable. Here is a clear case of the *andha-paramparā-nyāya*, as Daśanapura, mentioned as a seat of government of the Early Pallavas (who, it should be noted, ruled contemporaneously with the Vākāṭakas) and identified by scholars with modern Darsi in the Nellore District, had absolutely nothing to do with the Vākāṭakas. As a matter of fact, there is not the slightest evidence of Vākāṭaka rule in the Chattisgarh region of M. P., the ancient name of which was, moreover, Kosala (South Kosala) and never Daśārṇa. From numerous references to Daśārṇa in Indian literature,[2] it is clear that it was one of the old names of East Malwa and the adjoining region, with its capi-

1. See D. B. Diskalkar in *IHQ*, Vol. XX, 1944, pp. 367-68.
2. Cf., *e.g.*, *Meghadūta*, 24-25.

tal at Vidiśā (modern Besnagar near Bhilsa, Madhya Pradesh) and with the rivers Daśārṇā (modern Dhasan) and Vetravatī (modern Betwa) running through it.

In this connection, the attention of scholars may be drawn to another almost equally unjustifiable theory that the Vākāṭaka territory was known as Kuntala.[1] It is easy to show that the heart of the Vākāṭaka kingdom corresponded to the ancient *janapada* of Vidarbha (modern Berar and the adjoining regions) where most of the records of the family have been discovered. The founder of the Vākaṭaka family seems to have had something to do with East Malwa and the Vindhyan region and a few other members of the main branch of the family probably held sway over parts of Central India (especially Bundelkhand) after the eclipse of Gupta power in that area.[2] But the capital of a branch of the Vākāṭaka line was at the city of Vatsagulma, modern Basim in the Akola District, while the main branch of the family ruled from a place not very far away from modern Nāgpur.[3] The Vākāṭaka kings reigned in Vidarbha from the second half of the third century to the first half of the sixth. They were not only responsible for some of the magnificent caves at Ajaṇṭā; but it was apparently at their court that the celebrated Vaidarbhī *rīti* or the Berar style of Sanskrit composition originated and flourished and was recognised by the author of the *Kāvyādarśa* as the best style as early as the seventh century A.D. We have elsewhere suggested that the other important style known as the *Gauḍī rīti* has to be associated with the court of the dynasty of Gauḍa kings represented by Gopacandra and Śaśāṅka.[4]

In connection with the name Daśārṇa, it is interesting to note that the *Periplus of the Erythraean Sea*[5] place 'the region of Dosarene yielding the ivory known as Dosarenic' towards the east of Masalia (*i.e.* the country around Masulipatam in Andhra).

1. Cf. *The Successors of the Sātavāhanas*, 1939, p. 253, note.
2. Cf. *The Age of Imperial Unity*, p. 217.
3. See *The Classical Age*, pp. 181-82, 185.
4. Cf. above, pp. 125 ff. For the *Udīcya*, *Pratīcya*, *Dākṣiṇātya* and *Gauda* styles of composition, see also Bāṇa's *Harṣacarita*, I.7.
5. See Schoff's trans., p. 47.

Ptolemy mentions the city of Dosara[1] apparently as the metropolis of this land. He locates Dosara in the land of the Kokkonagai who lived to the west of the country watered by the mouths of the Ganges. Some scholars trace the Indian name Daśārṇa in Dosarene.[2] But it is impossible to locate the places mentioned by the Greek authors outside modern Orissa. It seems that *Dosara* is a modification of the Indian name *Tosala* (the same as *Tosalī* or *Tosalā*), identified with modern Dhauli in the Puri District. In early times, Tosalī was the capital of the Kaliṅga country.

II

A Rāṣṭrakūṭa ruler named Mānāṅka (who seems to have originally been a *rāṣṭrakūṭa* or subdivisional ruler) is known from the Undikavatika grant[3] of his great-grandson Abhimanyu who resided at Mānapura. He has been identified with king Mānāṅka, grandfather of Avidheya who issued the Pandurangapalli grant discovered in the neighbourhood of Kolhapur, There is reason to believe that the territories over which these rulers held sway lay in the Kolhapur region and the adjoining area of the South Marāṭhā country, and Mirashi may be right in identifying their capital with Man in the Satara District. Rāṣṭrakūṭa Bibhurāja of the Hingniberdi plates, Dejja-mahārāja of the Gokak plates, and Govindarāja, son of Śivarāja, who is known from the Naravana grant of 743 A.D. of the time of Vikramāditya II and seems to have been the subordinate ruler of a territory in the Satara-Ratnagiri region, may have been scions of this family.[4] The land ruled by this family seems to have been known as 'Māna's territory.'[5]

The Pandurangapalli charter[6] appears to describeMānāṅka,

1. *Geography*, VIII. 1. 77.
2. Cf. *JAS*, Letters, Vol. XVI, p. 266; Schoff, *The Periplus*, trans., p. 253.
3. *Ep. Ind.*, Vol. VIII, pp. 165 f.
4. Cf. *ibid.*, Vol. XXIX, pp. 174 ff.; Vol. XXI, pp. 289 ff.; *The Classical Age*, p. 199. Śvarāja's son Govindarāja is described as the grands of Nannappa in the Salem plates (*Ep. Ind.*, Vol. XXVII, pp. 145 ff.).
5. See below, pp. 193-94.
6. Cf. *ABORI*, Vol. XXV, pp. 36 ff.

who probably flourished in the latter half of the fifth century, as the conqueror of Vidarbha and Aśmaka and also as the chastiser (*praśāsitā*) of the Kuntalas. The Kuntala people under reference were no doubt the Kadambas of Vanavāsī (Banavāsi), whose territories comprised the North Karnara District and parts of Mysore, Belgaum and Dharwar. In the inscriptions of the Later Kadambas, the progenitor of the Kadamba family (sometimes called Mayūravarman, but in some cases Mukkana or Trilocana Kadamba) is represented as the ruler of the Kuntala country which is described as the land round the capital city of Banavāsi in the North Kanara District. The country of Vidarbha lay on both sides of the river Varadā (modern Wardha, a tributary of the Godāvarī) according to the *Mālavikāgnimitra*, and roughly corresponded to the major part of Berar and the western part of M. P. Its ancient capital was at the city of Kuṇḍina which has been identified with modern Kauṇḍinyapura on the Wardha in the Chandur Taluk of the Amaravati District of Berar. The city of Padmapura, which was the birth-place of the poet Bhavabhūti and was situated in the Vidarbha country in Dakṣiṇāpatha, has been rightly identified with modern Padampur near the Amgaon railway station in the Bhandara District of M. P. Vidarbha therefore included at least the Amaravati region in the west and the Bhandara area in the east. Epigraphic evidence shows that, about the time of Mānāṅka, the above land was entirely in the possession of kings belonging to the main branch of the Vākāṭaka family that had its headquarters near modern Nāgpur. Thus it will be seen that the Pandurangapalli grant seems to represent Mānāṅka of the south Marāṭhā country as having fought successfully with the Kadambas of Kuntala and the Vākāṭakas of Vidarbha.

But who were the Aśmakas, also mentioned in connection with the victorious campaigns of the Rāṣṭrakūṭa ruler? It appears to us that the Aśmakas under reference are no other than the Vākāṭakas of Vatsagulma which is mentioned separately from Vidarbha in the *Kāmasūtra* (V. 5. 33-34). The capital city of this branch of the Vākāṭaka family has been identified with modern Basim in the Akola District about the southern fringe of Berar; but their dominions certainly included the Ajaṇṭā region in the Aurangabad District and very probably also Nāndīkaṭa, identified with the Nander District, both in the northern part of the old

Hyderabad State. The Vākāṭakas of Vatsagulma therefore ruled over the southernmost region of Berar and the northern part of old Hyderabad, and the ancient Aśmaka country has been located by scholars precisely in this region.[1]

The *Pārāyaṇa*, incorporated in the *Suttanipāta*, speaks of a sage named Bāvari who was an inhabitant of Śrāvasti, but settled 'in the country of Aśmaka, in the vicinity of Mūlaka, on the bank of the Godāvarī.'[2] While describing the journey of Bāvari's disciples from the sage's hermitage in Aśmaka to a locality in Northern India, the same work says that the first place reached was Pratiṣṭhāna (modern Paiṭhan on the Godāvarī in the Aurangabad District) which was the capital of the Mūlaka country ; the men are said to have next reached Māhiṣmatī on the Narmadā, and then Ujjayinī outside the limits of Dakṣiṇāpatha. This shows that the Aśmaka country lay immediately to the south of Mūlaka which is the Paiṭhan region of the Aurangabad District. The ancient capital of the Aśmaka country was, according to the *Mahābhārata*, at the city of Paudanya which is known as *Potana* (*Podanna, Paudanya*) in the Pali literature . A variant of the form *Potana* is *Potali* which seems to be a mistake for *Potaṇa* or *Potaṇi* (cf. the striking similarity between the forms of the two letters *l* and *ṇ* at some stages of development), the latter being a possible corruption of *Paudanya* through another intermediate form *Podaniya*.[3] Raychaudhuri identifies Paudanya with modern Bodhan near the Godāvarī in the Nizamabad District abutting on the Nander District in the old Hyderabad State. In a narrow sense therefore the Aśmaka country may be identified with the Nander-Nizamabad region of Andhra Pradesh and the adjoining area. In ancient literature, however, Aśmaka is often represented as including Mūlaka, *i.e.* the Paiṭhan area of the Aurangabad District, and as abutting on Kaliṅga (roughly speaking, the coastal land between the Mahānadī and the Godāvarī), Vidarbha, Aparānta (the Northern Konkan) and Avanti, doubtless the celebrated Avanti-Dakṣiṇāpatha with its capital at Māhiṣmatī

1. Cf. *PHAI*, pp. 76, 21-22 ; *NHIP*, Vol. VI, p. 88; *JAHRS*, Vol. IX, iii, p. 1 ff.

2. See the Chapter on Gonarda (Ch. XIX) below.

3. Cf. *śākya=śākiya=śaki*; *mūlya =muliya=muli*; *ārogya=ārogyia=arogi*, etc. in my *Gram. Prak. Lang.*, p. 23.

on the Narmadā, identified with Mandhata in the Nimar District or Maheswar in the former Indore State. This wider sense must have been in the mind of Bhaṭṭasvāmin when he identified Aśmaka with Mahārāṣṭra, i. e. the Marāṭhā country. But the heart of this land appears to have been the northern districts of old Hyderabad including naturally the southernmost region of Berar. The heart of the ancient Aśmaka country thus seems to have corresponded to the dominions of the Vākāṭakas of Vatsagulma. Rāṣṭrakūṭa Mānāṅka therefore claimed victory not only over the Kadambas of Kuntala and the Vākāṭakas of Vidarbha, but also over the Vatsagulma branch of the Vākāṭaka dynasty of Aśmaka.

III

There have been some comments[1] on our views regarding the identification of the Kuntala and Aśmaka countries.[2] An attempt has been made to prove that 'Kuntala in ancient times did not comprise only the North Kanara District and parts of Mysore, Belgaon and Dharwar Districts as stated by Dr. Sircar, but that it extended much further to the north so as to include what we now call the Southern Marāṭhā Country.' Unfortunately it has not been noticed that practically the same view has been expressed by many writers including Fleet[3] and ourselves.[4] The fact is that where other scholars would locate, according to evidence, the Kuntala country proper in the heart of the Kannaḍiga area and distinguish between that land and the later empire (often called Kuntala) of the imperial Kuntala or Karṇāta (*i.e.* Kannaḍiga) dynasties, our critic would place Kuntala proper in the Marāṭhā country and would not distinguish the essential Kuntala country from the gigantic Kuntala or Kannaḍiga empire of later times. If, in the days of Warren Hastings, the Banaras District formed a part of Bengal, would it justify any one to identify Bengal with Banāras or to locate Bengal in Banaras ? The name Vaṅgāla (Bengal) originally indicated a small district in Southern Bengal; it was later

1. See V. V. Mirashi in *IHQ*, Vol. XXII, pp. 309-15.
2. Cf. *ibid.*, p. 233 ff.
3. *Bomb. Gaz.*, Vol. I, Part ii, p. 431.
4. *Suc. Sāt.*, p. 215. Cf. below, p. 246 and note 1.

used to signify the whole country comprising such ancient lands as Vaṅga, Samataṭa, Suhma, Tāmralipta, Gauḍa and Puṇḍravardhana. Can we locate the ancient Vaṅgāla country in any part of the later Vaṅgāla (Bengal) we like ?

Most of the questions raised in this connection were already answered in our *Successors of the Sātavāhanas* referred to above. It will be clear from the following quotations from that work :

"Cf. a record of A.D. 1077......: 'In the centre of that middle world is the Golden Mountain to the south of which is the Bhārata land in which, like the curls of the lady earth, shines the Kuntala country to which an ornament......is Banavāsi.' Some other inscriptions also prove that Kuntala was the district round Banavāsi."[1]

"Kuntala and Karṇāṭa are used as synonymous in the *Vikramāṅkadevacarita* by Bilhaṇa.[2] Vikramāditya VI has been called both *Kuntalendu* (or, *Kuntalendra*) and *Karṇāṭendu*...... Vaijayantī, identified with Banavāsi, has been described as a *tilaka* (that is to say, the capital) of the Karṇāṭa country in the Birur grant of Viṣṇuvarman......Karṇāṭa therefore signified the same territory as Kuntala or the country of which Kuntala formed a part......The separate mention of Kuntala, Karṇāṭa Banavāsi, Māhiṣaka (cf. Mahiṣa-visaya in a Kadamba grant), etc., in some of the traditional lists may possibly refer to the fact that these names originally, signified separate geographical units abutting on one another. Sometimes however one of them may have formed the part of another ; cf. the case of Tāmralipta

1. The tradition about Banavāsi having been the capital of the Kuntala country was remembered even after the foundation of the Kuntala-Karnāta (Kannadiga) empire. For a tenth century record mentioning the Kuntala king residing at *Vana āsa*, indicating the city as well as 'exile', see *Ep. Ind*, Vol. XXII, p. 132 It has to be remembered that, in the age in question the capital of the said empire was at Mālkhed in the Gulbarga District, Mysore.

2. Note that the Cālukyas of Badami (*Ep. Ind*, Vol. IV, p. 88, etc.) and Kalyāna (*ibid.*, Vol. V, p. 16, etc.) were regarded as kings of Kuntala, while the Cālukya army was called the Karnāṭaka *bala* (*Ind. Ant.*, Vol. XI, p. 112, etc.). In Cola records like the Kanyākumāri inscription (*Ep. Ind.*, Vol. XVIII, p. 27; vv 69, 76) the Cālukya enemies of Rājendra I are mentioned as lords of the Kuntalas, while the inscriptions of Vīrarājendra mention them as kings of the Karnāta family.

which is mentioned in literature as an independent state, as a part of Suhma and also as a part of Vaṅga; also the case of Taxila (Raychaudhuri, *Stud. Ind. Ant.*, pp. 186 f.). With the rise of Kanarese powers like the Cālukyas and the Rāṣṭrakūṭas, the name Karṇāṭa (sometimes also the name Kuntala) extended over a large part of the Western and Southern Deccan. In the *Kaliṅgattuparaṇi*, the Cālukyas have been descibed as the *Kuntalar*, 'lords of Kuntala' (literally, 'the Kuntalas,' indicating their Kanarese origin)....An inscription of Harihara II dated in Śaka 1307 ......says that Vijayanagara (modern Hampi in the Bellary District) belonged to the Kuntala *viṣaya* of the Karṇāta country."

Only a few words are required now to be added to the above. As Vidarbha and Vatsagulma are separately mentioned in the *Kāmasūtra*, our critic thinks that the former included the latter and that they are separately mentioned according to the Māṭhara-Kauṇḍinya *nyāya*. It will be clear from the extracts quoted from our book that we would partially apply the same *nyāya* to explain the separate mention of Vanavāsī and Kuntala in the *Mahābhārata* and the *Vāyu Purāṇa* on which the critic takes his stand. The *Daśakumāracarita* and the *Kāvyamīmāṁsā* were composed in an age when the name Kuntala was often applied to the Kannaḍiga empire of the Cālukyas and their successors,[1] which usually included the Kuntala country proper. The separate mention of Banavāsi indicating Kuntala proper and Kuntala signifying the Kannaḍiga empire in these works has nothing to do with the critic's location of the Kuntala country proper in the South Marāṭhā country.

The critic suggests that the root *śās* or *praśās* means governing when the object is a territory and chastising when the object is a living being and that, in the passage *śrīmat-Kuntalānāṁ praśāsitā*, the word *praśāsitā* means a ruler because the object of *praśās* in this case is not a living being but the Kuntala country. Unfortunately, he does not notice that even if the distinction in the meaning of the verb is conceded, there is absolutely no reason why *Kuntala* (like *Surāṣṭra* in verse 11 of the Junagadh inscription of Skandagupta) should be taken in the sense of the Kuntala country (an inanimate object) and not in that of the

1. See *Suc. Sāt.*, p. 216, note 1.

Kuntala people. There can certainly be no objection if *śrīmat-Kuntalānāṁ praśāsitā* is translated as 'the chastiser of the illustrious Kuntala people.'

We do not subscribe to the critic's identification of the Ṛṣika country with Khandesh. Sylvain Lévi rightly regards Ṛṣika as the southernmost country in Gautamīputra Śātakarṇi's empire, to the south of Aśmaka.[1] According to the reading of the Hathigumpha inscription preferred by Barua and ourselves, the city of Ṛṣikanagara (capital of the Ṛṣika country) was situated on the Kṛṣṇabenā (Kṛṣṇā). As regards the critic's identification of Aśmaka with the Ahmadabad and Bhir Districts, it is really impossible for us to understand why the Nander-Nizamabad region lying immediately to the east of that area could not be included. We consider Raychaudhuri's indentification of Paudanya, the Aśmaka capital, with Bodhan in the Nizamabad District as exceptionally satisfactory and suggest that even the Ahmadabad-Bhir area or parts of it may have been included in the Aśmaka country and in the dominions of the Vākāṭakas of Vatsagulma. Nothing more can be said in the present state of our knowledge. As however the Vākāṭakas of Vatsagulma are known to have ruled over the northern part of the old Hyderabad State, which is the ancient Aśmaka country proper according to many writers including Raychaudhuri and ourselves, they may have been regarded as the lords of Aśmaka. As regards the inclusion of Vatsagulma in Vidarbha, suggested by writers like Rājaśekhara (about the beginning of the tenth century), we may draw the critic's attention to what has been said about Tāmralipta and Takṣaśilā in the extracts quoted from our book. Vatsagulma, like those localities, appears to have been sometimes a separate state, sometimes a part of Aśmaka and sometimes a part of Vidarbha. It is well-known that Mūlaka or the land round Paiṭhan in the Aurangabad District was sometimes a separate country, but was often regarded as a part of Aśmaka.

IV

The country round the city of Mānapura, founded by Mānāṅka, appears to have been known in the early medieval

1. Cf. below ;Chapter XIX.

period as Māna-deśa which included modern Velāpur about eleven *kos* to the west of Pandharpur (possibly the same as Pāṇḍuraṅgapalli mentioned in the record of Mānāṅka's grandson Avidheya) in the Sholapur District of the South Marāṭhā country. Cf. *Māṇadeśa-saṁbaddha-Velāpura*, *Māṇadeśa-saṁbaddha-sarvādhikārī Brahmadeva-rāṇa*, etc., in the three Velāpur inscriptions (1300 and 1305 A.D.) of the Yādava king Rāmacandra of Devagiri in G.H. Khare's *Sources of the Medieval History of the Dekkan*, Vol. I, pp. 78-80; Vol. II, p. 9. In that case, the dominions of the Rāṣṭrakūṭas of Mānapura comprised parts of the Ratnagiri, Satara and Sholapur Districts of Bombay.

The name of this land reminds us of the former District of Mānbhūm (i.e. Mānabhūmi) in the south-western border of Bihar. As we have elsewhere[1] suggested, the Mānas who gave their name to this area probably ruled over a big kingdom covering parts of the Manbhum and Singhbhum regions of Bihar and of the Mayurbhanj District of Orissa and having its headquarters at Khijjiṅga (modern Khiching) in the northern part of the Mayurbhanj District.

1. See above, pp. 187-88.

Chapter XII

KĀMBOJA, PĀNCĀLA AND MĀLAVA

I

There has been considerable speculation on the location of the country of the well-known Kamboja people in the Uttarāpatha division of ancient Bhāratavarṣa. Wilson located it in the Ghaznī region to the south of Kabul,[1] and N. L. Dey in 'Afghanistan, at least its northern part'.[2] Kalhaṇa's *Rājataraṅgiṇī* (IV. 165-66) mentions the Kambojas along with the Tukhāras of the upper Oxus valley including Balkh and Badakhshan, and Stein identifies the Kamboja country with the eastern part of Afghanistan.[3] H. C. Raychaudhuri places this land in the areas extending from Rajauri to the south of Kashmir in the east to Kafiristan in Eastern Afghanistan in the west.[4] Among recent writers on the subject, Lassen's location of the Kambojas in the Ghalcha-speaking areas of the Pamirs has been accepted by many,[5] while C. A. Lewis has supported the location of the people in Kafiristan.[6]

The evidence in favour of the various suggestions lacks definiteness with the exception of the theory of Raychaudhuri who draws our attention to a passage in the *Mahābhārata* (VII. 4. 5), according to which Karṇa defeated the Kambojas at Rājapura. Raychaudhuri identifies this Rājapura with Rajauri mentioned as Ho-lo-she-pu-lo (Rājapura) by Hiuen-tsang.[7] Unfortunately the evidence loses its definite character if the expression *rājapura* in the *Mahābhārata* is understood in the sense of *rājadhānī* or the capital city of Kambojas. Even otherwise,

1. *JASB*, 1838, pp. 252, 267.
2. *Geog. Dict.*, s. v.
3. *Kalhaṇa's Rājataraṅgiṇī*, Vol. I, p. 136, notes.
4. *Pol. Hist. Anc. Ind.*, 1938, pp. 125 ff.
5. Cf. *Purāṇa*, Vol. V, No. 1, p. 172; see also Jayachandra, *Bhāratbhūmi aur uske Nivāsī*, Saṁvat 1987, pp. 297 ff.; Motichandra, *Geog. Econ. Stud.*, 1945, pp. 32 ff.; V. S. Agrawala, *India as known to Pāṇini*, 1953, pp. 48-49; etc.
6. *Purāṇa*, Vol. IV, No. 1, pp. 133 f.
7. Watters, *On Yuan Chwang's Travels in India*, Vol. I, p. 284.

the identification of Rājapura of the *Mahābhārata* with Rajauri cannot possibly be regarded as certain.

The most significant fact about the mention of the Kamboja people in the early literary and epigraphic records of India is that they are usually associated with the Yavanas and often with both the Yavanas and Gandhāras. Among the Rock Edicts of the great Maurya emperor Aśoka (third century B.C.), Rock Edict V has Yavana-Kamboja-Gandhāra,[1] while we have Yavana-Kamboja in Rock Edict XIII with the omission of Gandhāra.[2] The *Mahābhārata* (XII.207, 43) agrees with Aśoka's Rock Edict V in mentioning Yauna (Yavana)-Kamboja-Gandhāra.

The Assalāyanasutta of the Buddhist canonical work *Majjhimanikāya* (43. 1. 3) not only speaks of Yavana-Kamboja, but also says that the same social system prevailed in the western countries including the lands of the Yavanas and Kambojas. It is stated that in the areas in question, instead of the *catur-varṇa*, there were only two social grades, viz. the nobleman and the serf.[3] A similar information is supplied by Rock Edict XIII of Aśoka which says, "There is no country except the land of the Yavanas where these classes do not exist, viz. the Brāhmaṇas and the Śramaṇas."[4] According to the *Harivaṁśa* (1. 14. 16) and a number of the Purāṇas,[5] another custom peculiar to the Yavanas and Kambojas was that both had shaven heads.

The above facts suggest that the Kambojas were intimately associated with the Yavanas and the Gandhāras, particularly with the former, and it may be supposed that all three of them lived in contiguous areas of Uttarāpatha in the age of the Mauryas. But the Aśokan records speak sometimes of the Yavanas alone, sometimes of the Yavanas in the company of the Kambojas or of the Kambojas and the Gandhāras. There is no doubt that the inscriptions give prominence to the Yavanas

1. Hultzsch, *CII*, Vol. I, pp. 191-92.
2. *Ibid.*, p. 211.
3. योनकंबोजेसु अञ्ञेसु च पच्चंतिमेसु जनपदेसु द्वेव वण्णा अय्यो चेव दासो च। अय्यो हुत्वा दासो होति दासो हुत्वा अय्यो होतीति ।
4. Hultzsch, *op. cit.*, p. 209.
5. See, e. g., *Vāyu Purāṇa*, 88. 140; etc.

who were apparently the most important people of the north-western province of the Maurya empire. But whether the absence sometimes of Gandhāras and sometimes of both the Gandhāras and Kambojas from the references to the Yavanas suggests that the Kamboja an d Gandhāra lands were at times broadly included in the Yavana country cannot be determined from the above Aśokan epigraphs. In the age of the Indo-Greeks, as we know from coins and the *Milindapañha*, the Yavana country must have included Afghanistan in the west and Śākala (modern Sialkot in the Punjab).[1] Another fact to be remembered in this connection is that, before the rise of the Yavanas to prominence in Uttarāpatha, the most important peoples of that region were the Gandhāras and the Kambojas. This is indicated by the list of the sixteen Mahājanapadas flourishing in the age of the Buddha as enumerated in the *Aṅguttaranikāya* (VII.5.3 and 5), which mentions the Gandhāras and Kambojas but omits the Yavanas.

The Aṅkuravatthu section of the *Petavatthu* (verses 257-58) suggests that there was a direct caravan route between Dvārakā (Dvāravatī) in Kathiawar and the country of the Kambojas.[2] There is also no doubt that the Kamboja capital was directly connected with the capital of Gandhāra and of some other *Mahājanapadas* by an easy route. Now, of the three important peoples living in Uttarāpatha during the Maurya age, viz. the Yavanas, Kambojas and Gandhāras, the third is definitely known to have inhabited the modern Rawalpindi-Peshawar region of West Pakistan, since two of the most important Gandharan cities of the early period were Takṣaśilā in the Rawalpindi District and Puṣkarāvatī or Puṣkalāvatī in the Peshawar District.[3] This consideration seems to go against

1. Cf. Malalasekera, *Dictionary of Pali Proper Names*, s.v.
2. यस्स अत्थाय गच्छाम कंबोजं धनहारका ।
   यानं आरोपयित्वान खिप्पं गच्छाम द्वारकं ॥
3. गंधारविषये सिद्धे तयोः पुर्यौं महात्मनः ।
   तक्षस्य दिक्षु विख्याता रम्या तक्षशिला पुरी ।
   पुष्करस्यापि वीरस्य विख्याता पुष्करावती ॥ (*Vāyu Purāṇa*, 88. 189-90). See also *Rāmāyaṇa*, VII. 114. 11; H. C. Raychaudhuri, *op.cit.*, pp. 50 ff., pp. 124-25. Hiuen-tsang speaks of Puruṣapura (Peshawar) as the

the location of the Kamboja country in the Pamirs. Another important fact, generally forgotten, is that the territory called Bactria or Bactriana by the Greeks and Bāhlīka by the Indians formed no part of the Maurya empire. This land, bounded by the Hindu Kush in the south and east and by the Oxus in the north-east and having its headquarters at the site of modern Balkh in Northern Afghanistan, was a province of the Seleucid empire till the middle of the third century B.C., when it became an independent kingdom under Diodotus I. This fact suggests that the Pamirs lying beyond the Hindu Kush also lay outside the empire of Aśoka. It is therefore difficult to believe that Aśoka's Kamboja subjects lived in the Pamirs. We have now also to think of the light thrown on the problem by the recently discovered Kandahar inscriptions of Aśoka.

According to the *Mahāvaṁsa* (XXIX. 40), the chief city of the Yavana country was Alasanda or Alasandā which has been identified by some scholars with Alexandria founded by Alexander the Great near Kabul.[1] According to the *Milindapañha*,[2] the Dvīpa or Dwab of Alasanda was two hundred *yojanas* from Śākala, capital of the Yavana king Milinda (i. e. the Indo-Greek king Menander who flourished about the close of the second century B.C.), the distance between Śākala and Kashmir being only twelve Yojanas. The Kandahar Edict[3] of Aśoka has two versions, one of which is written in the Greek language and alphabet. There is no doubt that this part of the Edict was meant for Aśoka's Yavana or Greek subjects who lived in Afghanistan, particularly in the Kandahar region. Thus the concentration of the Greek population in the Maurya empire was in the said area. Since the name of Kandahar is derived from the Greek name Alexandria ('Alexandria among the Arachosians' founded by Alexander), it is also possible to think that Alasanda of the *Mahāvaṁsa* is identical with Kandahar. The Dwab of Alasanda may in that case be the land between the Helmund and Tarnak rivers in Southern Afghanistan.

capital of Gandhāra (Watters, *op.cit.*, pp. 198-99), while Al-Birūnī mentions Vaihand (Udabhāṇḍa or Und) as its capital (Sachau, *Alberuni's India*, Part I, p. 259).

1. Geiger, *Mahāvaṁsa*, p. 194. We have elsewhere suggested the location of an Alasanda near the mouth of the Indus. See below, Chapter XIV.
2. Ed. Trenckner, pp. 82-83.
3. *Ep. Ind.*, Vol. XXXIII, pp. 1 ff.

The other version of the Kandahar Edict is written in the Aramaic language and script which were adopted by the Achaemenian administration. Aramaic was also introduced in the north-western areas of the Indian sub-continent when they were conquered by Darius I about the close of the sixth century B. C., and it is well-known that the Kharoṣṭhī alphabet used in the Aśokan Edicts in the Peshawar and Hazara Districts, which are written in the Indian Prakrit language, is an Indian modification of the Aramaic script.

But who are the people for whom the Aramaic version of Aśoka's Kandahar Edict was intended ? Apparently they were Iranians living in the Maurya empire, and considering the close association of the Yavanas and Kambojas in Aśokan epigraphs, there can be no doubt that it was the Kambojas for whom the Aramaic version of the Kandhar Edict was meant. That the Kambojas were of Iranian extraction was long ago suggested.[1] They were settled in the Afghanistan region in Uttarāpatha. Their numbers were occasionally swelled by new migrants from Iran, especially during the age of the Achaemenians. Their main concentration was specially in Southern Afghanistan, where they appear to have lived side by side with the Yavanas. Of course, this does not prove that some of the Kambojas did not live elsewhere in the Maurya empire even during the reign of Aśoka or that some of them did not migrate to other areas at later dates.

We have seen how originally the Gandhāras and Kambojas were the most important peoples in Uttarāpatha and how, after Alexander's conquest, the Yavanas became more prominent in that area. Another interesting fact is that, with the advent of the Śakas on the scene, on the decline of the Indo-Greeks, they were often more closely associated with the Yavanas, and the joint mention of Śaka-Yavana became popular.[2] The culture of the Gandhāras and Kambojas appears to have been modified due to their contact with the Yavanas even in the Maurya age,

1. Cf. *JRAS*, 1912, pp. 255-57; *Ling. Surv. Ind.*, Vol. X, pp. 456-57.

2. See Patañjali's *Mahābhāṣya* on Pāṇini, II. 4. 10; *Milindapañha*, p. 327; also Śaka-Yavana-Pahlava in a Nasik inscription (*Sel. Ins.*, 1942, p. 197).

though later it underwent further modification as a result of the occupation of Uttarāpatha successively by the Śakas, Parthians and Kuṣāṇas. We also see that, in the later lists of the peoples of Uttarāpatha, such as those found in Varāhamihira's *Bṛhatsaṁhitā* and the Bhuvanakośa section of the Pūrāṇas, the Yavanas, Kambojas and Gandhāras are not usually mentioned in the same breath as in some of the older records, though two of them are sometimes placed side by side. Thus, while the *Manusmṛti* (X. 43) omits the Gandhāras and has the Kambojas, Yavanas and Śakas in close proximity, the Bhuvanakośa puts the Yavanas by the side of the Gandhāras, and the Kambojas elsewhere.[1]

It appears that the location of the Kamboja country in the Pamirs has been somewhat influenced by Kālidāsa's *Raghuvaṁśa* (IV. 60ff.) which describes how Raghu left the land of the Pārasīkas or Sassanians in the western region and subjugated the countries in the northern areas of Bhāratavarṣa. In this region, Raghu first defeated the Hūṇas on the banks of the Vaṁkṣu (Oxus), i. e. in the Bālhika country or Bactria, next subdued the Kambojas and then ascended the Himalayas. In the Himalayas, he met the Kirātas (probably in Nepal), defeated the Utsavasaṅketas and other hill tribes, accepted the services and presents of the Himalayan peoples like the Kinnaras, and finally reached the kingdom of Prāgjyotiṣa or Kāmarūpa in the valley of the Lauhitya or Brahma-putra. Unfortunately, the evidence is too vague for the definite location of the Kambojas in the Pamirs. It must also be admitted that the reference here is not to the age of the Mauryas whose empire, moreover, did not include the Pamirs.

II

It is well known that corrupt readings in the Purāṇic section on Geography created confusion in the minds even of the

1. Cf. above, p. 33, note 1; p. 34, note 4. According to legends envisaging ethnic relations between the Yavanas and Gandhāras, the Yavana, Gandhāra and Mleccha peoples were descendants respectively of Yayāti's sons Turvasu, Druhyu and Anu (*Matsya Purāṇa*, 34. 30; 48. 6 ff.). Gandhāra after whom the Gandhāra-viṣaya was named, is represented as the great-grandson of Druhyu who is sometimes also said to have been the progenitor of the Bhojas.

medieval writers. Thus the *Vāyu* and *Brahmāṇḍa Purāṇas* read *Andhra-Vākāḥ* (i.e. the Andhras and Vākas or Bākas) in the list of the East Indian peoples in the place of *Aṅgā Vaṅgāḥ* (i.e. the Aṅgas and Vaṅgas) in the corresponding text in several other Purāṇas,[1] and Yādavaprakāśa must have relied on such defective readings of the Purāṇas when he included, among the peoples of Eastern India, the Andhras, Vakas and Sālvas.[2] Familiarity with this type of corrupt texts formerly led us to suspect the corrrctness of a number of statements in the account of the 56 countries in and around the Indian sub-contiment, which is found in the late medieval work entitled *Śaktisaṅgama Tantra*.[3] But a re-examination became necessary in respect of a few of the cases, one of them relating to the country of Pāñcāla as described in the said work.

Verse 23 of the said section of the Tantra, referred to above, reads as follows :

> कुरुक्षेत्रात्पश्चिमे तु तथा चोत्तरभागतः ।
> इन्द्रप्रस्थान्महेशानि दशत्रियोजनान्तरम् ।
> पाञ्चालदेशो देवेशि सौन्दर्यगर्वभूषितः ।।

An alternative reading is *daśa-tri-yojan-ottaram* in the second line.

It is said that the Pāñcāla country lies to the west as well as to the north of Kuru-kṣetra (in the Karnal-Ambala region of the Eastern Punjab) at a distance of 30 *yojanas* from Indraprastha (in the Delhi region) or 30 *yojanas* to the north of Indraprastha. A *yojana* being regarded as equal to 8 or 9 miles, 30 *yojanas* in the above description would roughly indicate about 250 miles. Since however the ancient Pāñcāla country with its capitals at Ahicchatra and Kāmpilya in ancient times and at Kānyakubja in the early medieval age actually lay in the Bareilly-Farrukhabad region, i. e. to the east of the Eastern Punjab and the Delhi region, we were inclined to regard the Tantra text as faulty. But we have now noticed that the *Śaktisaṅgama Tantra* contains several passages indicating the location

1. Cf. above, p. 36, note 5.

2. See *Vaijayantikoṣa*, ed. Oppert, Bhūmi Section, Deśa Subsection, verse 32.

3. Cf. 1st ed. of the present work, pp. 68ff.

of the Pāñcāla country in the region of the Western Punjab and Southern Kashmir, and this consistency of the author appears to be rather significant.

Verse 48 describes the location of Kuru-deśa (with its ancient capitals at Hastināpura in the Meerut District and at Indraprastha near Delhi) of the present Delhi-Meerut region as follows :

हस्तिनापुरमारम्य कुरुक्षेत्राच्च दक्षिणे ।
पाञ्चालपूर्वभागेतु कुरुदेशः प्रकीर्तितः ॥

It says that the Kuru country extends from Hastināpura and lies to the south of Kuru-kṣetra and to the east of Pāñcāla. We have to note that the Pāñcāla country is located in verse 28 to the north and west of Kuru-kṣetra or the Karnal-Ambala region while verse 48 locates it in the land to the west of Kuru-deśa or the Delhi-Meerut region. Pāñcāla thus covered parts of Southern Kashmir and the Western Punjab.

That the Pāñcāla country of the *Śaktisaṅgama Tantra* is not identical with the ancient territory of that name lying in the Bareilly-Farrukhabad region is further suggested by a few other indications of the Tantra. Thus verse 24 says—

पाञ्चालदेशमारम्य म्लेच्छाद्दक्षिणपूर्वतः ।
काम्बोजदेशो देवेशि वाजिराशिपरायणः ॥

Here the Kāmboja country is located between Pāñcāla-deśa and the south-eastern border of the Mleccha land, i. e. the Muslim world, and it has to be noted in this connection that two Mleccha territories are described in the text, verse 30 speaking of Vakrānta (Makran) as *Mahāmleccha-parāyaṇa* and verses 31-32 of Khurāśana (Khurasan including Airāka or Iraq in its northern part) as *Mleccha-mārga-parāyaṇa*. The Kāmboja territory thus abutted on Pāñcāla in the east (about the Western Punjab) and on the south-eastern border of Khurāśana (somewhere in Southern Afghanistan) and this seems to be suggested also by verse 28 which runs—

काम्बोजदेशमारम्य महाम्लेच्छात्तु पूर्वके ।
बाह्लीकदेशो देवेशि अश्वोत्पत्तिपरायणः ॥

We know that Bāhlīka is the old Indian name of the Balkh region of Northern Afghanistan and that it lies to the east of Khorasan. The description of Bāhlīka-deśa as situated to the east of Mahāmleccha and extending up to the Kāmboja territory would suggest that Kāmboja lay roughly to the east and south of Bāhlīka (Northern Afghanistan) and to the west of Pāñcāla (Western Punjab and Southern Kashmir) and therefore comprised the Peshawar-Hazara region of West Pakistan and the Kafiristan-Kandahar region of Afghanistan including the tribal territory lying between the two areas. The Kāmbojas having been Iranian autochthons originally, the discovery of the Aramaic version of an Aśokan edict at Kandahar points to the existence of their most important settlement in Southern Afghanistan in the 3rd century B.C.[1] Whether Kāmboja included the Pamir region in the north cannot be determined; but its location between Pañcāla-deśa and the south-eastern border of the Mleccha country including the Khurasan region seems to go against the possibility.

Verse 26 of the *Śaktisaṅgama Tantra* should also be considered in this connection. It runs as follows :—

काम्बोजाद्दक्षभागे तु इन्द्रप्रस्थाच्च पश्चिमे ।
पाण्डुदेशो महेशानि महाशूरत्वकारकः ॥

It is said that the Pāṇḍu country lies to the south of Kāmboja and to the west of Indraprastha or the Delhi region. We have to note that, according to verse 48, Pāñcāla lay to the west of the Kuru country in the Delhi-Meerut region, while verse 26 says that Pāṇḍu-deśa lay to the west of Indraprastha in the Delhi region. It therefore appears that Pāṇḍu-deśa lay to the south of Pāñcāla in the region of the Northern Punjab and Southern Kashmir, i. e. in the Jaipur-Bikaner region of Northern Rajasthan. Under these circumstances, *Kāmbojād=dakṣa-bhāge tu* may be a mistake for *Pāñcālād=dakṣa-bhāge tu*. But, if *Kāmbojād=dakṣa-bhāge* is taken to be the correct reading, the Pāṇḍu country probably extended considerably towards the west and comprised Northern Sind and the adjoining region of the Western Punjab. The second suggestion seems to be supported by verse 53 which locates the Madra country (having its old capital at Sākala, modern Sialkot) between Virāṭa (in

1. See above, p. 199.

Jaipur) and Pāṇḍya (Pāṇḍu) in *pūrva-dakṣa-krama* (extending from east to south). It is further supported by Ptolemy's *Geography* (VII.1.46) which locates the Pandoouoi or Pāṇḍava people on the banks of the Bidaspes or Jhelum.

The application of the name Pāñcāla to an area considerably away from the old Pañcāla or Pāñcāla country in the Bareilly-Farrukhābād region in the *Śaktisaṅgama Tantra* seems to suggest that, by the late medieval period when the Tantra was compiled, Pāñcāla as the old name of a territory in U.P. was forgotten. But, at the same time, there must have been some basis for the same name being applied to the tract of land in the western Punjab and Southern Kashmir region. In this connection, it is interesting to note the application of the name Pīr Panjāl to a lofty range of hills forming the south-western boundry of Kashmir extending for about 40 miles from the Barāmūlā pass in the north-west to the Pīr Panjāl pass in the south-east as well as to the river of the same name rising in Kashmir in lat. 33° 30′, long. 74° 43′ and falling into the Jhelam in the Punjab in lat. 33° 16′, long 73° 38′. There is an interesting note on the name of the Pīr Panjāl from the erudite pen of Stein who has shown that the prefix *Pīr* is a late addition, that Panjāl is pronounced in Kashmīrī as *Pantsāl* and in Pāhāḍī as *Pañcāl* and that the ancient spelling of the name was Pañcāla since Śrīvara mentions the range as Pañcāladeva, while Kṣemendra speaks of the Pīr Panjāl pass as Pañcāladhārā.[1] Essentially therefore Pāñcāla was the name of the valleys of the hill range and river bearing the name Pīr Panjāl or Pañcāla.

The above discussion would show that the *Śaktisaṅgama Tantra* indicates the following location of the territories:

1. Kuru-deśa in the Delhi-Meerut region;
2. Kuru-kṣetra to the north of Kuru-deśa, in the Karnal-Ambala region;
3. Pāñcāla-deśa to the north of Kuru-kṣetra and to the west of Kuru-kṣetra and Kuru-deśa, in the region of the Western Punjab and Southern Kashmir;
4. Kāmboja-deśa extending from the Pāñcāla country in the region of the Western Punjab and Southern Kashmir, to the south-eastern border of the Mleccha

1. *Kalhaṇa's Rājataraṅgiṇī*, Vol. II, pp. 396-98.

country or Khurāśana (comprising, besides Iraq, etc., the north-eastern regions of Iran and the adjoining areas) in the Kandahar region of Southern Afghanistan, lying in the present Peshawar-Hazara region of West Pakistan and the Kafiristan-Kandahar region of Afghanistan;

5. Bāhlīka-deśa between the Kāmboja country in the east and south and the Mleccha land (Khurāśana) in the west, in Northern Afghanistan;
6. Pāṇḍu-deśa or Pāṇḍya-deśa to the west of Indraprastha or the Delhi region and to the south of Kāmboja, about the northern parts of Sind and the adjoining areas of the Western Punjab, abutting on the Madra country about the present Sialkot region of the Punjab.

III

The vast area between Bundelkhand in the east and Rajasthan in the west has been known as Malwa (Mālava) since medieval times. In the ancient period, the eastern part of this territory was called Ākara or Daśārṇa which had its capital at the city of Vidiśā, modern Besnagar on the Betwa (ancient Vetravatī) near Bhilsa in Madhya Pradesh. The Avanti country having its headquarters at Ujjayinī on the Siprā, also in Madhya Pradesh, was situated in the western part of Malwa. But when exactly the ancient Avanti and Ākara-Daśārṇa regions came to be known as Mālava (Malwa) has not yet been properly investigated. The territory could have been so called only after its occupation by the Mālava people.

In the eighth decade of the fourth century B.C. when Alexander the Great of Macedon invaded the north-western regions of Bhāratavarṣa, the Mālavas, called Malloi by the Greeks, are known to have been living in the land lying to the north of the confluence of the Ravi and the Chenab and were probably confederated with the Kṣudrakas who inhabited the Montgomery District of West Pakistan. From the said area, the Mālavas, or at least a large section of the tribe, migrated to the Jaipur-Tonk region of Rajasthan. This movement may have begun during the Indo-Greek occupation of the Punjab, but seems to have continued down to the Scythian conquest of that territory.[1]

Thousands of Mālava coins were discovered at the village of Nagar (ancient Mālavanagara) near Uniyara in the Tonk District. Nagar must have been the chief city of the Mālava people, though their political influence soon spread over wide areas of Rajasthan. This is indicated by the discovery of a number of inscriptions, bearing dates in the Kṛta or Mālava era, in different parts of the State.[2] There is little doubt that the old Avanti and Ākara-Daśārṇa regions came to be known as Mālava due to their contact with these Mālavas. But historians do not appear to have any clear idea as to when the name Mālava became popular in the sense of the territory now called Malwa. That is why whenever the name Mālava is noticed in the epigraphic and literary records of the post-Gupta age, it is generally regarded as identical with modern Malwa.[3] But there is evidence to show that this is erroneous.

About the beginning of the seventh century A.D., Bāṇabhaṭṭa's *Harṣacarita* speaks of the Mālavas as friends of the Gauḍas of West Bengal and as enemies of the Puṣyabhūtis of Thanesar and the Maukharis of Kanauj.[4] The Chinese pilgrim Hiuen-tsang, who travelled in India in the second quarter of the same century, aslo gives an account of the Mo-la-p'o or Mālava country.[5] The Mālavas are again mentioned by Ravikīrti in the Aihole inscription of 634 A.D. as having been defeated by the celebrated Cālukya king Pulakeśin II of Badami.[6] But these contemporary authorities do not speak of the same Mālava country or people.

What Bāṇabhaṭṭa meant by 'Mālava' is clearly indicated by his *Kādambarī*.[7] In this work, Vidiśā on the Vetravatī is represented as the capital of Mālava while Ujjayinī on the Siprā is described as the chief city of the Avantī (Avanti)

1. *The Age of Imperial Unity*, p. 163.
2. *Ibid.*, pp. 164-65.
3. Cf. *The Classical Age*, pp. 98, 105-06; *The Age of Imperial Kanauj*, pp. 9, 24, 26; etc.
4. Tripathi, *History of Kanauj*, pp. 51, 65 ff.
5. Watters, *On Yuan Chwang's Travels in India*, Vol. II, pp. 242 ff.
6. *Ep. Ind.*, Vol. VI, p. 6, verse 22.
7. Ed. H. Siddhāntavāgīśa, pp. 19 and 183. ; cf. Sircar, *Ancient Malwa and the Vikramāditya Tradition*, p. 10

country. This shows that East Malwa had already become famous as Mālava, though West Malwa still retained its ancient name Avanti. That this nomenclature of East and West Malwa was not totally forgotten in much later times is known from Yaśodhara's commentary on Vātsyāyana's *Kāmasūtra*, entitled *Jayamaṅgalā*, which was composed in the thirteenth century, as well as from the *Śaktisaṅgama Tantra* of a still later date. Yaśodhara explains *Mālavī* (the Mālava girl) as 'born in Eastern Mālava' while *Āvantikā* (the girl of Avanti) is explained by him as 'born in the land of Ujjayinī' and 'the girl of Western Mālava'.[1] The *Śaktisaṅgama Tantra* likewise applies the names Mālava and Avanti respectively to East and West Malwa.[2]

Hiuen-tsang also distinguishes between Mo-la-p'o (Mālava) and the Wu-she-yen-na (Ujjayinī) territory. But he locates Mālava not in East Malwa as done by Bāṇabhaṭṭa, but in the valley of the river Mo-ha, *i.e.* the Mahī in Gujarat, and further says that Kheṭa (modern Kaira) and Ānandapura (modern Vadnagar) both in the present Gujarat State, formed parts of Mālava. Thus, while Bāṇabhaṭṭa's Mālava lay to the east of Avanti (Ujjayinī), Hiuen-tsang places Mālava to the west of the Ujjayinī region.

After the extirpation of the Śakas of Western India by Candragupta II Vikramāditya about the close of the fourth century, East and West Malwa were being ruled respectively by the Later Guptas and the Aulikaras as vassals of the Guptas. Both these ruling families appear to have belonged to the Mālava clan, though the clan-name attached itself only to the former probably because two contiguous kingdoms could have hardly borne the same name.

The Aihole inscription seems to represent the Mālavas as the neighbours of the Lāṭas who had their headquarters at Navasārikā (modern Nausari in the Surat District of South Gujarat) and the Gurjaras who ruled from the city of Nāndīpurī in the present Broach District in the same neighbourhood. These Mālavas therefore appear to have lived in Hiuen-tsang's Mālava in the Gujarat region and not in East Malwa which

1. Cf. VI. 5. 22 and 24 with Yaśodhara's commentary thereon.

2. See above, pp. 98-99.

is called Mālava by Bāṇabhaṭṭa. In later records, Kakka, viceroy of the Rāṣṭrakūṭa emperor Govinda III (794-814 A.D.), claims to have been stationed in the Lāṭa country (South Gujarat) for the purpose of checking Gurjara-Pratihāra encroachment into Mālava.[1] This Mālava is apparently Hiuen-tsang's and not Bāṇabhaṭṭa's, since East Malwa could hardly have been defended by an officer stationed at a distance in Southern Gujarat.

In West Malwa, the Hūṇas subdued the Aulikara feudatories of the Guptas about the close of the fifth century, though Yaśodharman of the Aulikara dynasty re-established the fortunes of the family by defeating the Hūṇa king Mihirakula about 532 A.D.[2] A few decades later, the Kalacuris of the Narmadā valley extended their power over Malwa. A charter of Kalacuri Śaṅkaragaṇa was issued from Ujjayinī in 595 A.D. while his son Buddharāja issued a grant from Vaidiśa (Vidiśā) in 608 A.D.[3] But Hiuen-tsang states that Mālava in Gujarat formed a part of the dominions of the Maitraka king Śīlāditya Dharmāditya (605-09 A. D.) of Valabhī (modern Vala near Bhavnagar in Kathiawar), while Śīlāditya's successor Kharagraha I is known to have issued his Virdi plates from Ujjayinī in 617 A.D.[4] We also know from the *Harṣacarita* and the Puṣyabhūti inscriptions that the Mālava (East Malwa) king Devagupta was defeated by king Rājyavardhana of Thanesar about 605 A.D.[5] It is not improbable that, about the beginning of the seventh century, the Puṣyabhūtis and Kalacuris were allied against a combination of the Maitrakas and Later Guptas. Sometime later, however, the Puṣyabhūti king Harṣavardhana seems to have occupied both East and West Malwa and compelled the Maitraka king Dhruvasena II Bālāditya (629-43 A.D.) to become his subordinate ally.[6] Dhruvasena II is known to have granted land in the district called Mālavaka (Mālava) apparently in the Mahī valley.[7]

1. *Ind. Ant.*, Vol. VII, p. 163.
2. Cf. *Select Inscriptions*, pp. 386 ff., 394 (verse 43), 395 (verse 6).
3. Bhandarkar's List of Inscriptions, Nos. 1206-07.
4. Gadre, *Important Inscriptions from the Baroda State*, Vol. I, pp. 7 ff. The *Āryamañjuśrīmūlakalpa* (v. 586) includes Ujjayanī in Śīlāditya's kingdom.
5. Tripathi, *op. cit.*, pp. 64 ff., 72.
6. *Proc. A.I.O. Conf.*, Vol. XII, p. 525.
7. Bhandarkar, *op. cit.*, Nos. 1346-47.

The above sketch of the history of Malwa does not throw any light on the problem regarding the application of the name Mālava to ancient Avanti. More important in this respect is the evidence furnished by Rājaśekhara's *Kāvyamīmāṁsā* composed in the earlier half of the tenth century A.D. and the Muslim historian Balādhurī who wrote his *Kitāb Futūh al Buldān* a century earlier. Among the territories of Western India, Rājasekhara mentions 'Avanti, Vaidiśa, Surāshṭra, Mālava, Arbuda, Bhṛugukaccha and others'.[1] The list distinguishes Mālava from Avanti (West Malwa) and Vaidiśa (East Malva) and seems to locate the Mālava country between Saurāshṭra (Kathiawar) and Arbuda (Mount Abu). This Mālava is no doubt Hiuen-tsang's Mālava in the Mahī valley. Balādhurī likewise distinguishes Ujjainī from Mālava when he says that Junayd, the Arab governor of Sind, raided about 725 A.D. certain West Indian localities including Uzain (Ujjayinī) and al-Malibah (Mālava).[2] It appears therefore that Mālava did not become popular as the name of West Malwa even in the first half of the tenth century.

Till the middle of the tenth century, the Paramāras appear to have been ruling in the Mahī valley where Hiuen-tsang locates the Mālava country. In 949 A.D., the Paramāra king Harṣa Sīyaka issued his Harsola plates[3] from a camp on the Mahī at the request of the ruler of Kheṭaka (modern Kaira), who was apparently his feudatory. The same Paramāra king claims to have defeated the Rāṣṭrakūṭa monarch Khoṭṭiga (967-72 A.D.) of Mānyakheṭa.[4] Scholars generally trace the same achievement of the Paramāras in a statement of Dhanapāla in his *Pāiyalacchī* (972-73 A.D.) according to which the Mālavas devastated the city of Mānyakheṭa.[5] It seems therefore that these Paramāras were regarded as Mālavas. Whether they represented a sect of the ancient Mālava clan or were called Mālava as rulers of the Mālava country in the Mahī valley is difficult to determine, though the first alternative seems to be more probable.

1. *Kāvyamīmāṁsā*, G.O.S., p. 9.
2. Ray, *Dynastic History of Northern India*, Vol. I, p. 9.
3. Bhandarkar, *op. cit.*, No. 64.
4. *Ep. Ind.*, Vol. I, pp. 135, 137 (verse 12).
5. Ray, *op. cit.*, Vol. II, pp. 850-51.

Vākpati Muñja, son of Harṣa Sīyaka, issued a charter from Ujjayinī in 975 A.D.[1] His successors ruled over East and West Malwa for several centuries with the centres of their power in the city of Dhārā (modern Dhar) and the fort of Maṇḍapa (modern Mandu), both in West Malwa. The Paramāra kings including Vākpati and Bhoja (1000-55 A. D.) became famous as the lords of Mālava.[2] In the eleventh century, the Paramāra contemporary of Cālukya Someśvara I (1043-68 A.D.) is called *Mālavendra* in Bilhaṇa's *Vikramāṅkadevacarita* (III.67) while Yādavaprakāśa's *Vaijayantī* regards Mālava and Avanti as identical.[3] The popularity of the name Mālava in the sense of West Malwa thus appears to date from the Paramāra occupation of the territory in the latter half of the tenth century A.D.

IV

Many are the cases in which the same names are found to be borne by more localities than one. Thus, in the geography of modern India, we have at least three places called Pāṭnā, viz. (1) headquarters of the Patna District of Bihar, situated at lat. 25° 37′, long. 85° 12′ 31″; (2) headquarters of the former Patna State, now merged in Orissa, standing at lat. 20° 36′, long. 13° 9′; and (3) town in the Belgaum District of Maharashtra, 20 miles to the west of Belgaum, at lat. 15° 52′, long. 75° 18″. It is well known that the name Pāṭnā, applied to these places, is derived from the Sanskrit word *pattana* meaning 'a township'. Similar is the case of the several places called Navagrāma (Nowgong), literally 'a new village or habitation'. There are some places bearing names ending in the words *purī*, *nagara*, etc., meaning 'a city or town', which are really the latter part of geographical names like Puruṣottamapurī and Rājapurī, Pāṭaliputranagara (Śrīnagara) and Mālavanagara, etc.

1. Bhandarkar, *op. cit.*, No. 84. Sīyaka's father Vairisiṁha probably ruled Dhārā as a feudatory of his Rāṣṭrakūṭa overlord. See *A.R. Ep.*, 1957-58, p. 2.

2. Cf. the fourteenth century work *Prabandhacintāmaṇi*, Tawney's trans., pp. 16, 29-30, 32, 36, 44, etc.

3. See Oppert's ed., p. 36. In the twelfth century. Hemacandra's *Abhidhānacintāmaṇi* (verse 956) supports the identification. This work gives Avantī as another name of Ujjayinī (verse 976), though the variants *Avanti* and *Avantī* are both known to have been used to indicate the country and its capital.

There are also cases in which the name of one place is purposely applied to another. Thus the Brāhmaṇas of Pāṭaliputra appear to have given the name Pāṭaliputra to modern Cuddalore (headquarters of the South Arcot District of Tamil Nadu) where they must have settled in considerable numbers.

Besides cases like the above, the same name is sometimes found to be borne by different localities in different parts of the world, although their origin must have been different. The people of India, to whom the city of Delhi and the State of Bihar are so familiar, need not be surprised if Delhi was the name of the chief town of the Delaware County in the State of New York, USA, situated on the northern bank of the Mohawk branch of the Delaware river, 70 miles west of Albany, and if Bihar was the name of a County of Upper Hungary, which bordered on Transylvania, was intersected by the Korosh river and had its headquarters at Groswarden. Likewise the name of the well-known East Asiatic country of Korea was borne by one of our princely States, the headquarters of which (also called Korea, lat. 23° 6′, long. 82° 26′) lay 153 miles north-west of Sambalpur (Orissa) and 135 miles south-west of Sherghati (Gaya District, Bihar).

It is of course not common for people to make a confusion between two localities of the same name, but belonging to different countries. However, our attention was recently drawn to a case falling under the same category of confusion.

The ancient city of Ujjayinī (Pali-Prakrit—Ujjenī, Ujeni) on the Siprā river was the capital of the Avanti (West Malwa) country. Ujjayinī (now called Ujjain), at present the headquarters of the District of that name in the western region of Madhya Pradesh, is famous for its temple of the god Mahākāla (Śiva) to which pious pilgrims flock from different parts of India even today. There was another town of the name Ujjayinī in Ceylon, about which Malasekara's *Dictionary of Pali Proper Names* (Vol. I, p. 345) has the following note : "a city in Ceylon founded by Vijaya's minister Accutagāmi", and the reader is referred to the *Dīpavaṁsa*, IX.36, and *Mahāvaṁsa*, VII. 46. In the same context, Malalasekera also refers to a third place of the same name, viz. the *nigama* or township of Ujjayinī which is known from the *Buddhavaṁsa* commentary.

It is unfortunate that Ujjayinī, capital of Avanti in India, has been confused with the Ceylonese Ujjayinī, said to have been founded by Accutagāmi, not only at two places of B.C. Law's *Historical Geography of Ancient India* (published about 1954), pp. 52 and 332, but also in the chapter contributed by him to *The Age of Imperial Unity* (*History and Culture of the Indian People*, Vol. II, 1951), p. 14. At the three places referred to above we have the following three statements:

I. "The capital of Avanti which was one of the sixteen great *janapadas* was Ujjayinī which was built by Accutagāmi."

II. "Ujjayinī which was the capital of Avanti......was built by Accutagāmi according to the *Dīpavaṁsa* (p. 57)."

III. "Ujjayinī, the capital of Avanti, was built by Accutagāmi."

We are glad to note that the said confusion is not found in Law's earlier publication entitled *Ujjayinī in Ancient India*, Gwalior, 1944.

CHAPTER XIII

GAṄGĀ AND THE GANGARIDAE

Classical writers represent the Nanda kings of Magadha primarily as rulers of the peoples known to them as the Gangaridae and the Prasii. The second of the two names is rightly taken to be the Greek plural form of the Indian word *Prācya* meaning the inhabitants of the Eastern Division of India.[1] It is well known that Bhāratavarṣa was divided into five countries, *viz*. Prācya or Eastern India, Pāścātya or Western India, Uttarāpatha or the Himalayas with the north-western region and the adjoining areas of Central Asia, Madhyadeśa or the central part of Northern India, and Dākṣiṇātya or the peninsula of the Deccan.[2] The Prācya country or the land of the Prācya people comprised Bihar and Bengal together with the eastern part of the U. P. and the eastern fringe of Orissa. The western limit of this land was the Kālaka-vana (probably near Allahabad) according to the authors of the Sūtra works,[3] Prayāga or

. See above, p. 38, note 2.

2. *Ibid*., pp. 30 ff. The Himalayan and Vindhyan regions were sometimes added to these. See above, pp. 42-46 (Sections VI-VII.). Dākṣiṇātya was also called Dakṣiṇāpatha and Pāścātya also Aparānta.

3. Cf. *Baudhāyana Dharmasūtra*, I. 2. 14-16 :

प्रागदर्शनात्प्रत्यक्कालकवनाद्दक्षिणेन हिमवन्तमुदक् पारियात्रमेतदार्यावर्त्तम् । तस्मिन् य आचारः स प्रमाणम् । गंगायमुनयोरंतरमित्येके । अथात्र भाल्लविनो गाथामुदाहरन्ति—

पश्चात्सिन्धुर्विसरणी सूर्यस्योदयनं पुरः ।
यावत्कृष्णो विधावति तावद्धि ब्रह्मवर्चसम् ॥

But according to Śaṅkhalikhita : प्राक् सिन्धुसौवीराद्दक्षिणेन हिमवतः पश्चात्कांपिल्यादुदक्पारियात्रादनवद्यं ब्रह्मवर्चसम् and according to Paiṭhinasi : आ हिमवत आ च कुमार्याः सिन्धुवैतरणी नदी सूर्यस्योदयनं पुरः । यावदा कृष्णमृगो विचरति तत्र धर्मश्चतुष्पादो भवति. See Rangaswami Aiyangar, *Rājadharma*, p. 60. It should be noticed that what was originally called Āryāvarta was later named Madhyadeśa while in later times Āryāvarta included the whole of Northern India. This is related to the gradual spread of Aryanism.

Allahabad according to the *Manusmṛti*[1] and Vārāṇasī or Banāras according to the *Kāvyamīmāṁsā*[2] composed by Rājaśekhara about the beginning of the tenth century A.D.[3] Prāgjyotiṣa or Assam, often regarded by ancient writers as a Himalayan country, formed no part of the dominions of the Nandas and Mauryas of Magadha. In the seventh century A.D., the Chinese pilgrim Hiuen-tsang noticed *Stūpas* attributed to the Maurya king Aśoka (who is said to have built 84000 *Stūpas* throughout his empire) in all parts of Bengal, but not a single one of them anywhere in the dominions of Bhāskaravarman, king of Assam. The country of Kaliṅga comprising wide areas of Orissa was usually regarded in early times as a land belonging to Dākṣiṇātya or the Southern Division of India.[4]

The people called Gangaridae is usually located in Bengal. They therefore represented one of the various tribal elements constituting the so-called Prācya people. The separate mention of the Gangaridae and the Prasii (who apparently included the Gangaridae) indicates the importance of the former among different elements of the Prācya people under the Nandas (who may have belonged to that tribe) in the fourth century B. C. The name *Gangaridae* is sometimes

1 See II, 22-23 :

आ समुद्रात्तु वै पूर्वादासमुदात्तु पश्चिमात् ।
तयोरेवान्तरं गिर्योरार्यावर्त्तं विदुर्बुधाः ।।
हिमवद्विन्ध्ययोर्मध्यं यत्प्राग्विनशनादपि ।
प्रत्यगेव प्रयागाच्च मध्यदेशः प्रकीर्त्तितः ।

2. *G.O.S.* ed., pp. 93-94 : तत्र वाराणस्याः परतः पूर्वदेशः. . माहिष्मत्याः परतो दक्षिणापथः. . . देवसभायाः परतः पश्चाद्देशः. . . पृथूदकात्परत उत्तरापथः ।

3. In their eagerness to include the land of the Buddha's birth and activities in Madhyadeśa, the Buddhist writers give its boundaries as Kajaṅgala (Kānkjol near Rājmahāl in the Santal Parganas District on the eastern fringe of Bihar) in the east and beyond it, Mahāsāla; the river Salāvatī in the south-east, the town of Setakannika in the south; the town and district of Thūna in the west; and Mount Usīradhvaja in the north (*Mahāvagga*, V. 12. 13). According to the *Divyāvadāna* (ed. Cowell and Neil, p. 22), the sacred land was bounded by Puṇḍravardhana (Mahāsthān in the Bogra District of North Bengal) in the east and, beyond it, Mount Puṇḍralakṣa; by the town of Śarāvatī in the south and beyond it the river Śarāvatī; by the Brāhmaṇa villages Sthūṇa and Upasthūṇa in the west, and by Usira-giri in the north. The countries outside the limits were called *pratyanta*. According to Weber (*Hist. Ind. Lit.*, p. 115, note), Kāmpilya (Kampil in the Farrukhābād District, U. P.) was originally the eastern limit of Madhyadeśa.

4. See above, p. 39, note 2.

taken as a Greek corruption of Sanskrit *Gaṅgā-rāṣṭra*, *Gaṅgā-rāḍha* or *Gaṅgā-hṛdaya*. The unsoundness of these suggestions is clearly demonstrated by the fact that the word *Gangaridae* in the plural is derived from the base *Gange* or *Ganges*[1] as Achaemenidae (the Achaemenians) from *Achaemenes* and *Sassanidae* (the Sasanians) from *Sassan*.[2] The name of the people is therefore actually Gange or Ganges and not Gangaridae which simply means the Ganges or Gangian people. That the correct form of the name of this people was Ganges is certain since, as will be shown below, it is found in the *Periplus of the Erythraean Sea* as the name of its habitat. The particular area of Bengal that was inhabited by the Gangaridae (Ganges people or Gangians) is indicated not only by the *Periplus of the Erythraean Sea* (*circa* 80 A.D.) but very definitely also by the *Geography* of Ptolemy (*circa* 145 A.D.).

Ptolemy[3] refers to the five estuaries of the river Ganges as (1) the westernmost mouth called Kambyson (144° 30′, 18° 15′), (2) the second mouth called Mega (145° 45′, 18° 30′), (3) the third mouth called Kamberikon (146° 20′, 18° 40′), (4) the fourth mouth called Pseudostomon (147° 40′, 18° 30′), and (5) the fifth mouth called Antebole (148° 30′, 18° 15′). Although owing to the changes in their courses, the identification of these mouths of the Ganges with her present estuaries may not be beyond doubt, Ptolemy's estimate of the distance between the easternmost and westernmost of them to be four degrees of latitude seems to suggest that the area washed by the estuaries of the Ganges in the second century A.D. was practically the same as it is today, that is to say, the land between the Bhāgīrathī or Hooghly river in the west and the Padmā-Meghnā estuary in the east. Elsewhere in his work,[4] the Greek geographer says, "All the region about the mouths of the Ganges is occupied by the Gangaridae with the following city—Gange, the royal city (i.e. capital of the country)—146°, 19° 15′." It is clear from the above account that the people called Gangian or Ganges received their name from

1. The letter *r* in the plural form is an augment.
2. See above, pp. 171 ff.
3. *Geog.*, VII. i.18.
4. VII. i. 81.

Gange, the chief city of their country, just as the people of Attika were known as Athenian from the capital city of Athens. In India, names of cities like Puṇḍravardhana, Karṇasuvarṇa or Tāmralipta usually also indicated the districts round them and, moreover, the names of a country and its people were almost always the same. Thus, according to Ptolemy, the Ganges people or Gangians inhabited the delta of Southern Bengal watered by the lower spurs of the river Ganges. In the second century A.D., they appear to have been an autonomous people with their capital at the city of Gange which, as its name indicates, was situated on the river Ganges. The latitude and longitude of the city as given by the geographer, although they can hardly be relied on, would suggest its location not far from the confluence of the Ganges and the sea. It is difficult to say whether the Maroundae of Ptolemy,[1] whose country abutted on that of the Gangaridae and lay to the east of the Ganges, were the Puṇḍras who of North Bengal or the Muruṇḍas of Bihar.

It is very interesting to note that the *Periplus*[2] applies the name Ganges not only to a river but also to a country and its capital both located about the principal mouth of the river. The author, who was a Greek navigator and merchant, says that while sailing towards the east in the Bay of Bengal 'with the ocean to the right and the shore remaining beyond to the left', the country called Ganges comes into view : "there is a river near it called the Ganges. On its bank is a market-town which has the same name as the river, Ganges. Through this place are brought malabathrum and Gangetic spikenard and pearls, and muslins of the finest sort which are called Gangetic. It is said that there are gold mines near these places and there is a gold coin which is called *kaltis*." The reference to the Gangetic muslin no doubt reminds us of the worldwide celebrity of the muslin manufactured at Dacca in the medieval period and also of the *dukūla*, *kauśika*, *patrorṇa* and *prāvara*, mentioned in the Sabhā-parvan of the *Mahābhārata* as presents made to Yudhiṣṭhira by the Vaṅgas, Kaliṅgas, Tāmraliptas and Puṇḍras of Eastern India.[3] It is also interesting that the city of Ganges, situated

1. VII. ii. 14.
2. Para. 63.
3. Cf. Motichandra, *Geog. Econ. Stud.*, pp. 112 ff. See also above, pp. 172 ff.

on the river of that name (apparently indicating the principal mouth of the Ganges) has been called a market-town; but the fact that the town bore the same name as the country round it shows that, although it was a market-town in the eyes of a foreign merchant, the city of Ganges was the metropolis of the country called Ganges. It is probable that, soon after the dissolution of the Maurya empire, the Gangian people of Lower Bengal formed a powerful state with their headquarters at the city of Gange or Ganges. As, however, the name of the people was apparently derived from their chief city, the Gangian city seems to have existed as early as the days of the Nandas and Mauryas of Magadha.

But a people called Gaṅgā, Gāṅga or Gāṅgeya inhabiting Lower Bengal and having their capital at a city called Gaṅgā (Greek *Gange* or *Ganges*) is not known from ancient Indian literature. This powerful people, known to the foreign writers from the fourth century B.C. down at least to the second century A.D., was apparently known to the Indians by a different name. Curiously enough Kālidāsa, who lived in the fourth and fifth centuries A.D., locates, in connection with the mythological *dig-vijaya* of Raghu, the celebrated Vaṅga people, known from ancient Indian literature, exactly in the same region where the Gangaridae or Gangians are placed by the Classical writers. In Canto IV of Kālidāsa's *Raghuvaṁśa*, Raghu is said to have reached the shores of the Bay of Bengal (verse 34) and to have accepted the submission of the Suhmas (verse 35) and subdued the Vaṅgas (verses 36-37). The Suhma country, later called Rāḍha, lay on both banks of the Ajay river in the modern Burdwan region of South-West Bengal. The *Raghuvaṁśa* further says that, after having conquered the Vaṅgas, the Ikṣvāku hero crossed the Kapiśā (the modern Kaśāi river in the Midnapur District, identified by Raychaudhuri with Ptolemy's Kambyson, the westernmost mouth of the Ganges) and, through the country of the Utkalas (inhabiting parts of the Midnapur and Balasore Districts), reached Kaliṅga in the Cuttack-Puri-Gangam region (verse 38). It seems that the Kaśāi river in the Midnapur District was once regarded as the demarcating line between the Vaṅga and Utkala countries and that the Vaṅgas occupied the delta of Lower Bengal as far as the Kaśāi in the west. This is possibly supported by Ptolemy's *Geography* (if Raychaudhuri's indentification of the Kambyson is accepted) and certainly by the Jain *Prajñāpanā* which speaks of Tāmralipti or Tāmralipta (near Tamluk in the Midnapur District) as a locality belonging to the land of the Vaṅgas.[1] But far more important in determining the home of the Vaṅga people is verse 36 of Kālidāsa's text describing Raghu's victory over the Vaṅgas :

> वंगानुत्खाय तरसा नेता नौसाधनोद्यतान् ।
> निचखान जयस्तम्भान् गंगास्रोतोऽन्तरेषु सः ।।

The mouths of the Ganges, referred to as *Gaṅgā-srotas*, were near about the Bay of Bengal mentioned earlier in verse 34. The passage *Gaṅgā-sroto-'ntareṣu* reminds us of *stan-āntare* (i.e. in the space between the breasts) in the passage *na mṛṇāla-sūtraṁ racitaṁ stan-āntare*. Mallinātha rightly explains *Gaṅgā-sroto-'ntareṣu* as *Gaṅgāyāḥ srotasāṁ pravāhāṇām = antareṣu dvīpeṣu*.[2] The verse therefore means to say that, after having totally routed the Vaṅga people who were fighting from their boats, Raghu established columns of victory in the land interesected by the mouths of the Ganges, which was apparently the country inhabited by that people. This shows beyond doubt that, in the age of Kālidāsa (i.e. the fourth and fifth centuries A.D.), the Vaṅgas inhabited the deltaic region of Southern Bengal and that they were no other than the Gangaridae or Gangians who, according to the Classical writers of the first and second centuries A.D., lived in the region about the mouths of the Ganges and had their chief city near about the confluence of the Ganges and the sea.

The location of the city of Gaṅgā, capital of the Gangians or Vaṅgas in the vicinity of the confluence of the Gaṅgā and the Sāgara (sea) suggests that it was no other than the celebrated holy city of Gaṅgāsāgara or Gaṅgāsāgarasaṅgama mentioned in Indian literature. As a personal name like Satyabhāmā is known to have been abbreviated into both Satyā and Bhāmā, a locality like Puruṣottamapurī could be and has actually been celebrated both as Puruṣottama and as Purī. The city of Gaṅgāsāgara likewise seems to have been known both as Gaṅgā

1. See Raychaudhuri, *Stud. Ind. Ant.*, p. 186.
2. These points have also been discussed above, pp. 171ff.

(at least to foreigners of the early period) and as Sāgara (still preserved in the name of the Sāgar or Gaṅgāsāgar Island).

The ancient city of Gaṅgāsāgara at the eastern extremity of the Cakravarti-kṣetra and of Bhāratavarṣa, Kumārīdvīpa and Āryāvarta became one of the famous all-India *tīrthas* at least before the rise of the Guptas in the fourth century A.D., if not in much earlier times.[1] The popularity of pilgrimage to holy places is noticed in the records of Aśoka (*circa* 269-232 B.C.) who visited several Buddhist *tīrthas* including Sambodhi (Mahābodhi or Bodhigayā, modern Bodhgayā)[2] and of the Hinduised Śaka king Ṛṣabhadatta (119-23 A.D.) who visited Puṣkara in Rājputānā and made arrangements in various *tīrthas* in Western India for the benefit of the pilgrims.[3] The *Baudhāyana Dharmasūtra*[3] (*circa* 5th century B.C.) discouraged travels in the lands of the Āraṭṭas, Kāraskaras, Puṇḍras, Sauvīras, Vaṅgas, Kaliṅgas and Prāṇūnas, as it prescribed purification of the travellers by the performance of the Punastoma or the Sarvapṛṣṭhā.[4] But later authorities allowed travels in the Aṅga, Vaṅga, Kaliṅga, Surāṣṭra and Magadha countries if they were undertaken for the purpose of pilgrimage.[5] The Tīrthayātrā section of the Vanaparvan of the *Mahābhārata*, composed earlier than the Gupta

1. Since pilgrimage to holy places is a non-Aryan institution later adopted by the Aryans, the antiquity of Gaṅgāsāgara may be pre-Aryan.
2. See *Select Inscriptions*, 1942, pp. 28, 70-71.
3. *Ibid.*, pp. 161-62.
4. I. 2. 14-16 :

अवन्तयोऽङ्गमगधाः सुराष्ट्रा दक्षिणापथाः ।
उपावृत्सिन्धुसौवीरा एते संकीर्णयोनयः ॥

आरट्टान्कारस्करान्पुण्ड्रान्सौवीरान्वंगान्कलिगान्प्राणूनानिति च गत्वा पुनस्तोमेन यजेत सर्वपृष्ठया वा । अथाप्युदाहरन्ति—

पद्भ्यां स कुरुते पापं यः कलिगान्प्रपद्यते ।
ऋषयो निष्कृतिं तस्य प्राहुर्वैश्वानरं हविः ॥

The epithet *saṅkīrṇa-yoni* applied to some peoples suggests that their culture was an admixture of Aryan and Dravidian elements. The same work (I. ii. 1-4) accepts cross-cousin marriage as an established custom among the Dākṣiṇātyas (*mātula-pitṛsvasṛ-duhitṛ-gamanaṁ Dākṣiṇātyānām*).

5. Bagchi, *Pre-Aryan and Pre-Dravidian in India*, p. 74:

अंगवंगकलिंगेषु सौराष्ट्रमगधेषु च ।
तीर्थयात्रां विना गच्छन् पुनः संस्कारमर्हति ॥

age, refers to Gaṅgāsāgara as a very great *tīrtha*.[1] The *Śabda-kalpadruma* quotes the following verses from the *Kūrma Purāṇa*, one of the older *Purāṇas* known to Al-Bīrūnī :

गंगायां ज्ञानतो मृत्वा मुक्तिमाप्नोति मानवः ।
अज्ञानाद् ब्रह्मलोकं च याति नास्त्यत्र संशयः ।।
गंगायां च जले मोक्षो वाराणस्यां जले स्थले ।
अन्तरिक्षे च गंगायां गंगासागरसंगमे ।।

It is said that a person dying in the waters of any part of the Ganges gets salvation ; at Banāras salvation reaches him even if he dies on the banks of the river ; but at Gaṅgāsāgara salvation is guaranteed whether one dies in the waters or on the land or in the air. The *Viṣṇusaṁhitā*, a post-Gupta work in its present form, recognises the importance of this *tīrtha* in regard to the performance of funerary rites.[2] That Gaṅgā-sāgara was well known in North-Western India in the eleventh century A.D. is suggested by the reference in Al-Bīrūni's work on India (*circa* 1030 A.D.) to this city 'where the Ganges flows into the sea.'[3]

Due to several factors, the chief amongst them being the change of the main course of the Gaṅgā through the Padmā as well as the impotence of the descendants of the valiant Vaṅgas

The verse is ascribed to the *Ādi Purāṇa* in the *Vīromitrodaya*, *Saṁskāra-prakāśa*, Chowkhamba, p. 546. Cf. *Mahābhārata*, VIII. 45. 14-15, according to which, the *śāśvata-dharma* is known by certain proples including the Kaliṅgas :

कुरवः सहपाञ्चालाः शाल्वा मत्स्याः सनैमिषाः ।
कोशलाः काशयोऽङ्गाश्च कालिंगा मागधास्तथा ।
चेदयश्च महाभागा धर्मं जानन्ति शाश्वतम् ।।

1. III, 85, 4-5 :

गंगायास्तत्र राजेन्द्र सागरस्य च संगमे ।
अश्वमेधं दशगुणं प्रवदन्ति मनीषिणः ।।
गंगायास्त्वपरं पारं प्राप्य यः स्नाति मानवः ।
त्रिरात्रमुषितो राजन् सर्वपापैः प्रमुच्यते ।।

2. Chapter 85. 28 :

गंगाद्वारे प्रयागे च गंगासागरसंगमे ।
सततं नैमिषारण्ये वाराणस्यां विशेषतः ।।

3. Sachau, *Alberuni's India*, Vol. I, pp. 201, 261.

of old as guardians of the sacred place, the all-India importance of the Gaṅgāsāgara *tīrtha* gradually waned and its enviable position as the greatest *tīrtha* in Eastern India was ultimately usurped by the medieval Purī *tīrtha* in Orissa. As a local *tīrtha* , however, it preserved its popularity in Eastern India throughout the medieval period in spite of the depredations of Magh and Portuguese pirates in Lower Bengal. The practice of dedicating children to the goddess Gaṅgā in the waters of the confluence at Gaṅgāsāgara prevalent in that age in South-West Bengal and the neighbourhood is well known to all students of Indian history. Medieval Bengali literature refers to the belief that a person offering to die in the waters of Gaṅgāsāgara with a solemn desire was sure to get it fulfilled in his next life. Caṇḍīdāsa's *Śrīkṛṣṇakīrtana* (about the 16th century) has the passage :

सागरसंगम गिया गायेर मास काटिया
आपना मगर भोज दिया ।
सागरसंगमजले त्यजिव मो कलेवरे.

In the sixteenth century, Jayānanda, a poet of the Burdwan region, wrote in his *Caitanyamaṅgala* :

आर रमणी बले गंगासागरे मरिब.

In the same century, *Kavikaṅkaṇa* Mukundarāma put the following words in the mouth of Śrīmanta, one of the heroes of his *Caṇḍīmaṅgala* :

खण्डिया विधिर थाम्य सागरे करिब काम्य
पूजा करि सङ्केतमाधव ।
भुञ्जिया संसारसुख देखिब बापेर मुख
पुनरपि हइया मानव ।।

Śrīmanta is also represented as saying :

यदि वा पितार सङ्गे नहे दरशन ।
कामना करिया मोर सागरे मरण ।।

Elsewhere in the same work, one of the jealous ladies, who were comparing their own husbands unfavourably with Gaurī' bridegroom (Śiva), is found saying :

आर युवती बले आमि मन्दार याव ।
कामना करिया गिया सागरे मरिब ।।[1]

1. *Vide* C. C. Banerji, *Caṇḍīmaṅgalabodhinī*, pp. 822, 857. *Sāgara* has been used here to mean *Gaṅgāsāgara*.

A number of medieval Sanskrit works such as the *Gaṅgā-vākyāvalī,* attributed to the Maithila writer Vidyāpati (*circa* 1375-1450), and the *Tīrthacintāmaṇi* of Maithila Vācaspati-miśra (*circa* 1425-80), which is divided into five Prakāśas, *viz.* Prayāga, Puruṣottama (Puri), Gaṅgā, Gayā and Vārāṇasī and deals with topics like the purpose of pilgrimage, its preliminaries, rites to be performed at the *tīrthas,* subsidiary holy places at the principal *tīrthas,* etc., make prominent mention of the Gaṅgāsāgara *tīrtha.* The most valuable information about the religious and monastic establishment at Gaṅgāsāgara, one of the greatest and oldest Indian *tīrthas,* is however found in the *Tīrtha-sāra* composed by the Bengali writer Bhavadeva Nyāyālaṅkāra in Śaka 1651 (1729 A.D). This work shows that, as late as the first half of the eighteenth century, considerable parts of the old city existed and that people flocked to the *tīrtha* with a desire to touch the feet of Pitāmaha (Brahman), to have a bath in the waters of the Mandākinī (Gaṅgā on her way to Pātāla) and to worship the images of Mādhava (Viṣṇu) named after an ancient local king of that name, Amaraśaṅkara (Śiva), Kapila (founder of the Sāṅkhya system of Indian philosophy, mythologicaly represented as a sage indirectly responsible for the digging up of the beds of the oceans by the sons of Sagara as well as for the descent of the Gaṅgā on the earth and regarded as an *Avatāra* of Viṣṇu), Skanda-Kārttikeya and Hari (Viṣṇu).[1]

The city of Gaṅgāsāgara seems to have been often encroached upon by the sea and affected by the occasional changes in the surface of the land caused by the silting up of one or other of the mouths of the Ganges and by the creation of new mouths.

1. Cf. *IHQ*, Vol. XXII, p. 128 : माधवनामा नृप आसीत्तस्य देशस्य । तेन स्थापितो माधवनामा विष्णुस्तत्प्रतिमा । तदधिष्ठानचिह्नेन तीर्थस्य परिचयः कृतः ... तत्र कपिलनाममुनिप्रतिमास्ति ।... तत्र मन्दाकिनी पातालगा गंगा तिष्ठति । ... कार्त्तिकेयप्रतिमा तत्रास्ति ।.... हरिप्रतिमा तिष्ठति ।...

शंकरः कपिलश्चैव श्वेतद्वीपपतिस्तथा ।
द्वीपे साधारणः स्कन्दः पंचमश्च महोदधिः ।।

इति स्कान्दे । शंकरः अमरशंकरः । श्वेतद्वीपपतिर्माधवः ।

कपिलस्य मुखं दृष्ट्वा स्पृष्ट्वा पैतामहं पदम् ।
तरगंगोदकं पीत्वा पुनर्जन्म न विद्यते ।।

The same reason seems to have been at the root of the rise, in different ages in the neighbourhood, of such market towns as the ancient Tāmralipta, the medieval Saptagrāma and Hooghly and the modern Calcutta. But some of the old temples and images at Gaṅgāsāgara survived as late as 1842 A.D. when the remaining part of the ancient city was finally and completely washed away by the waves of the sea.[1] Just a year before the remnants of the last temple were swallowed up, a very interesting account of the 'Mela at Ganga Saugor' appeared in the *Friend of India*, Vol. VII, 1841, p. 70 : "...a mere sandbank, about a mile in length and about a quarter mile broad, of crescent form with the wide sea opening in front and the back covered by a dense jungle. At one corner stands the solitary temple of the celebrated Sanyasi Cupil Mani (Kapilamuni)...The temple is the last remnant of what has evidently been a large monastic institution for devotees, the ruins of which may be walked over at low water. These ruins show that the buildings must have been very extensive as well as massy..." The Kapila temple is said to have been 'built of Concha stone brought from Orissa' and was in a precarious condition 'soon to moulder away.' It was inhabited by a number of mendicants, some of whom owed allegiance to a pontiff of distant Jaipur in Rajasthan. There were two stone images in the temple, one of Kapilamuni and the other of 'Mahdeb' (either Mādhava, i.e. Viṣṇu, or less probably Mahādeva, i.e. Śiva). The correspondent of the *Friend of India* noticed an inscription in the temple, about the letters of which he remarks, "...yet are they like neither to things in heaven above or in the earth beneath, nor could the Śanyasis themselves decipher the hieroglyphics." The epigraph was supposed to contain a date which was believed to correspond to 430 or 437 A.D.[2] The above shows clearly that the

1. *JASB*, 1850, p. 538; *IHQ*, *loc. cit.*

2. *JASB*, *loc. cit.*, *IHQ*, *op. cit.*, p. 129. Yule and Burnell, in the *Hobson-Jobson*, s.v. *Saugor*, *Saugor Island*, observe, "It is said once to have been populous, but in 1688 (the date is clearly wrong) to have been swept by a cyclone-wave. It is now a dense jungle haunted by tigers." Amongst the quotations in this work from earlier European authors, Hedges speaks of his visit in 1683 to the 'Pagodas at Sagor'. He further says, "James Price assured me that about 40 years since, when the Island called Gonga Saugor was inhabited, the Raja of the Island gathered yearly rent out of it to the amount of 2 lacks of rupees." In 1705, Luillier spoke of the temple

sea washed away the most important sacred place in Eastern India, associated with the names of an ancient Vaṅga king called Mādhava and the sage Kapila, just as Dvārakā, a West Indian *tīrtha* of all-India reputation, had been swallowed up by the waves of the Arabian Sea. But whereas an entirely new city, founded in the neighbourhood, was endowed by the people of the Pāścātya division of India with the name and sanctity of ancient Dvārakā, the inhabitants of the Bengal delta did practically nothing to preserve the sacred memory of Gaṅgāsāgara. The real purpose of pilgrimage to holy places in all the four extremities of India was to travel over the length and breadth of the whole land in order to know it properly and to feel the fundamental unity of the various elements in the Indian population. With the absence of Gaṅgāsāgara, there is no ancient *tīrtha* of all-India recognition in the eastern border of the country to attract pilgrims from other parts of India. This seems to be detrimental to the cause of the unity of the Indian people. The East Indians should think of rebuilding the great *tīrtha* at the junction of the Gaṅgā and the Sāgara in all its ancient glory.

at 'Sagore' and of two famous ascetics residing there, while in 1727 Hamilton says, "...among the Pagans, the Island Sagor is accounted holy, and great number of Jougies (asectics) go yearly thither in the months of November and December to worship and wash in salt water, though many of them fall sacrifices to the hungry tigers." The account of Hedges suggests that the Island was once severely hit by a cyclone about the middle of the seventeenth century.

Chapter XIV

ARIAKA, ALASANDA AND BARBARA

I

The *Periplus of the Erythraean Sea*, composed in the eighth decade of the first century A.D.,[1] says, "Beyond the Gulf of Baraka (Dvārakā in Kathiawar) is that of Barygaza (Bharukaccha, Bharoch or Broach on the mouth of the Narmadā in Southern Gujarat) and the coast of the country of Ariaka which is the beginning of the kingdom of Mambarus (apparently a Śaka king) and of all India. That part of it (the kingdom of Mambarus) lying inland and adjoining Skythia (Śakasthāna in the lower valley of the Indus) is called Abiria (Ābhīra country), but the coast is called Syrastrene (Surāṣṭra or Kathiawar)."[2]

There is difference among scholars on the identification of the land called Ariaka. Lassen took it to be a mistake for *Larika* standing for Lāṭa (the Nausari-Broach region of South Gujarat), while Bhagwanlal Indraji considered it to be the same as Aparāntikā (Aparānta or the Northern Konkan about the Thana District).[3] The separate mention of both Ariake and Larike in the Geography of Ptolemy (composed about the middle of the second century A.D.) on the basis of material

1. Cf. Smith, *EHI*, 1924, p. 245, note 1. The reference in the *Periplus* (Schoff's trans., p. 42) to 'Ozene, formerly a royal capital' shows that the work had been composed before Caṣṭana (Tiastenes) made Ozene (Ujjayinī in West Malwa) his capital in the second quarter of the 2nd century A. D. since it appears to have remained the capital of Caṣṭana's descendants till their overthrow by Candragupta II about the close of the fourth century A.D. See Sircar, *Studies in Indian Coins*, pp. 115ff.

2. Schoff's trans., p. 39. The *Periplus* also says, "In these places there remain even in the present time signs of the expedition of Alexander, such as ancient shrines, walls of forts and great wells." These antiquities, probably seen by the author himself, may be associated with Alexander's stay in the land of the lower Indus for some months waiting for the Etesian winds for the dispatch of a part of his forces under Nearkhus by the sea before his own departure from India with the remainder of the army by land in the year 324 B.C.

3. Schoff, *op. cit.*, pp. 174-75; Smith, *op. cit.*, pp. 108-09.

collected a little earlier, however, suggests that *Ariaka* may not stand for *Larika*. The author of the *Periplus*, it appears, included Larika or Lāṭa in Ariaka which has to be identified with Aparānta, since Suppara (Sūrpāraka, the capital of Aparānta) was located in Ariaka as we shall see below.

Ptolemy first gives the places in Syrastrene (Kathiawar), next those in Larike (Southern Gujarat) and then those of Ariake which he calls 'Ariake of the Sadenoi'.[1] In the land of Larike, he locates the mouth of the river Mophis (Mahī) and, on the Gulf of Barygaza in the same country (as we shall see later), the mouth of the Namades (Narmadā) and a locality called Nausaripa (Navasārikā, modern Nausari in the Surat District). Larike is therefore undoubtedly the same as Lāṭa. On the other hand, the geographer locates in the coast of Ariake the places called Souppara (Sūrpāraka, modern Sopara in the Thana District) and Semyla (modern Chaul in the same District, 23 miles to the south of Bombay).[2] This Ariake is no doubt identical with Aparānta or the Thana region (Northern Konkan).

The *Periplus* seems to include Ariaka, comprising Ptolemy's Larike or Lāṭa, in the kingdom of Mambarus who had his capital at Minnagara near Barygaza. Thus about 80 A.D., both Lāṭa and Aparānta appear to have formed parts of the Śaka kingdom of Western India. The same state continued till the time of the Kṣaharāta-Śaka ruler Nahapāna (119-24 A.D.) who ruled over wide areas of Western India as far as the Nasik-Poona region of Maharashtra in the south. There is an early Jain tradition that Nahapāna had his capital at Bharukaccha which was repeatedly attacked by the Śātavāhana king of Pratiṣṭhāna (i.e. Gautamīputra Śātakarṇi, c. 106-30 A.D.), who ultimately conquered it.[3] The partial destruction of Bharukaccha as a result of this struggle may have led the next Śaka ruler Caṣṭana (Tiastenes) to make Ujjayinī (Ozene) his headquarters, as we learn from Ptolemy's Geography.

1. *La Geographie de Ptolemee, L'Inde* (VII. 1-4), ed. Renou, pp. 3-4 (Section 6).

2. Hippokoura (116° 45′—19° 30′) is described elsewhere (*op.cit.*, p. 36, Section 83) as the capital of Baleokouros, usually identified with Vilivāyakura of the Kolhapur coins.

3. Cf. *JBORS*, Vol. XVI, p. 288.

It will be seen that, while the *Periplus* includes Ariaka in the kingdom of Mambarus, Ptolemy mentions Ariake as the land of the Sadenoi and apparently includes Larike in the kingdom of Caṣṭana. A second section in Ptolemy's Geography says that, in the country of Larike, the market town of Barygaza lay to the west of the river Namades (Narmadā) while, among the towns lying to the east of that river, there was Ozene (Ujjayinī), the capital of Tiastenes (Caṣṭana).[1]

Who were the Sadenoi occupying Ariake when the Kārdamaka-Śaka ruler Caṣṭana was holding Larike about the forties of the second century A.D.? That they were no other than the Śātavāhanas appears to be clear from a third section of Ptolemy's Geography wherein Baithana (Pratiṣṭhāna on the Godāvarī in the Aurangabad District of Maharashtra), capital of Siriptolemaios (Śrī-Pulumāvi who was the son of Gautamīputra Śātakarṇi and ruled in c. 130-59 A.D.), and the town of Tagara (modern Ter in the Osmanabad District of Maharashtra) are located in the same country to the west of the river Bendas.[2]

The Śātavāhana occupation of Aparānta before the Śaka conquest of Western India is referred to in the *Periplus* which says. "The market-towns of this region are, in order, after Barygaza: Suppara (Sūrpāraka, Sopārā) and the city of Kalliena (Kalyāṇ in the Thana District, 33 miles north-east of Bombay) which in the time of the elder Saraganus (an early Śātavāhana king named Śātakarṇi) became a lawful market-town; but since it came into the possession of Sandanes (probably a governor or predecessor of Mambarus), the port is much obstructed, and Greek ships landing there may chance to be taken to Barygaza under guard."[3]

About 124 A.D., the later Śātavāhana king Gautamīputra Śātakarṇi (c. 106-30 A.D.) overthrew the Kṣaharāta-Śaka ruler Nahapāna and not only occupied the Nasik-Poona region and Northern Konkan (Aparānta), but even succeeded in annexing

1. Renou, *op.cit.*, pp. 28-29 (Section 63). Although Larike thus seems to have comprised the Ujjayinī region, Nasika given here as a town in Larike cannot be the modern Nasik in Maharashtra especially because the Kārdamaka-Śakas could not recover the Nasik-Poona region from the Śātavāhanas.

2. *Ibid.*, p. 35 (Section 82).

3. Schoff, *op.cit.*, p. 43. Kalyāṇa is located in Aparānta in a Kanheri inscription (Lüders' List of Inscriptions, No. 1013).

Malwa (Ākara and Avanti) and Kathiawar (Surāṣṭra).[1] But Gautamīputra is known to have lost much of the conquered territory to the Kārdamaka-Śakas before his death about 130 A.D. In that year, the dominions of Kārdamaka Caṣṭana and his grandson Rudradāman (who ruled jointly with his grandfather as a sub-king) included Kutch, while the Junagadh inscription (150 A.D.) states that Rudradāman twice defeated Śātakarṇi (Gautamīputra Śātakarṇi) the lord of Dakṣiṇāpatha, though the latter was not extirpated because of the closeness of relations between the two rulers. The inscription further says that Aparānta and the other districts in the northern part of Gautamīputra's empire were then in the possession of the Kārdamaka-Śakas.[2]

As regards the closeness of relations between Gautamīputra Śātakarṇi and Rudradāman, we know that the former's younger son Vāsiṣṭhīputra Śātakarṇi married the latter's daughter.[3] The reference to two wars between Rudradāman (who was originally a sub-king under his grandfather Caṣṭana) and Gautamīputra may suggest that there were two stages in the reconquest of the northern districts of Gautamīputra's kingdom by the Kārdamaka-Śakas. Probably the Malwa-Kathiawar area was conquered before the conquest of Aparānta and the neighbouring regions. The absence of any mention of Lāṭa in the list of territories in the possession either of Gautamīputra or of Rudradāman appears to suggest that it was included in the administrative unit of Aparānta as indicated by the statement of the *Periplus* referred to above.

The question now is whether the whole of Aparānta was reconquered by the Kārdamaka-Śakas as suggested by the Junagadh inscription or parts of it still remained in the hands of Gautamīputra and his successors as is indicated by Ptolemy's Geography. It has to be noted that epigraphic records of the Kārdamakas are conspicuous by their absence in the Nasik-Poona region and the Northern Konkan, while a number of inscriptions of the successors of Gautamīputra Śātakarṇi have been discovered in the said areas. Thus we have the Nasik and

1. See *The Age of Imperial Unity*, ed. Majumdar, p. 201.
2. *Ibid.*, p. 183.
3. *Loc. cit.*

Karle (Poona District) inscriptions of Vāsiṣṭhīputra Pulumāvi (c. 130-59 A.D.), the Kanheri (Thana District) inscription of Vāsiṣṭhīputra Śātakarṇi (c. 159-66 A.D.), the Nasik and Kanheri inscriptions of Gautamīputra Yajña Śātakarṇi (c. 178-202 A.D.), etc., etc.[1] Formerly we had Śaka type silver coins of Yajña Śātakarṇi, which were apparently issued for circulation in areas previously under the Śakas and one of which was found at the base of the ruined Buddhist establishment at Sopara. This led to the belief that Yajña conquered Aparānta from the Kārdamaka-Śakas.[2] But coins of the same type issued by Vāsiṣṭhīputra Pulumāvi and Vāsiṣṭhīputra Śātakarṇi have also been recently discovered.[3] Thus the Nasik, Poona and Thana Districts appear to have remained in the possession of Gautamīputra and his successors. Ptolemy's representation of Ariake as the land of the Sadenoi (Śātavāhanas) is therefore justifiable, while his inclusion of Larike (comprised in Ariaka according to the author of the *Periplus* who, as we have seen, seems to be supported by the Śātavāhana and Śaka inscriptions) apparently in the kingdom of Tiastenes of Ozene shows that the northern part of the Ariaka of the *Periplus* (i.e. Larike or Lāṭa) was reoccupied by the Śakas.

According to some scholars, Ptolemy distinguishes Ariake of the Sadenoi from Ariake of the Pirates, comprising the coast country towards the south stretching as far down as the country of Lymirike (Damirika of the *Periplus*, i.e. Tamiḻaka or Draviḍa) which seems to have included the land of the Keralaputra (the king of Kerala).[4] This extension of Aparānta in the south seems to be supported by Kālidāsa's *Raghuvaṁśa* (IV.54-58) wherein the river Muralā is probably mentioned as the boundary between Kerala and Aparānta.[5]

1. See Rapson's *Catalogue of Coins*, pp. l-liii; *Ep. Ind.*, Vol. XXXV, p. 25?; Sircar, *Studies in Indian Coins*, pp. 107ff.

2. Rapson, *op.cit.*, p. cxxi.

3. *Ep. Ind.*, Vol. XXXV, pp. 247 ff.

4. S. N. Majumdar Sastri, *McCrindle's Ancient India as described by Ptolemy*, p. 45. Cf. the *Periplus* locating Tyndis, one of the first markets of Damirika, in the kingdom of Kerobothra (Schoff, *op.cit.*, p. 44).

5. Mallinātha, the South Indian commentator of the *Raghuvaṁśa*, regards the Muralā as a river in the Kerala country, while the South Indian product entitled *Avantisundarīkathā* speaks of the sports of the Aparānta elephants in the waters of the Muralā (cf. Sircar, *Select Inscriptions*, p. 453, note 1).

II

A large number of cities and towns in the vast area subdued by Alexander the Great were called *Alexandria* after the mighty conqueror. An Indian form of the name *Alexandria* is *Alasanda* found in the *Milindapañha* and other works. It is sometimes called the *nagara* (city or town) of the Yavanas (the Greeks probably of Bactria and India) and sometimes a *dvīpa* (a territory between two rivers or an island).[1] Since however, there were several Alexandrias even in the north-western areas of India and in Afghanistan (both comprised in the Uttarāpatha division of Bhāratavarṣa), there has been difference of opinion among scholars as regards the idetification of Alasanda mentioned in Indian literature.

Among the Alexandrias of Uttarāpatha, mention may be made of—(1) Alexandria-under-the-Caucasus (Hindukush), identified with modern Charikar or Opian near Kabul; (2) Alexandria-among-the-Arachosians, identified with modern Kandahar; (3) Boucephala Alexandria on the western bank of the Jhelum; (4) Nicia (Alexandria) on the eastern bank of the Jhelum; (5) Alexandria at the confluence of the Chenab and the Indus; (6) Sodrian Alexandria below the confluence of the Punjab rivers; and (7) several cities and towns built by Alexander the Great in the land watered by the mouths of the river Indus.[2] The most important Alexandria outside the above zone is of course the celebrated city of that name in Egypt.

About the Alexandrian towns in the lower valley of the Indus, Cunningham says, "In the time of Alexander, the only place of note in the Delta was Patala; but he is said to have founded several towns himself during his long stay in Lower Sind, waiting for the Etesian winds to start his fleet. Unfortunately, the historians have omitted to give the names of these places. Justin alone notes that on his return up the Indus, he built the city of Barce....Ptolemy has preserved the names of

1. Cf. Malalasekera, *Dictionary of Pali Proper Names*, s. v. Alasanda.

2. See *Cambridge History of India*, Vol. I, pp. 376 ff.; Smith, *Early History of India*, 1924, pp. 52, 72, 110 ff.; Raychaudhuri, *Political History of Ancient India*, 1938, p. 212.

several places, as Barbara, Sousikana, Bonis and Kolaka, of which the first is most probably the same as 'Barbariké emporium' of the *Periplus* and perhaps also the same as the Barce of Justin."[1] According to Ptolemy, Patala and Barbarei stood on islands in the Lower Indus.[2]

It is interesting to note in this connection that a people called Varvara or Barbara is well known from ancient and medieval Indian literature. Pargiter says, "The Varvaras are generally mentioned in conjunction with the Śakas or Yavanas (Sabhā-P., XXXI, 1199; Vana-P., CCLIII, 15254; Śānti-P., CCVII 7560-61, etc.); and from the allusions, it appears that they were mainly a western or north-western race."[3] The Purāṇas also mention the Varvaras along with the Śakas or the Śakas and the Yavanas,[4] while in late medieval works, they are not only associated with Sauvīra (Lower Indus valley to the east of the river) and Lāṭa (Southern Gujarat), but are also mentioned between Lāṭa and Saindhava (Sind or the Lower Indus valley to the West of the river) and placed between two localities called Māyāpura and Saptaśṛṅga.[5] Since these Varvaras apparently lived in the lower valley of the Indus, there seems to be no difficulty in associating them with the city of Barbara probably founded by Alexander the Great in the same area.

As regards the identification of Alasanda mentioned in Indian literature, the general tendency among scholars now is to regard it as the same as Alexandria-under-the-Caucasus near Kabul.[6] Of course the recent discovery of an edict of Aśoka in Greek and Aramaic (meant apparently for the Yavana and Kamboja subjects of the Maurya emperor) at Kandahar has made it possible to think that Alexandria-among-the-Arachosians was the headquarters of the Yavana-Kamboja province in Aśoka's empire.[7] But Sylvain Lévi has emphatically dec-

1. *Ancient Geography of India*, ed. Sastri, p. 319.
2. VII.i.59; cf. Smith on Alexander's haven (*op. cit.*, pp. 110-11).
3. *Mārkaṇḍeya Purāṇa*, trans., p. 319, note 7.
4. See above, p. 30, note 1; p. 67, note 9.
5. *Ibid.*, pp. 71, 79-80, 106.
6. See, e.g., *Cambridge History of India*, Vol. I, p. 348, etc.
7. *Purāṇa*, Vol. V, pp. 251 ff.

lared that Alasanda of the Indian literature is identical with Alexandria in Egypt.[1]

Lévi's theory is primarily based on the evidence of the *Milindapañha* which mentions Alasanda in three different passages. In the first of these passages, the Yavana king Milinda (Menander) says that his *jāta-bhūmi* (land of birth) was Alasanda-dvīpa which was 200 *yojanas* from Śākala (Sialkot in the Punjab), the capital of the Indo-Greek king.[2] Lévi points out that the Chinese translations of the *Milindapañha* give the distance between Alasanda and Śākala as 2000 *yojanas* which would suit a distant locality like Alexandria in Egypt. But the distance of 200 *yojanas* given in the Pāli text cannot possibly be discarded without sufficient reason, and Lévi's argument on this point is not strong enough as we shall presently see.

The second passage in the *Milindapañha* mentions Alasanda in the following list of territories—Śaka-Yavana, Cīna-Kirāta, Alasanda-Nikumba, Kāśi-Kosala, Kaśmira and Gandhāra.[3] The most interesting feature of this list is that Alasanda is mentioned here separately from the lands of the Śakas and the Yavanas. This seems to suggest that Alasanda, associated with the Yavanas, lay outside the Indian Yavana country which included Śākala according to the *Milindapañha*. This fact, however, does not definitely suggest its identification with Alexandria in Egypt.

More important is the third passage saying how a seafaring merchant could visit the following lands in course of his voyage—Vaṅga, Takkola, Cīna, Sauvīra, Surāṣṭra, Alasanda, Kola-pattana and Suvarṇabhūmi.[4] Since Alasanda is here represented as situated on the sea-shore, Lévi thinks that it is certainly identical with Alexandria in Egypt. But there is another passage in the *Milindapañha* which seems to go clearly against the above identification.

1. *Indian Historical Quarterly*, Vol. XII, pp. 124 ff. But see the views of Sastri in Cunningham, *op. cit.*, pp. 692 ff.

2. Ed. Trenckner, p. 82.

3. *Ibid.*, p. 331.

4. *Ibid.*, p. 359. The identification of Vaṅga with the deltaic region of South Bengal is not less satisfactory than that with the Banka Island.

After declaring that Alasanda-dvīpa was the land of his birth, Milinda states that his *jāta-nagara* (the township where he was born) was Kalasīgrāma which bears an Indian name and could not have been situated in the Egyptian Alexandria or in its neighbourhood.[1] There is moreover some evidence to show that Alasanda was situated near about the mouths of the river Indus.

The *Kauṭilīya Arthaśāstra* mentions two kinds of *pravāla* (coral), viz. *alakandaka* (v. l. *alakandraka, alakṣandraka*) and *vaivarṇika* (v.l. *vaivalguka*),[2] and Lévi is certainly right in regarding the name Alakanda, from which *alakandaka* is derived, as identical with Alasanda. Lévi has also drawn our attention to Bhaṭṭasvāmin's commentary on the *Arthaśāstra*, which explains *alakandraka* as follows—*Alakandro Varvara-kūle samudraikadeśaḥ; tatra jātam = alakandrakam*, "Alakandra is a part of the sea near the shore in Varvara; the coral produced there is called *alakandraka*." This suggests that Alakanda = Alasanda was a locality on the sea-shore in the Varvara country which, as we have seen, lay near about the mouths of the Indus. This small territory seems to have been subject to Scythia or Indo-Scythia in the age of the *Periplus* and Ptolemy's *Geography*.

Lévi, however, says, "The coast of Barbara, Barbaria of the *Periplus*, is the coast of Somali on the Gulf of Aden where the port of Berbera still preserves the ancient name."[3] In his opinion therefore, Alakanda = Alasanda has to be identified with Alexandria in Egypt. But the distance between Alexandria and Berbera being about 2000 miles, there is really no sense in locating the former on the coast of the latter. Moreover, Varvara in the Lower Indus Valley was well-known to the Indians till the late medieval period, so that a fear of confusion would have prevented Bhaṭṭasvāmin from applying the same name to the Berbera coast nearly 2000 miles away. In these circumstances, if the Chinese versions of the *Milindapañha* locate the birth place of Mi-lan (Milinda) on the sea-

1. *Ibid.*, p. 83. The existence of some Brāhmaṇas in the population of the Egyptian Alexandria in Roman times can hardly explain the nomenclature of Kalasīgrāma.

2. Ed. Shamasastri, p. 78; Lévi, *loc. cit.*

3. *Op. cit.*, p. 129.

shore", as Lévi points out,[1] Kalasīgrāma in Alasanda-dvīpa may have been a place in Lower Sind.

In this connection, Lévi further draws our attention to the fact that, according to the *Periplus*, the Mediterranean coral was imported to the Indian ports (including Barbaricum) which did not export coral outside, while Pliny also refers to the export of coral by the Gallic people to India. But this apparently refers to the *vaivarṇika* coral of the *Arthaśāstra*, which Bhaṭṭasvāmin explains as follows—*Yavana-dvīpe Vaivarṇa-nāmo samudr-aika-deśaḥ*; *tatrabhavaṁ vaivarṇikam*, "a part of the sea in Yavana-dvīpa is called Vivarṇa; coral produced there is called *vaivarṇika*."[2] If then the *alakandraka* coral was the product of the Mediterranean, as Lévi thinks, where could the *vaivarṇika* coral of Yavana-dvīpa have been fished ? Yavana-dvīpa here seems to indicate the Roman world. The *Garuḍa Purāṇa* (Ch. 80) also mentions the following places as the source of coral—Kerala and other (Indian) territories as well as the Nisaka, Romaka (the Roman empire) and Devaka countries. If coral, possibly fished near Barbaricum, was not exported outside, the reason may be that it had a greater demand in India itself.

Lévi points out that early Jain literature mentions *alisaṁdaya* or *alisaṁdaga* as the same as or as a variety of *cavalaka* or chick pea and regards the name as derived from Alexandria in Egypt.[3] But the grain seems to have been an Indian produce (probably growing in the Lower Indus valley), and there is perhaps no reason to associate it with Alexandria.

A tradition regarding the existence of the Yavanas in the Lower Indus valley in the age of the Guptas seems to be known to Kālidāsa's *Raghuvaṁśa*, Canto IV. Verse 60 says that Raghu started from Aparānta (Northern Konkan), by the land-route, to Persia to conquer the Pārasikas. The preference of the land-route to the sea-route is explained by verse 61 which states that Raghu wanted to frighten the Yavana girls as it were. The

1. *Op. cit.*, p. 126.
2. *Ibid.*, pp. 129-30.
3. *Ibid.*, pp. 127-28.

stanza suggests that, in going to Persia from the Northern Konkan, Raghu had to cross the country of the Yavanas in the lower valley of the Indus, with whom he had no occasion to fight.[1] It is tempting to associate these Yavanas with the Varvara country in which Alasanda was situated.

1. Sircar, *The Successors of the Sātavāhanas in the Lower Deccan*, pp. 325-26; olso *Studies in the Society and Administration of Ancinet and Medieval India*, Vol. 1—*Society*, pp. 71-72.

Chapter XV

VĀHLIKA, VĀHIKA AND GURJARA

I

Elsewhere we have tried to show that very often in inscriptions and literary works we get, in connection with the description of a powerful king's *dig-vijaya* (conquest of the four quarters, or of 'the whole earth'), the traditional boundaries of the conventional *cakravarti-kṣetra*.[1] These boundaries are generally : (1) the Himālaya, Badarikāśrama on the Gandhamādana, the country near the Vaṁkṣu (Oxus), or the like, in the north; (2) the Udaya mountain (mythical), the Eastern Sea (Bay of Bengal), the Lauhitya, Vaṅga, or the like, in the east; (3) the Malaya mountain, the Mahendra mountain, the Southern Sea (Indian Ocean), Ceylon, or the like, in the south; and (4) the Asta mountain (mythical), the Mandara mountain (mythical), the Western Sea (Arabian Sea), the Pārasika country, or the like, in the west. In that connection, we have also tried to prove that the Meharauli pillar inscription describing the *dig-vijaya* of a king named Candra (i.e. Candragupta II) in the verse :

यस्योद्वर्त्तयतः प्रतीपमुरसा शत्रून्समेत्यागतान्
वङ्गेष्वाहववर्तिनोऽभिलिखिता खड्गेन कीर्त्तिर्भुजे ।
तीर्त्वा सप्त मुखानि येन समरे सिन्धोर्ज्जिता वाह्लिका
यस्याद्याप्यधिवास्यते जलनिधिर्वीर्य्यानिलैर्दक्षिणः ॥[2]

gives the following boundaries of the *Cakravarti-kṣetra* : (1) Vāhlika (spelt also with *b* for *v* and *ī* for *i*) in the north; (2) Vaṅga in the east ; (3) Southern Sea in the south and (4) the seven mouths of the Indus (all falling into the Western or Arabian Sea) in the west. The Vāhlika-Bāhlika-Vāhlīka Bāhlīka country is certainly the modern Balkh region on the Oxus in the northern part of Afghanistan.[3]

1. See above, pp. 8 ff.
2. *Corp. Ins. Ind.*, Vol. III, p. 141, verse 1; *JRASB*, Letters, Vol. V, 1939, pp. 407 ff.; above, p. 10.
3. See above, p. 32, mote 6.

According to a celebrated historian,[1] the country of the Vāhlikas mentioned in the Meharauli pillar inscription, which speaks of the pillar being raised on the Viṣṇupada, has to be placed on the Beas (Vipāśā) on the strength of a verse found in the *Rāmāyaṇa* as follows :

ययुर्मध्येन वाह्लीकान्सुदामानं च पर्वतम् ।
विष्णोः पदं प्रेक्षमाणा विपाशां चापि शाल्मलीम् ॥[2]

"They went through the Vālhīka country to Mount Sudāman, viewing Viṣṇupada and also the Vipāśā and the Śālmalī."

And the following conclusion has been drawn on the basis of it: "It will be seen from this verse that Viṣṇupada, Vipāśā and Śālmalī, if not even Sudāman, were all in the Vāhlīka country and close to one another........these places were in the close proximity of the Vipāśā which, we know, is the modern Beas, where it is joined by another river, the Śālmalī."[3]

The suggestion that the country watered by the Beas was called Vāhlīka or Bāhlīka is however certainly wrong. Numerous passages can be quoted from the epic, Purāṇic and Classical Sanskrit literature to prove that the Punjab = Pañcanada, 'the land of the Five Rivers' (one of the five being the Beas), was in ancient times called the Vāhīka country, and there is absolutely no doubt that the reading *Vāhlīkān* in the passage quoted from the *Rāmāyaṇa* is a mistake for *Vāhīkān*. A flood of light on the ancient Vāhīka country and its people is thrown by several chapters of the Karṇa-parvan of the *Mahābhārata*. We have therein (Ch. 44) the following verses regarding the geography of this country :

पञ्चानां सिन्धुषष्ठानां नदीनां येऽन्तरा स्थिताः ।
तान्धर्मबाह्यानशुचीन्वाहीकानपि वर्जयेत् ॥ ७

* * *

साकलं नाम नगरमापगा नाम निम्नगा ।
जार्तिका नाम वाहीकास्तेषां वृत्तं सुनिन्दितम् ॥ १०

* * *

1. See D. R. Bhandarkar in *Journ Andh. Hist. Res. Soc.*, Vol. X, pp. 86 ff.
2. *Rāmāyaṇa*, II. 68, verses 18-19.
3. *Op. cit.*, p. 87.

पञ्च नद्यो वहन्त्येता यत्र पीलुवनान्युत ।
शतद्रुश्च विपाशा च तृतीयैरावती तथा ॥ ३१
चन्द्रभागा वितस्ता च सिन्धुषष्ठा बहिर्गिरेः ।
आरट्टा नाम ते देशा नष्टधर्मा न तान्व्रजेत् ॥ ३२

* * *

पञ्चनद्यो वहनन्त्येता यत्र निःसृत्य पर्वतात् ।
आरट्टा नाम वाहीका न तेष्वार्यो द्व्यहं वसेत् ॥ ४०

The above verses and several others from the Karṇa-parvan (Chs. 44-45) prove that the kingdom (or the people) over which Śalya ruled was generally referred to as Pañcananda or Vāhīka and specifically as Madraka, Jārtika and Āraṭṭa. In other words, the Madrakas, Jārtikas and Āraṭṭas were generally designated as Pañcanada and Vāhīka. The Madraka or Madra people lived in the land around Sākala (modern Sialkot) while the Jārtikas are the modern Jāṭs. There is however one verse which suggests that Vāhīka was originally the name of a country or people on the Beas (Vipāśā) :

वहिश्च नाम हीकश्च विपाशायां पिशाचकौ ।
तयोरपत्यं वाहीका नैषा सृष्टिः प्रजापतेः ॥[1]

"In the Vipāśā, there were two Piśācas named Vahi and Hīka ; their descendants are the Vāhīkas who are not the creation of Prajāpati."

The geography of this well-known Vāhīka country has been discussed by a number of scholars. Reference may be made to Cunningham's *Ancient Geography of India*, Calcutta, 1924, pp. 247, 686-87, and V. S. Agrawala's 'Patañjali and the Vāhīka-grāmas' in the *Indian Culture*, Vol. VI, 1939, pp. 129 ff. In view of the voluminous evidence regarding the existence of a Vāhīka country near the Vipāśā, no Vāhlīka (which is no doubt a mis-lection for *Vāhīka*) country can be located in the same region on the strength of a solitary passage of the *Rāmāyaṇa*.

If the Vāhīkas were originally a people living on the Beas, we have to explain how the sense of the word *Vāhīka* expanded so as to indicate all the tribes living in Pañcanada, i.e. the Punjab. It is interesting in this connection to note that the *Mahābhārata* sometimes uses the terms Vāhīka, Madra, Jārtika,

1. *Mahābhārata*, Karṇa-parvan, Ch. 44, verse 10.

Āraṭṭa and Pañcananda synonymously. It therefore appears that the lands of the tribes which lived close to one another in the Punjab became in course of time moulded into a big kingdom under the powerful kings of Śākala (Sialkot). As Vāhīka was beyond Kurukṣetra and therefore lay outside the boundary of Brahmāvarta,[1] its analogical connection with the word *bahis* may have been another cause of the expansion of its geographical sense.

II

Furdoonjee D. J. Paruck refers to a drachm which he originally attributed[2] to the Sassanian emperor Hormazd II (303-10 A.D.), but was later inclined, after further study, to assign to Hormazd I (272-73 A.D.).[3] In the legend of the coin in question, Paruck reads the names of three territories forming parts of the Sassanian empire as Inde, Iradate and Harezi, which he identifies respectively with Sindh, Mūltān and Rājpūtānā. Whatever, however, be the value of the reading and interpretation of the names *Inde* and *Iradate*, there is absolutely no doubt that the suggestion regarding the third name is wrong.

Paruck says, "The reading of the word in the second line on the upper part of the reverse remained for a long time completely illusive ; but now I am able to propose the reading HREZI. According to all the early Arab geographers, the old name of Rājpūtānā was Haraz. It is probable that the original form of this name was Harez, as on this drachm."[4] He refers to Cunningham's *Ancient Geography of India*, Calcutta, 1924, p. 358, and quotes the following passage in support of the identification of Harez and Rājpūtānā : "The name of the country is somewhat doubtful as the unpointed Arabic characters may be read as Haraz or Hazar and Kharaz or Khazar as well as Jurz or Juzr. But fortunately there is no uncertainty about its position

1. *Ibid.*, VII. 41. 6.
2. Cf. *Revue Archeologique*, 1930, pp. 234 ff.
3. *JNSI*, Vol. I, pp. 58-71.
4. *Loc. cit.*, pp. 61-62.

which is determined to be Rājpūtānā by several concurring circumstances......"

It is a matter of regret, however, that Cunningham's views have not been quoted faithfully and completely. For, immediately before the above passage, he says, "All the early Arab geographers speak of a kingdom named Jurz or Juzr which from its position would appear to be the same as the Kiu-che-lo of Hwen Thsang." Cunningham further says, "Edrisi, quoting Ibn Khordādbeh, states that Juzr or Huzr was the hereditary title of the king as well as the name of the country. This statement confirms my identification of Juzr with Guzr or Gujar, which is a very numerous tribe whose name is attached to many important places in North-West India and the Punjab and more especially to the great peninsula of Gujarāt."

It will be seen that, according to Cunningham, the ancient name of Rājpūtānā was Gujar or Gurjara (Gūrjara)[1] and not Harez as stated by Paruck. It is a matter of satisfaction that the suggestion of the great Indologist, made on insufficient data as early as 1871 A.D., has later been proved definitely by unquestionable epigraphic evidence. The editor of the Calcutta edition of Cunningham's work, consulted by Paruck, remarks, "The inscriptions of the 9th century also applied the name Gurjara-rāṣṭra to the country north of Ajmer and the Sambhar Lake.[2] Its capital Pi-lo-mo-lo is modern Bhinmāl, 25° N. and 76° 16′ E., a town in the Jashwantpur District of Jodhpur.[3] Bhinmāl or Bhillamāla, the plateau of the Bhils, was the capital of the Gurjara-Pratīhāras who transferred their headquarters to Kanauj." The statement, although it confuses the Gurjara-Pratīhāra empire with the Gurjara country and is not strictly accurate, is sufficient to disprove Paruck's contention.

That Hiuen-tsang's *Kiu-che-lo* and its capital *Pi-lo-mo-lo* stand respectively for *Gurjara* and *Bhillamāla* is supported by the Daulatpura inscription[4] of 843 A.D., according to which the Gurjara-Pratīhāra emperor Vatsarāja (*circa* 775-800 A.D.) granted the village of Śivā in the Deṇḍavānaka *viṣaya* (modern

1. In South Indian inscriptions, the name is generally spelt *Gūrjara*.
2. *Imp. Gaz. Ind.*, Vol. XII, p. 350.
3. *Ibid.*, Vol. IV, p. 111.
4. *Ep. Ind.*, Vol. V, p. 211 ff.

Didwana in the former Jodhpur State, Rajasthan) of the Gurjaratrā *bhūmi*. In the days of the early Arab travellers, therefore, the name of the Jodhpur region of Rajasthan was Gurjaratrā or Gujarāt, a name derived from that of the Gurjara people. That, however, this country included the former Alwar and Bharatpur States in Eastern Rajasthan is suggested by the celebrated Arab writer Al-Bīrūnī who composed his work on India about 1030 A.D. Al-Bīrūnī refers to 'Bazāna, the capital of Gujarāt' and places it 25 *farsakh* (about 92 miles) to the north of the kingdom of Mewār (capital Jattaraur, *i.e.* Citrakūṭa or Chitor) and 15 *farsakh* (about 55 miles) from Rājaurī lying on the route towards the south-west from Kanoj.[1] Rājaurī (*i.e.* Rājyapurī) is no doubt identical with Rājyapura (modern Rājorgaḍh in the former Alwar State of Rajasthan) which was the capital of *Mahārājādhirāja* Mathanadeva of the Gurjara-Pratīhāra lineage according to an inscription of 960 A.D.[2] Al-Bīrūnī further says, "The distance between Mathurā and Kanoj is the same as that between Kanoj and Bazāna, *viz.* 28 *farsakh* (about 103 miles)."[3] The above indications show almost beyond doubt that Bazāna is no other than the modern Bayānā in the former Bharatpur State of Rajasthan. It may be pointed out that Bazāna seems to have ceased to be the capital of 'Gujarāt' in the days of Al-Bīrūnī, as he says, "This town is called Nārāyan by our people. After it had fallen into decay the inhabitants migrated to another place called Jadūra."[4] The name Nārāyaṇa was apparently derived from a celebrated temple of that god at Bayānā. As regards Jadūra, Sachau says, "This reading is uncertain. Perhaps all the signs of the Arabic text are the name of a place."[5] It, however, seems that the reading intended is Rājaurī, i.e. Rājyapurī or Rajyapura which is the same as the modern Rājor. It may further be pointed out that Al-Bīrūnī's *Canon Masudicus* gives the latitudes and longitudes of both Bazāna (or Nārāyaṇa) and Bhillamāla, longitudes being reckoned from the coast of the Atlantic and Bagdad

1. Cf. Sachau, *Alberuni's India*, Vol. I, p. 202; Sircar, *Cosm. Geog. E. Ind. Lit.*, p. 176.
2. *Ep. Ind.*, Vol. III, p. 266.
3. Sachau, *loc. cit.*; Sircar, *loc. cit.*
4. *Loc. cit.*
5. Sachau, *op. cit.*, Vol. II, p. 319.

being located at 70°.[1] As, however, the longitude recorded for Bazāna is palpably wrong, it hardly helps us in locating the city.

The above discussion will show that Harez was never the name of Rājpūtānā. According to the combined testimony of inscriptions and of Hiuen-tsang (7th century) and Al-Bīrūnī (11th century), the capital of Gurjaratrā or the Gurjara country in Rājpūtānā was at first at Bhinmāl in the former Jodhpur State, next it was at Bayānā in the former Bharatpur State, and afterwards at Rājorgaḍh in the former Alwar State. The Arab geographers, therefore, speak of Gurjara and not of Harez.

1. *Loc. cit.*

CHAPTER XVI

MALAYA AND MAHIṢA

I

There has been some criticism of the view that 'Malaya is the southern end of the Western Ghāṭs'. The critic says, "The Purāṇas do not help us in locating the Malaya mountain."[1] But his views appear to go against the evidence supplied by the Purāṇic sections on Indian geography.

It is well known that the *Mārkaṇḍeya*, *Vāyu*, *Kūrma*, *Matsya*, *Vāmana* and *Brahmāṇḍa Purāṇas* contain a long list and the *Viṣṇu*, *Brahma* and *Śiva Purāṇas* a short list of the Indian rivers. According to both these lists, the rivers Kṛtamālā and Tāmraparṇī issue from the Malaya mountain.[2] Even the shortest list of rivers found in the *Agni Purāṇa* speaks of the Kṛtamālā as rising from the Malaya. Of the two rivers mentioned above, Tāmraparṇī, still known by practically that name, flows from the southern part of the Western Ghāṭs to the Bay of Bengal. The celebrated ports of Korkai and Kayal stood on its banks. The identification of the Kṛtamālā with the modern Vaigai running by Madurai or Dakṣiṇa-Mathurā is corroborated by the *Caitanyacaritāmṛta*, an old biography of the Bengali Vaiṣṇava saint Caitanya (1485-1534 A.D.). It is clearly stated in this work that Caitanya took his bath in the waters of the holy Kṛtamālā at Dakṣiṇa-Mathurā in the course of his tours of pilgrimage in South India.[3]

The evidence of the Purāṇas definitely locates the Kul-āçala or Kula-parvata (literally, a mountain associated with certain tribes) called Malaya, which was the source of the Tāmraparṇī and Kṛtamālā (Vaigai) rivers running through the Pāṇḍya country, at the extreme southern end of the Western Ghāṭs.[4] The name of this mountain no doubt lies at the root of the names

1. See C. Kunhan Raja in *IHQ*, Vol. XXII, pp. 223-27.
2. See above, p. 60, note 3.
3. See *loc. cit.*; also p. 64, notes 4-5.
4. For the identification of the Malaya, see Raychaudhuri, *Stud. Ind. Ant.*, pp. 109 ff.; Pargiter, *Mark. Pur.*, trans., pp. 285, 303-04; Dey, *Geog. Dict.*, s. v. See also above, p. 60, notes 3 and 6.

Malayāḷam and Malabar; but reference has been made to certain 'conflicting evidences' that would suggest the location of the Malaya elsewhere.[1]

As a matter of fact however the Kula or tribal mountain called Malaya cannot be located in any other place excepting the southern end of the Western Ghāṭs although there were probably other hills called Malaya which may have given rise to the 'conflicting evidences'. We know that the hilly region of Central Ceylon was known by the name Malaya and one of the heirs of the Ceylonese kings was often styled *Malaya-rāja*. A Viṣṇukuṇḍin inscription speaks of Prince Mādhavavarman II, who had his headquarters at Amarapura (i. e. Amarāvatī near Vijayavāḍa), as *Trikūṭa-malay-ādhipati*, probably indicating 'governor of the province of Trikūṭa-malaya'. Since the name *Malaya* is derived from the Dravidian word *malai* meaning 'hill', it is not improbable that this word affixed to the end of the names of certain South Indian hills was often Sanskritised as *malaya* and thus gave rise to a confusion. The tendency to Sanskritisation is still noticed in the name Madra-maṇḍala applied to the Madras territory (in spite of the fact that the ancient Madra people lived in the Punjab with their capital at Śākala, i. e. modern Sialkot), and we can understand the working of the mind of a Sanskritist if he would refer to Tiru-malai and Anna-malai respectively as Śrī-malaya (or Tri-malaya) and Anna-malaya. But there is absolutely no doubt as to the location of the celebrated Malaya Kul-ācala (Kula-parvata) at the southern end of the Western Ghāṭs.

II

A famous historian has recently discovered the existence of several hitherto unknown kings and ruling families of ancient India such as Mahiṣa[2] (or Mahiṣya)[3] and Sebaka.[4] These dynastic names, derived according to him from those of the progenitors of the families in question, are said to have been found on certain coins of about the second or third century A. D. A

1. Cf. C. Kunhan Raja in *IHQ*, Vol. XXIII, p. 69.
2. V. V. Mirashi in *JNSI*, Vol. XI, pp. 1 ff.
3. *IHQ*, Vol. XXII, pp. 24 ff.
4. *JNSI*, Vol. VIII, pp. 197 f.

proper examination of the coins has convinced us that the dynastic or tribal name *Mahiṣa* does not occur in the legend.[1] But assuming that the reading of the name of the Mahiṣas is justified, a word may be said about the location of the country associated with a ruling family or a people called Mahiṣa.

The historian has quoted passages from the *Rāmāyaṇa*, the *Mahābhārata* and the Purāṇas to show that the Mahiṣa country 'was situated in [the] Southern Deccan'.[2] He further says, "The royal name Mahiṣa (or Mahiṣya) was probably derived from the country of Māhiṣaka or Mahiṣa-maṇḍala where this family was ruling. In my previous article on this Śaka king Māna, I followed Rapson in supporting that Māhiṣaka was the country round Māhiṣmatī, modern Oṅkār Māndhātā in the Nemāḍ District of the Central Provinces. Since then the provenance of these coins has become known. As stated before, they were found in the excavations at Kondapur and Maski in the Hyderabad State. The country under the rule of the Mahiṣa dynasty seems therefore to have been the southern portion of the Hyderabad State."[3] It may be pointed out in this connection that there is definite epigraphic evidence regarding the existence of a tract of land associated with a people or a royal family called Mahiṣa in the old Mysore State which is not very far from 'the southern portion of the Hyderabad State'.

An inscription of the Kadamba king Viṣṇuvarman I (about the end of the fifth century)[4] was found in a village in the Tumkur Taluk of the Tumkur District in the northern part of the old Mysore State. This is the Hebbata grant published in the *Annual Report of the Mysore Archaeological Department* for 1925.[5] The grant was issued by Kadamba Viṣṇuvarman I in his fifth regnal year, when the king was staying at the *adhiṣṭhāna* (meaning 'a city') of Kūḍalūr. By this grant, Viṣṇuvarman I created an *agrahāra* of the village called Herbbaṭa situated in the Sāṭṭipalli-Jāripāṭa division of the Mahiṣa *viṣaya* (district) and offered it with *dakṣiṇā* and libation of water, in accordance with the

1. See *Ep. Ind.*, Vol. XXXV, pp. 69 ff.; Sircar, *Stud. Ind. Coins*, pp. 126 ff.
2. *JNSI*, vol. XI, p. 4.
3. *Loc. cit.*
4. See *The Successors of the Sātavāhanas*, p. 393.
5. See *op. cit.*, p. 98, Plate X.

brahmadeya-nyāya, to a Yajurvedīya Brāhmaṇa belonging to the Badira family or clan and to the Gautama *gotra*. It seems clear from this inscription that the present Tumkur area in the Mysore State was known in the fifth century A.D. as the Mahiṣa territory which formed a part of the dominions of the Early Kadamba kings. While discussing the evidence of this interesting inscription, we observed elsewhere, "The name of the Mahiṣa *viṣaya* (cf. Māhiṣaka of the Purāṇic lists) is evidently the source from which the present Mysore (Mahiṣūr) has derived its name."[1] It seems to us that the original name of Mysore was derived from the words *Mahiṣa* and *ūr* and signified 'the city of the Mahiṣas'. Whatever however be the value of this suggestion, the Hebbata grant of the Kadamba king Viṣṇuvarman leaves no room for doubt that the northern part of the old Mysore State was known in ancient times as the Mahiṣa country. Whether this country included, towards the north, the present Kondapur and Maski areas of old Hyderabad, at least in a particular period of history, can hardly be determined in the present state of our knowledge. But it is by no means improbable. We know that, in ancient records, the extent of a country is often differently indicated. Thus Tāmraplipta is mentioned in literature sometimes as an independent state, sometimes as a part of Suhma and sometimes as a part of Vaṅga. Similarly, Taxila (Takṣaśilā) is sometimes represented as an independent kingdom, sometimes as a part of Gandhāra and sometimes as a part of Uraśā (roughly the modern Hazara District of the North-West Frontier region of West Pakistan).[2] This is no doubt due to the fact, also supported by modern history, that a small kingdom retained its old name even when it was expanded to cover wide areas. It may be recalled in this connection that, with the rise of Kanarese powers like the Cālūkyas, Rāṣṭrakūṭas and the houses of Vijayanagara, the name Karṇāṭa and Kuntala extend-

1. *The Successors of the Sātavāhanas*, p. 293; also p. 216, note.

2. See *ibid.*, p. 216, note; also H. C. Raychaudhuri, *Studies in Indian Antiquities*, pp. 186 f. Raychaudhuri further points out how the city of Uragapura sometimes belonged to the Cola, but sometimes to the Pāṇḍya, kingdom. Similarly Koṭivarṣa (the modern Dinajpur District) was sometimes included in the Puṇḍravardhana *bhukti*, but sometimes in Rāḍha. See also above, pp. 191-92.

ed over wide regions outside the Kannaḍa-speaking area.[1] Since the English East India Company extended its power over extensive areas of Northern India from their base in Bengal, the Company's dominions in North India, including the present Bihar and the major part of Uttar Pradesh, were for a long time known as the Presidency of Bengal.

Sometimes, however, it is seen that a tract of land, later annexed to a kingdom, retains the name of the latter even when the heart of the kingdom itself comes in a later epoch to be known by a different name. We have referred above to the more recent case of Karṇātaka in the Tamil land, which owes its origin to the expansion of the Vijayanagara empire outside the original Karṇāṭa country. The ancient names Andhrāpatha near the mouths of the Kṛṣṇā, Śātavāhanīya-rāṣṭra or Śātavāhanīy-āhāra in the present Bellary District and Nalavāḍi in the present Bellary-Anantapur region may possibly offer older instances of the same type, as the hearts of the Śātavāhana (Andhra) and Nala countries probably lay elsewhere. But, in the absence of any clear evidence to the contrary, it is possibly better, in the present state of our knowledge, to believe that the heart of the ancient Mahiṣa country lay in the old Mysore State.

The name Mahiṣa, applied to a people and their land, must have originally derived from the totem of the said people. It is thus possible to think that there were more settlements of the Mahiṣa people in different parts of the country than one. Māhiṣmatī on the Narmadā seems also to owe its name to the Mahiṣas. But there is as yet no genuine evidence (excepting wrongly interpreted legends of a few coins) to show that the southern part of the former Hyderabad State was known as the Mahiṣa country.

1. Note how the name Karṇāṭaka (Carnatic) came to be applied to the dominions of the Nawabs of Arcot in the Tamil country. See above, pp. 190 ff.; cf. pp. 82 (verse 16), 96-97. Cf. also the case of Gandhāra (Raychaudhuri, *PHAI*, 1938, p. 202).

Chapter XVII

KRMILĀ

I

The Monghyr plate of king Devapāla (*circa* 810-50 A.D.) of the Pāla dynasty of Bengal and Bihar was discovered as early as 1780. It was first published in 1788 in the *Asiatic Researches*, Vol. I, pp. 13 ff. The inscription was re-edited by F. Kielhorn in the *Indian Antiquary*, Vol. XXI, 1892, pp. 254 ff.; by A. K. Maitreya in the *Gauḍalekhamālā*, B. S. 1319 (1913 A.D.), pp. 33 ff. ; and by L. D. Barnett in the *Epigraphia Indica*, Vol. XVIII, 1925-1926, pp. 304 ff.

The grant was issued by the *Paramasaugata-Parameśvara-Paramabhaṭṭāraka-Mahārājādhirāja* Devapāladeva in the 33rd year of his reign from his *jaya-skandhāvāra* at Mudga-giri, *i.e.* modern Monghyr, the headquarters of a District of that name in Bihar. By this charter, the Pāla king granted the village called Meṣikā-grāma in favour of a Brāhmaṇa named Vihekarāta-miśra. The village was situated in the *viṣaya* or district of Kṛmilā which formed a part of the *bhukti* or province of Śrīnagara. From the inscriptions of the Pālas, we come to know of the existence of two *bhuktis* or provinces in Bihar, viz., Śrīnagara-bhukti and Tīra-bhukti.[1] The word *tīra* refers to the banks of the Ganges. Tīra-bhukti is the same as the modern Tirhut Division and apparently indicated parts of Bihar lying to the north of that river. The expression *Śrī-nagara* meaning 'the illustrious city,' *i.e.*, the city *par excellence*, referred to the celebrated ancient city of Pāṭaliputra (of which the modern representative is Pāṭnā derived from Sanskrit *pattana* or township) and the Śrīnagara-bhukti no doubt included the districts of South Bihar having their administrative headquarters at the above city.[2] Of the *viṣayas* or districts forming the Tīra-bhukti in the age of the Pālas, we know of the Hodreya and Kakṣa *viṣayas*, and, of those forming the

1. *History of Bengal*, Vol. I, Dacca University, p. 273.

2. Cf. the commentary on Vātsyāyana's *Kāmasūtra*, VI. 5. 30: *Nāgarikā iti Pāṭaliputrikāḥ*, etc.

Śrīnagara-bhukti, such districts as Gayā-viṣaya, Rājagṛha-viṣaya and Kṛmilā-viṣaya are known from inscriptions.[1] Of these *viṣayas*, those of Gayā and Rājagṛha can be easily identified as the tracts of land round respectively the modern towns of Gayā and Rājgir (*i.e.*, Rājagṛha), the latter being now situated in the administrative division known as the Patna District of Bihar. But the exact situation of Kakṣa *viṣaya* in the Tīra-bhukti and the Kṛmilā-viṣaya in the Śrīnagara-bhukti could not be satisfactorily determined. Recently we discovered certain new inscriptions of the Pāla period, which throw interesting light on the location of the Kṛmilā-viṣaya.

About the beginning of January 1950, I visited, in course of a search for inscriptions, certain villages in the neighbourhood of the Kiul and Luckeesarai railway stations on the former East Indian Railway (now Eastern Railway) about the western fringe of the Monghyr District of Bihar. Long ago, Alexander Cunningham made an epigraphical survey of many villages of South Bihar and the interesting results of that survey are recorded in his celebrated Reports. But it is doubtful if it was possible for him to visit all the villages. Moreover, images both inscribed and uninscribed, are being discovered every year in various villages of that area at the point of the cultivator's ploughshare and the workman's spade and a large number of them, discovered after Cunningham's survey, have accumulated in many villages. There is no doubt that many of such images have often been carried away from the find-spots by interested persons; but the epigraphic survey conducted by me only in a few villages convinced me fully that very good results may still be obtained if the images (unfortunately broken in most cases), scattered over almost all old villages in South Bihar, are made the object of a careful search in the line initiated by Cunningham more than eighty years back, but not seriously continued afterwards.

There is a village called Valgūdar (often said to be Vargūjar), near Rajauna and Chauki, on the side of the railway line between the Luckeesarai and Mankatha railway stations. I visited the village on the 9th of January and discovered no less

1. Ray, *Dynastic History of Northern India*, Vol. I, pp. 274, 400. For the Hodreya-viṣaya, see *Ep. Ind.*, Vol. XXIX, p. 50.

than three interesting inscriptions. A stone pedestal of a lost image in a locality, called Saṅgat owing to its being a Sikh religious establishment in the village, was found to bear a very important inscription dated both in the 18th regnal year of king Madanpāla of the Pāla dynasty and in the Śaka year 1083. The importance of the date of this record has been discussed elsewhere; but the fact it records is no less interesting. It says that two Vaiṣṇava brothers of a Brāhmaṇa family established an image of the god Nārāyaṇa at Krimilā. The second inscription discovered and examined by me at Valgūdar was found on a broken image of a goddess with a child on her lap and with a lion as her emblem, lying in the veranda of the Kutchery of Babu Dalipnarayan Sinha who was a zamindar of Bhagalpur. It may be mentioned here that images representing the Devī with a child seem to have been very popular in all parts of South Bihar. We have noticed such images in many villages. One such image is now in the Asutosh Museum of Indian Art of the University of Calcutta, and it bears an inscription of the time of Rāmapāla. The find-spot of this image also seems to us to be some village of South Bihar. A similar image at Rajauna near Valgūdar is known, from the inscription on it, to have been called Puṇḍeśvarī. It is possible that this rural deity, apparently not uncon nected with the conception of Pārvatī with Skanda on her lap as indicated by the theme of Kālidāsa's *Kumārasambhava*, and of the Buddhist Hārītī, was worshipped under different local names in various places.[1] The inscription on the Valgūdar image of the Devī with a child says that it was installed by a person named Nṛkaṭṭa at the *adhiṣṭhāna* or city of Kṛmilā. It may be noticed that the spelling of the name in this record is the same as in the Monghyr inscription of Devapāla, although in the

1. A similar image, but with four arms and without the lion emblem, was found near Dacca and has been tentatively identified by Bhattasali (*Iconography*, pp. 63 ff.; Plate XX) with Hārītī. Another form of the same primitive Mother-goddess with a child on her lap was conceived with a snake-canopy over her head and was worshipped both in Bihar and Bengal. I found one such image on the bank of the Saṁsārpokhrī at Luckeesarai. In Bengal, this deity later came to be identified with the snake-goddess Manasā (cf. *ibid.*, pp. 212 ff.; *His. Beng.*, *op. cit.*, pp. 460 f.); but that she was originally worshipped under different local names is suggested by the Marāil (Dinajpur District) image of the goddess known to have been called Bhaṭṭiṇī Maṭṭuvā.

Valgūdar inscription of Madanapāla's time it is written slightly differently. The word *kṛmilā* means 'a fruitful woman' or 'a place full of worms' and the word *kṛmi* forming its first part is sometimes also written as *krimi*. Our second inscription does not refer to the reigning monarch. Its palaeography however seems to suggest a date earlier than the days of Madanapāla. The third inscription that I discovered at the same village was on a stone pedestal of a lost image now being used by people as a platform for washing feet in the compound of the house of Babu Kesav Sinha. The inscription it bears says that the image in question was installed at the *adhiṣṭhāna* or city of Kṛmilā during the rule of Dharmapāla whose reign is now assigned to *circa* 770-810 A. D.[1] The regnal year is not referred to.[2]

Now the above three inscriptions, all discovered in the small village of Valgūdar, show beyond doubt that the city of Kṛmilā or Krimilā, headquarters of the *viṣaya* or district of that name forming a part of the Śrīnagara-bhukti within the dominions of the Pālas, stood either on the very site of that village or on a site, parts of which are now occupied by the village. There is a stone slab representing the twelve Ādityas and containing an inscription dated in the fifth regnal year of the Pāla king Śūrapāla, probably the first king of that name who reigned about the middle of the ninth century, at the neighbouring village of Rajauna. This inscription also says that the slab was installed at Kṛmilā, and it appears that the slab had been originally found at Valgūdar, but was later carried to Rajauna. It is however not improbable that the site of the modern village of Rajauna was also within the bounds of the ancient city of Kṛmilā, although we are not quite sure about that. In any case, if the ancient city of Kṛmilā has to be identified with the present village of Valgūdar near Luckeesarai in the western fringe of the Monghyr District, there is no difficulty in locating the *viṣaya* or district of that name in the tract of land around that village, that is to say, roughly in the western part of the Monghyr District, to the south of the Ganges between the land round Pāṭnā and that round Monghyr. It is quite probable that these two regions centering round Pāṭnā

1. *History of Bengal*, *op. cit.*, p. 177.

2. The inscriptions have been published in *Ep. Ind.*, Vol. XXVIII, pp. 137 ff.

(*i. e.* ancient Śrīnagara) and Monghyr (ancient Mudgagiri) formed separate *viṣayas* within the Śrīnagara-bhukti and were known as the Śrīnagara-viṣaya and Mudgagiri-viṣaya in the age of the Pālas. But on this point nothing can be said definitely until further evidence is forthcoming.

The *viṣaya* of Kṛmilā or Krimilā is also mentioned in the Nālandā plate[1] of Samudragupta who flourished in the fourth century A. D. As however the charter is spurious and seems to have been forged a few centuries after Samudragupta's time, it may or may not prove the existence of a *viṣaya* and therefore of a city of the above name in the fourth century. But that both existed before the Pāla occupation of Bihar may be suggested by the mention of the *viṣaya* on certain old seals found at Nālandā.[2] The village of Kavāla in the Krimilā *viṣaya* known from one such seal appears to be no other than the present Kawāli not far from Valgūdar.

II

There is a tradition recorded in the *Harivaṁśa* (I. 31. 24-28) and the *Vāyu Purāṇa* (99. 18-22).[3] The *Harivaṁśa* text of the verses runs as follows :

Uśīnarasya patnyas=tu pañca rājarṣi-vaṁśajāḥ |
Nṛgā Kṛmī Navā Darvā pañcamī ca Dṛṣadvatī || 1
Uśīnarasya putrās=tu pañca tāsu kul-odvahāḥ |
tapasā vai sumahatā jātā vṛddhasya Bhārata || 2
Nṛgāyās=tu Nṛgaḥ putraḥ Kṛmyāṁ Kṛmir=ajāyata |
Navāyās=tu Navaḥ putro Darvāyāḥ Suvrato='bhavat || 3
Dṛṣadvatyās=tu saṁjajñe Śibir=Auśīnaro nṛpaḥ |
Śibes=tu Śibayas=tāta Yodheyās=tu Nṛgasya ha || 4
Navasya Navarāṣṭran=tu Kṛmes=tu Kṛmilā purī |
Suvratasya tath=Ambaṣṭhā.......... || 5

The *Vāyu Purāṇa* offers the following variant readings: *Mṛgā* for *Nṛgā* in verse 1; *jāta-vṛddhāś=ca dhārmikāḥ* for *jātā vṛddhasya Bhārata* in verse 2; and the text of verses 3-4 as follows:

Mṛgāyās=tu Mṛgaḥ putraḥ Navāyā Nava eva tu |
Kṛmyāḥ Kṛmis=tu Darvāyāḥ Suvrato nāma dhārmikaḥ || 3

1. *Select Inscriptions*, p. 263.
2. *Mem. A.S.I.*, No. 66, pp. 34, 54.
3. For a corrupt text referring to the tradition, see *Matsya Purāṇa*, 48. 16 ff.; cf. *Bhāgavata Purāṇa*, IX. 23. 2 ff.

Dṛṣadvatī-sutaś = c = āpi Śibir[1] *= Auśīnaro dvijāḥ* ।
Śibeḥ[2] *Śivapuraṁ khyātaṁ Yaudheyan = tu Mṛgasya tu* ॥ 4

According to this tradition, king Uśīnara of the Puru dynasty had five queens, viz., (1) Nṛgā or Mṛgā, (2) Kṛmī, (3) Navā, (4) Darvā, and (5) Dṛṣadvatī, who gave birth respectively to the following five sons, viz. (1) Nṛga (or Mṛga), (2) Kṛmi, (3) Nava, (4) Suvrata and (5) Śibi (or Śivi) Auśīnara. Śibi was the lord of the Śibi people or of the city of Śivapura, while Nṛga (or Mṛga) was the ruler of the Yodheyas or of Yaudheyapura. The other three sons of Uśīnara, viz. Nava, Kṛmi and Suvrata, were the lords respectively of Navarāṣṭra, Kṛmilāpurī and Ambaṣṭhāpurī.

Although the above tradition does not help us in locating the city of Kṛmilā, the Purāṇic writers appear to have believed in its hoary antiquity. Early Buddhist works not only regards Kṛmilā as a very old city, but locates it on the Ganges in the area witnessing the Buddha's activities.

In the Buddhist canonical works in Pali, the name of the city of Kṛmilā is met with in the forms *Kimilā*, *Kimmilā* and *Kimbilā*, while the name of an inhabitant of the city is given as Kimila, Kimmila, or Kimbila. It may be pointed out that Malalasekera's *Dictionary of Pali Proper Names*[3] recognises the spellings *Kimilā* or *Kimbilā* and *Kimila* or *Kimbila*, but prefers the forms *Kimbilā* and *Kimbila*. J. Kashyap's edition of the Pali texts, however, quotes all the three variants of the two names, though it prefers *Kimilā* and *Kimila*.[4] We have of course to admit that all the three variants of both the names are correct. *Kimilā* is the normally expceted Pali form for Sanskrit *Kṛmilā* (or. *Krimilā*). But the form *Kimmilā* is derived through the modified pronunciation of the Sanskrit name as *Kirmilā, while Kimbilā is derived through the intermediate form Kimmilā. The same is the case with the Pali forms of the personal name Kṛmila (or Krimila).

Two Suttas, entitled the *Kimilāsutta* and *Kimilasutta*, were

1. Or *Śivi*°.
2. Or *Śiveḥ*.
3. Vol. I, pp. 604-06.
4. *Suttapiṭaka-Aṅguttaranikāya*, I, pp. 55, 217; II, p. 486; *Saṁyutta-nikāya-Saḷāyatanavagga*, p. 164; *Mahāvagga*, p. 274; cf. *Vinayapiṭaka-Cullavagga*, p. 281; *Mahāvagga*, p. 481.

preached by the Buddha when he was camping at the *veḷu-vana* or *niceḷu-vana* at the city of Kimilā = Kimmilā = Kimbilā (*i.e.* Kṛmilā or Krimilā) which is stated to have been situated on the bank of the Ganges. The river is now at a short distance from the villages of Valgūdar and Rajauna, on the site of which the ancient city stood. It is said that the city existed in the time of Kāśyapa Buddha Daśabala who was the twentyfourth Buddha (the third of the present aeon called the Bhadra-kalpa),[1] the twentyfifth and last of the Buddhas being Gautama (*c.* 566-486 B.C.). One of the celestial palaces seen by Nimi when he visited heaven is said to have belonged to a Deva who had originally been a pious inhabitant of Kṛmilā.[2] The Buddha is said to have been born in a previous birth as Nimi who later succeeded his father on the throne of Mithilā.

It appears that the early Buddhists believed in the existence of the city of Kṛmilā before the days of the Buddha. The only indication regarding the location of the city in the Buddhist works is of course that it lay on the bank of the Ganges. But it is clear that it lay in the area which was the sphere of the Buddha's activities. It is well known that, according to early Buddhist works, this area comprised the six great cities of (1) Campā,[3] capital of the Aṅga country in the Monghyr-Bhagalpur region in East Bihar which was annexed to the kingdom of Magadha (Patna-Gaya region of South Bihar) about the close of the sixth century B. C.; (2) Rājagṛha (modern Rajgir in the Patna District), the older capital of Magadha before it was transferred to Pāṭaliputra; (3-4) Śrāvasti (modern Set-Mahet on the borders of the Gonda and Bahraich Districts of U. P.) and Sāketa (near Ayodhyā in the Fyzabad District, U.P.), both in the kingdom of Kosala,[4] (5) Kauśāmbī (modern Kosam on the Jamuna, about thirtyfive miles to the west of Allahabad, U. P.), the capital of the Vatsa country annexed to Avanti after the middle of the sixth century B.C.; and (6) Vārāṇasī (U.P.), the capital of the kingdom of

1. Malalasekera, *op. cit.*, s. v. *Kimbila* and *Kassapa.*

2. *Ibid.*, s. v. *Nimi.*

3. Campā has been located near Bhagalpur. See above, p. 36, note 5; p. 90.

4. Śrāvastī was the capital of the Kosala king Prasenajit who was a contemporary of the Buddha.

Kāśī which was annexed to the Kosala kingdom about the close of the sixth century B.C.[1] Thus the sphere of the Buddha's activities covered Bihar and the eastern part of U.P. and the city of Kṛmilā was no doubt situated in this area.[2] The location of the city at the site of the modern villages of Valgūdar and Rajauna in the western fringe of the Monghyr District, to the south of the Ganges, does not militate against Buddhist evidence.

1. Kashyap, *Suttanipāta-Dīghanikāya-Mahāvagga*, pp. 113, 130. At p. 175 of the same work, reference is made to the following big cities of the whole of India : (1) Dantapura of the Kaliṅgas; (2) Potana of the Aśmakas; (3) Māhiṣmatī of the Avantis; (4) Roruka of the Sauvīras; (5) Mithilā of the Videhas; (6) Campā of the Aṅgas; and (7) Vārāṇasī of the Kāśis. This list seems to refer to an earlier time when the kingdoms of Magadha and Kosala were not very prominent.

2. With the spread of Buddhism, new traditions developed in order to prove that the Buddha visited areas far away from the said region, e.g. Khotan in Central Asia. Cf. below, p. 274.

CHAPTER XVIII

CAURĀSĪ

Caurāsī as a geographical unit indicating a group of 84 villages is known from different parts of India. There is a group of villages called Bhināikī Caurāsī near Ajmer in Rajasthan, and it is probably mentioned as Caturaśīti in the fragmentary Barli inscription[1] of the first century B. C. Caurāsī is also a Pargana in the Mirzapur District of U.P., while Corāsī (Caurāsī) is a Subdivision of the Surat District of Bombay. There is a village called Caurāsī in the Sadar Subdivision of the Purī District of Orissa. This village likewise seems to have. been originally the headquarters of a geographical and administrative unit consisting of 84 villages. That it had some importance in the past is possibly indicated by its yielding of a copper-plate inscription of the Bhauma-Kara monarch Śivakara II who flourished about the end of the ninth century A.D.[2] It seems that there are other areas similarly called Caurāsī in different parts of the country. *Caturaśīti*, the Sanskrit form of *Caurāsī*, is found in a number of ancient Indian records.

A copper-plate inscription, dated Śaka 734 (812-13 A.D.), of the Rāṣṭrakūṭa chief Kakka II Suvarṇavarṣa of Lāṭa or S. Gujarāt was published in the *Indian Antiquary*, Vol. XII, pp. 156 ff. It records the grant of the village of Vaḍapadraka in the Aṅkoṭṭaka eighty-four (*Aṅkoṭṭaka-caturaśīti-antarggata-Vaḍapadrak-ābhidhāna-grāma*). Aṅkoṭṭaka is modern Akoṭā, a suburb of Baroda (Vaḍapadraka). The Set-Mahet plate[3] of the Gāhaḍavāla king Govindracandra, dated Vikrama Saṁvat 1186 (1130 A.D.), records the grant of certain villages in the *Pattalā* or district called Vādā (or, Vājā) *Caturaśīti*. Another grant[4] of Madanapāla dated Vikrama 1164 (1108 A.D.), similarly speaks

1. *JBRS*, Vol. XL, Part i, pp. 3-4.
2. Cf. *JBORS*, Vol. XIV, pp. 292 ff.
3. *Ep. Ind.*, Vol. XI, pp. 20 ff.
4. *Ibid.*, Vol. XXXIII, p. 176. Cf. Ratnapura-caturāsikā and Caturuttara-caturāsikā in Gujarat (Majumdar, *Chaulukyas of Gujarat*, pp. 211-12).

of the gift of a village in Mānighapura-caturāsikā, in which *Caturāsikā* is no doubt the same as *Caturaśītikā* or *Caurāsī*. It should, however, not be supposed that only the word 'eighty-four' is associated with the names of districts. We have numerous other numbers similarly associated with geographical units.[1] A *Pratijāgaraṇaka* or Pargana within the dominions of the Paramāra kings of Malwa is called Saptāśiti-pratijāgaraṇaka (*i.e*, Pargana consisting of eighty-seven villages) in an inscription of Vikrama 1331.[2] But such numbers as 87 were not as popular as 84 in the context of geographical names. The popularity of constituting an administrative unit of 84 villages may be connected with the conception of 'the typical clan-chief's estate of 84 villages'.[3]

In the inscriptions of the Kannaḍa-speaking area, territorial units are often mentioned along with a number exactly as in Bhināikī Caurāsī and Aṅkoṭṭaka-caturaśīti; e.g., Gaṅgavāḍi ninety-six thousands, Raṭṭappāḍi seven and a half lakhs, Banavāsi twelve thousands, Alande one thousand, Purigere three hundreds, etc., etc. Similar instances are sometimes found also in the records of other parts of the country. Although there is possibly some exaggeration in some of these cases speaking of very big numbers, there is no doubt that the reference in such cases is to villages which may have been often very small or even nominal, As regards territorial units with smaller numbers, we may mention Nirugundagi twelve, Puriddha ten, Tamba six, Koṇḍavati two, etc., etc.

Considering the abnornally high numbers mentioned with some of the names and the fact that it is palpably impossible

1. Cf. *e. g.*, the *Pattalā* named Rudamaua-bayālisi in Govindacandra's Kamauli plate of Vikrama 1190 (1133 A. D.), published in *Ep. Ind.*, Vol. IV, pp. 111-12. In this name *bayālisī* means 'forty-two'. An inscription (*Ind. Ant.*, Vol. XVIII., pp. 344 ff.) of Paramāra Udayādiya, dated Vikrama 1229 (1171 A.D.), mentions the area around Bhilsa in the former Gwalior State as Bhāillasvāmi-mahādvādaśaka-maṇḍala (*i. e.*, the district called Bhāillasvāmin consisting of twelve subdivisions). It also mentions a subdistrict consisting of sixty-four villages as the Bhṛṅgārikā-catuḥṣaṣṭi *pathaka* (cf. *Ep. Ind.* Vol. XXX, pp. 210-11). A grant of the Somavaṁśī king Mahāśivagupta Yayāti I (*c.* 970-1000 A. D.) of Orissa mentions a district called Saṇṇavatī, *i. e.* ninety-six (No. 27 of *A. R. Ep.*, 1052-53, App. A).

2. *Ep. Ind.*, Vol. XXXII, p. 140.

3. Cf. Ghoshal, *The Agrarian System in Ancient India*, p. 66.

to believe that the small district of Gaṅgavāḍi could actually claim more than only a fraction of 96000 'villages' if a village is understood in its modern sense, we were formerly reluctant to take the word *grāma*, even when specifically mentioned in such cases, in the sense of 'a village'; cf. the passage *trayāṇāṁ Mahārāṣṭrakāṇāṁ navanavati-sahasra-grāma-bhājām* occurring in the celebrated Aihole inscription[1] of Pulakeśin II and referring to the three Mahārāṣṭras as having no less than 99000 *grāmas*. Thus in the name Veṅgīpura-viṣaya-grāma-sahasra, sometimes called Veṅgī-sahasra, we were inclined to interpret the word *grāma* as 'what is the subject of an assessment.'[2] This appeared to be justified by the fact that the Malkapuram inscription[3] of Śaka 1183 (1261 A. D.) refers to a grant of no less than three lakhs of *grāmas* in Ḍāhala-maṇḍala (the modern Jubalpur area) lying between the Ganges and the Narmadā by the Kalacuri king Yuvarāja (I or II, probably the latter) in favour of a Śaiva saint named Sadbhāvaśambhu,[4] as it is impossible to believe that even the whole kingdom of the Kalacuri king had nearly 20,000 villages in the modern sense of the term. There is, however, some evidence to show that the word *grāma* was used in such cases in the sense of a village whatever its size and nature may have been.

In the Aland (ancient *Alande*) inscription of *Yuvarāja* Mallikārjuna, published in the *Epigraphia Indica*, Vol. XXVIII, pp. 31 ff., Alande itself is mentioned as the *modala-vāḍa* or 'foremost village' of Alande-sāsira or 'Alande one thousand'. This makes it clear that the geographical and administrative unit of Alande consisted of one thousand *vāḍas* or villages, of which the village of Alande was the foremost; that is to say, it was the headquarters of the district in question. It has been reported to us that the people of Kuknūr even now quote an old list, traditionally handed down, of the thirty villages which formed the ancient administrative unit called 'Kuknūr thirty' and that all these villages can be traced in the

1. Cf. *ibid.*, Vol. VI, pp. 1 ff.
2. Cf. *The Successors of the Sātavāhanas*, p. 400.
3. *JAHRS*, Vol. IV, pp. 156-57.
4. Cf. भागीरथीनर्मदयोर्मध्यं डाहलमंडलम् .... तस्मै निस्पृहचेतसे कलचुरिक्ष्मापालचूडामणिर्ग्रामाणां नृपतिर्भिक्षां त्रिलक्षीं ददौ ।

area around Kuknūr even to-day. It is also interesting to note that an inscription mentions Kuknūr as the foremost (*i.e.* headquarters) of thirty *grāmas*.[1]

The conclusion seems to be supported by a section of the *Skanda Purāṇa*[2] apparently interpolated into the original work sometime in the medieval period.[3] This section says how the *nava-khaṇḍa* Bhārata, *i.e.* Bhārata-varṣa consisting of nine divisions, was subdivided into seventy-two *vibhedas* or subdivisions, and quotes a list of the seventy two (actually, seventy-five) countries together with the number of *grāmas* contained in each one of them. The list is introduced by the verse :

तेषां नामानि ग्रामांश्च पत्तनानि च फाल्गुन ।
वेलाकूलानि संख्यां च वक्ष्यामि तव तत्त्वतः ॥

and ends with the following passage :

द्वासप्ततिरमी देशा ग्रामसंख्याः प्रकीर्तिताः ।

The second of the two passages is followed by the statement that the entire Bhārata-khaṇḍa had 967,200,000 *pattanas* (towns) and 36,000 *velākūlas* (harbours). It can hardly be doubted that the word *grāma* has been used in this context in the sense of 'a village'. But the fact that the list speaks only of *grāmas* and not of *pattanas*, etc., seems to suggest that, for the purpose of enumeration, the latter also were counted as *grāmas*. The list further shows that the traditional or conventional number of *grāmas* in a country as given in it could hardly have been true if the word *grāma* is taken exactly in its modern sense.

We quote below the list of countries together with the number of villages in each one of them.

Number	*Name of Country*	*Number of Grāmas*
1.	Nivṛt *maṇḍala*	4 crores
2.	Bālāka *deśa*	$2\frac{1}{3}$,,
3.	Khurasāhaṇaka (Khorasan)	$1\frac{1}{4}$,,
4.	Aṁdhala or Amala (sic—Andhra)	4 lakhs
5.	Nepāla ·	1 lakh
6.	Kānyakubja	36 lakhs

1. त्रिंशद्ग्रामाधिपतिश्रीकुक्कुनूरु (*JBBRAS*, Vol. XII, 1876, p. 43).
2. Māheśvara-khaṇḍa, Kumārikā-khaṇḍa, Ch. 39, verses 127 ff.; cf. A. B. L. Awasthi, *Stud. Skand. Pur.*, pp. 24 ff., for variant readings.
3. Cf. *JRASB*, Letters, Vol. XIV, 1948, p. 25.

7.	Gājaṇaka or Mājaṇaka (i.e. Ghaznī or Ghazna)	72 lakhs
8.	Gauḍa *deśa*	18 „
9.	Kāmarūpa	9 „
10.	Ḍāhala or Veda (sic—Cedi)	9 „
11.	Kāṁtipura	9 „
12.	Mācīpura (sic—Kāṁcīpura)	9 „
13.	Oḍḍiyāna *deśa* (in the Swat valley)	9 or 7 lakhs
14.	Jālandhara	9 „
15.	Lohapura (i.e. Lahore)	9 „
16.	Pāṁbīpura	7 „
17.	Raṭarāja (i.e. Raṭṭa-rājya, the Rāṣṭrakūṭa kingdom)	7 „
18.	Harīāla (sic—Hariyāṇa, the Delhi region)	5 „
19.	Draḍa or Draḍasya (sic—Draviḍa) *viṣaya*	$3\frac{1}{2}$ „
20.	Vaṁbhaṇavāhaka (i.e. Brāhmaṇavāha or Bahmanābād in the Hyderabad region of Sind)	$3\frac{1}{2}$ „
21.	Nīlapura	21 thousands
22.	Amala or Mala *viṣaya* (redundant; see No. 4)	1 lakh
23.	Nareṁdu (sic—Varendra) *deśa* (see No. 72 below)	$1\frac{1}{4}$ lakhs
24.	Tilaṅga *deśa* (Telengana)	$1\frac{1}{2}$ „
25.	Mālava	118,092
26.	Sayaṁbhara (Śākambhara) *deśa*	$1\frac{1}{4}$ lakhs
27.	Mevāḍa (Medapāṭa)	$1\frac{1}{4}$ „
28.	Vāguri	80 thousands
29.	Gujarātra (sic—Gurjaratrā, probably modern Gujarāt)	70 „
30.	Pāṁḍu (Pāṇḍya) *viṣaya*	70 „
31.	Jahāhuti (Jejābhukti, the Khajurāho region of Bundelkhand)	42 „
32.	Kāśmīra	68 „
33.	Kauṁkaṇa (Konkan—North?)	63 „
34.	Laghu-Kauṁkaṇa (Smaller Konkan)	1422
35.	Sindhu	20 thousands
36.	Kaccha *maṇḍala*	1422
37.	Saurāṣṭra (Kathiawar)	55 thousands
38.	Lāḍa (Lāṭa, the Nauṣari-Bharoch region of Gujarāt)	21 „

39.	Atisindhu	10	thousands
40.	Aśvamukha	10	,,
41.	Ekapāda	10	,,
42.	Sūryamukha or Śūrpanakha (sic—Sūrpāraka, the Sopara region in the Thana District, Bombay)	10	,,
43.	Ekabāhu *deśa*	10	,,
44.	Saṁjāyu *deśa* (Sañjān in the Thana District ?)	10	,,
45.	Śiva (Śibi) *deśa*	10	,,
46.	Kālahayaṁjaya (Kālañjara ?)	10	,,
47.	Liṅgodbhava *deśa*	10	,,
48.	Bhadra	10	,,
49.	Devabhadra	10	,,
50.	Cāṭa	36	,,
51.	Virāṭa (the Jaipur-Alwar-Bharatpur region of Rajasthan)	36	,,
52.	Yamakoṭi (mythical city placed 90° east of the meridian of Laṅkā)	36	,,
53.	Romaka *deśa*	18	crores
54.	Tomara	$1\frac{3}{4}$	lakhs
55.	Karṇāṭa	$1\frac{1}{4}$	,,
56.	Jāṅgala	$1\frac{1}{4}$	,,
57.	Strī-rājya[1]	5	,,
58.	Pulastya or Mulastya *viṣaya*	10	,,
59.	Kāmboja	10	lakhs
60.	Kośala (North or South Kosala, possibly the latter in the Raipur-Bilaspur-Sambalpur region in Madhya Pradesh and Orissa)	10	,,

1. Strīrājya seems to have been a woman-dominated State in the Himalayan region. It may be noted that, in every Bihal (*vihāra*) of the Newars of Nepal, a young girl (*Kumārī*) having no scar on her body is worshipped as the living form of Kālī or Durgā. There is also a similar Kumārī for the whole kingdom, who is periodically elected from among the girls of the Vanra (priest) community during the last night of the *navarātri* festival after a trying test. The Newars believe that the valley of Nepal belongs to the Kumārī and therefore every year the king has to receive from her a fresh mandate for ruling the country. The Kumārī is replaced by another before she is found to be approaching her first menstruation. See Gopal Singh Nepali, *The Newars*, Bombay, reviewed in the *Swarajya*, March 26, 1966, p. 27.

61.	Bāhlika	4 lakhs
62.	Laṅkā *deśa* (see Siṁhala, No.67)	36 thousands
63.	Kuru *deśa*	64 thousands
64.	Kirāta-vijayojaya (sic—Kirāta *deśa*)	1¼ lakhs
65.	Vidarbhā (Vidarbha)	5 ,,
66.	Vardhamāna	14 thousands
67.	Siṁhala *dvīpa* (Laṅkā, No. 62)	10 ,,
68.	Pāṁḍu *deśa* (probably in the Punjab; see No. 30 above)	36 ,,
69.	Bhayānaka	1 lakh
70.	Māgadha (Magadha) *deśa*	66 thousands
71.	Pāṁgu or Pāṁḍu *deśa* (redundant; see Nos. 30 and 68)	60 ,,
72.	Vareṁdu (redundant; see No. 23 above)	30 ,,
73.	Mūlasthāna (i.e. Mūltān)	25 ,,
74.	Yavana	40 ,,
75.	Pakṣabāhu or Yakṣabāhu	4 ,,

It may be pointed out that the text of the above Purāṇic section is not free from errors.[1] Some of the names show considerable Prakrit influence[2] while many of them are doubtful. There is however epigraphic support in favour of some of the traditional numbers of *grāmas* quoted in the list. Thus the Śākambhara or Śākambharī country (No. 26) is actually known to have been otherwise called *Sapādalakṣa* or 'one and a quarter lakhs'.[3] But Kosala (No. 60) is stated to have contained 6 lacs and 96 villages in a Kuruspal inscription[4] while a well-known tradition gives the number of villages in Kashmir (No. 32) as 66,063.[5]

1. Nos. 22, 71 and 72 are redundant. The list also omits many well-known tracts.

2. *E. g.*, Nos. 20, 26, 27, 38.

3. Cf. No. 26 above and Bhandarkar's List of Ins., Nos. 279, 356. For another Sapādalakṣa country in the Nizamabad-Karimnagar region of Andhra Pradesh, see *Jour. Or. Res.*, Vol. XVIII, Part i, p. 40. There was similarly a Pādonalakṣa (i. e. 'one lakh minus one quarter') country about the Shivapuri District in the former Gwalior State (*Ep. Ind.*, Vol. XXXIII, p. 68).

4. Hiralal's List of Inscriptions, No. 273 (p. 161).

5. Stein, *Kalhaṇa's Rājataraṅgiṇi*, Vol. II, pp. 438-39. The Muslim writers give it as 1 lakh or 10,000.

The traditional numbers of villages in particular countries are also referred to in some other works. Vinayacandra's *Kāvyaśikṣā*, e. g., mentions Surāṣṭra 9 thousands, Lāṭa-deśa 21 thousands, Gurjara-deśa 70 thousands, Pārata 70 thousands, Ḍāhala 9 lakhs, Mālava 9 lakhs and 92, Kānyakubja 46 lakhs, etc.[1] According to Rashīduddīn and Wassāf, Gujarat comprised 80,000 or 70,000 villages and towns, Siwalik (cf. No. 26) 125,000 and Malwa (No. 25) 1,893,000. Hemacandra says that there were 1800 villages in Arbuda (Abu) while Forbes refers to a tradition that Caulukya Kumārapāla's kingdom as well as Ajmer each contained 1 lakh villages. The *Tabaqāt-i-Nāṣirī* gives 70,000 as the number of villages in Gondwana.[2]

It should be pointed out that sometimes numbers mentioned with the name of a tract do not refer to villages, but groups of them, although a distinction is made in such cases. The Udaypur (former Gwalior State, now in Madhya Pradesh) inscription of 1173 A.D. speaks of a village in Bhṛṅgārikā-catuḥṣaṣṭi-pathaka which formed a part of Bhāilasvāmi-mahā-dvādaśaka-maṇḍala.[3] It appears that the *pathaka* or subdistrict called 'Bhṛṅgārikā 64' consisted of sixty-four villages while the *maṇḍala* or district called 'Bhāilasvāmin great-12' consisted of twelve subdistricts. The speciality of the second case has been indicated in the record by the introduction of the word *mahat* (great) prefixed to the number.

Geographical names of this kind were more popular in the Kannaḍa-speaking area than elsewhere in India. There is evidence to show that, in the said region, the idea of coining similar names was not unknown even in the fourth century A.D.[4]

1. Rājaśekhara's *Kāvyamīmāṁsā*, G. O. S. ed., pp. 248-49. Cf. Nos. 6, 10, 25, 29, 37 and 38 of the *Skanda Purāṇa* list quoted above.
2. See P. Niyogi, *The Economic History of Northern India*, pp. 3ff.
3. *Ind. Ant.*, Vol. XVIII, pp. 344 ff.; see above, p. 257, note 1.
4. Cf. *The Successors of the Sātavāhanas*, p. 250 (Sahalāṭavī-grāmāhāra, probably consisting of twelve villages), and p. 305 (Tagare-mahāgrāma consisting of twenty-four *pallis*) referring to certain inscriptions of the Early Kadambas. On the whole question, see Fleet in *JRAS*, 1912, pp. 707 ff., and V. S. Agrawala in *Jadunath Sarkar Com. Vol.*, II, pp. 14ff. Our former interpretation of *grāma* as a unit of revenue assessment (elaborated by Agrawala as 'one plough measure of land assessed at one silver Kārṣāpaṇa') cannot be true in respect of many of the cases specified above. Of course the number of such revenue units or income in coins may have been confused with that of the villages of a territory in the cases quoting very big numbers.

CHAPTER XIX

GONARDA[1]

The *Pārāyaṇa* incorporated in the Pali *Suttanipāta* is one of the most ancient Buddhist works. The parallel writings, attested by frequent mention, have disappeared. But the Pali text, guaranteed as it is by two commentaries, viz. the *Mahāniddesa* and the *Cullaniddesa* which too are very ancient, has an incontestable value. In the account which sets forth the introduction to the collection, the Brāhmaṇa Bāvari, emigrated from the 'charming city of the Kosalas' (*i.e.* Śrāvasti), comes to settle himself 'in the country of Assaka, in the vicinity of Muḷaka,[2] on the banks of the Godhāvarī'. He sends his disciples in a mission to the Buddha at his place of birth, i.e. Sāvatthi of the Kosalas. The poet sums up in three verses (1011-1013) the steps in their route : "Patiṭṭhāna of Muḷaka; then the city of Māhissatī ; also Ujjenī and Gonaddha; Vedisā; Vana-savhaya; Kosambī ; and also Sāketa; and the big city of Sāvatthī, Setavya, Kapilavatthu; and the city of Kusīnārā ; and Pāvā; Bhoga-nagara; the Magadhan city of Vesālī and the Pāsāṇaka Cetiya."

The itinerary deserves the honour of a careful study. Here, however, I shall only occupy myself with the intermediate step between Ujjenī and Vedisā, two perfectly definite localities. The one is still now Ujjain to the north of Indore, lat. 23° 11′ 10″ N., and long. 75° 51′ 45″ E.; the other is Besnagar very near Bhilsa, lat. 23° 31′ 35″ N., and long. 77° 55′ 39″ E. The *Pārāyaṇa* places the city of Gonaddha between these two points.

The catalogue of the Yakṣas in the *Mahāmāyūrī*[3] follows an exactly identical order : "at Avanti the Yakṣa is Priyadarśana; at Gomardana Śikhaṇḍin; at Vaidiśa Añjalipriya." Avanti is another name of Ujjayinī; *Vaidiśa* is the Sanskrit form of

1. [This is a translation of an article in French, entitled *Gonarda, le berceau du Gonardīya*, by Sylvain Lêvi, which was published in the *Sir Asutosh Mookerjee Silver Jubilee Volumes*, Vol. III (Orientalia), Part ii, pp. 197-205.]

2. [The name may also be spelt as *Mūlaka.*]

3. *Journ. Asiat.*, 1915, I, p. 43, verse 19.

Pali *Vedisa.* The name of the intermediate locality is fluctuating in the tradition of the manuscripts. I have reproduced the reading of the Mss O and H in the text; but D reads *Gonardane.* Of the three Chinese versions, S transcribes *Kiu-kia-t'o-na,* which supposes an original *Gogardana* ; Y translates *You-hi,* 'Bull-joy', which restores *Gonandana;* A translates *You-ts'oei,* 'Bull-compress', that is, *Gomardana.* The Tibetan translator has followed the same text (*Ba-lan 'joms,* 'Bull-compress'). The evidence of the *Suttanipāta* comes to confirm the reading *Gonardana,* because it is evident that, in both cases , the question is of the same locality. The Pali *Gonaddha* may be restored, without difficulty, to Sanskrit *Gonarda.* The aspiration, introduced subsidiarily in the Pali form, is a phenomenon which is nothing exceptional. In that very introduction to the *Pārāyaṇa,* we have already also met with the Sanskrit name of the Godāvarī, modified by aspiration of the dental in the interior of the word as *Godhāvarī.* A list of analogous cases will be found, *e.g.*, in the *Pali* by Geiger, § 40 and § 62, and, for the Prakrits in general, in the *Grammatik der Prakrit Schprachen* by Pischel, § 207-209. In proper names, the phenomenon seems due generally to an erroneous interpretation; *e.g.*, *Khanda,* 'the god Skanda', owes its aspiration to a confusion with *skandha,* 'the shoulder'; *Erāpatha = Airāvata,* 'the divine elephant', has suffered the contamination of *patha,* 'way' (as, in Sanskrit also, it has later suffered the contamination of *patra,* 'leaf', in becoming *Elāpatra*). One would have believed to recognise the word *godhā,* 'big lizard', in the first syllables of the name of the Godāvarī. Sanskrit *go-narda* (or *go-nardana* by developed suffixation) clearly signifies 'the lowing of the cow'. Passed to Prakrit under the form *gonadda* (or *go-naddana*), the word becomes unintelligible. The root *nard* seems to have submitted in Prakrit to the root *nad* from which it hardly differentiates itself. Moreover, the word *go* quite naturally suggested the word *naddha,* 'attached, bound'.

Whatever be the explanation, the equivalence Gonarda = Gonaddha is certain. The name of Gonarda is indissolubly connected with the memory of Patañjali, 'the Gonardian', *Gonardīya.* A constant tradition attested by Kaiyaṭa, by the author of the *Trikāṇḍaśeṣa,* and by Hemacandra, identifies the personage designated by the name of *Gonardīya* in the *Mahābhāṣya* with the author of the *Mahābhāṣya.* Kielhorn, it is true, vigourously contested the value of that tradition[1] and maintained that the Gonardīya was the author of the Kārikā in verse, utilised and cited by Patañjali. The authority of Kielhorn in questions concerning the *Mahābhāṣya* surely deserves the highest consideration. But his interpretation is not in requisite contradiction of the Indian tradition. Patañjali might refer in the *Mahābhāṣya* to an anterior work which he had composed, designating himself with an appellation of impersonal character, derived from his place of birth. The geographical situation of Gonarda always agrees marvellously with the rare indications, which may be deduced from the *Mahābhāṣya*, for fixing the date of Patañjali. The two essential facts are : (1) the mention of Puṣyamitra, of his court (*sabhā*), and of his sacrifice (*iha Puṣyamitraṁ yāja-yāmaḥ*); and (2) the mention of the Greek conquests in India (*aruṇad=Yavanaḥ Sāketam, aruṇad=Yavano Madhyamikām*). Now, Gonarda is the nearest step to Vidiśā, which, according to the evidence of the accounts followed by Kālidāsa in the *Mālavi-kāgnimitra*, was the capital where the son of Puṣyamitra resided in the capacity of a viceroy. And Vidiśā was also in close connections with Greek politics. The column of Besnagar, on the site of the ancient Vidiśā, preserves the memory of a Greek ambassador (*Yona-dūta*), Heliodorous, sent by the Greek king Antialkidas, to the Indian king Kāśīputra Bhāgabhadra. Moreover, the geographical horizon of the *Mahābhāṣya* sets itself harmoniously around the Gonarda-Vidiśā region as the centre. Excepting some general designations of territories, *e.g.*, Vidarbha, Videha, Cola, Kerala, etc., which do not implicate any direct and personal knowledge, the names of localities, cities or towns form themselves in a sort of triangle, of which the base goes from Pāṭaliputra to the Punjab, and of which the apex reaches the lower Narmadā about Māhiṣmatī, a place figuring in the itinerary of the *Pārāyaṇa* as an intermediate step which the disciples of Bāvari, who started from Pratiṣṭhāna, reached on their way to Ujjayinī. Likewise the *Mahābhāṣya* has :

उज्जयिन्याः प्रस्थितो माहिष्मत्यां सूर्यमुद्गमयति ।[2]

A difficulty seems to oppose directly the location of Gonarda that I propose. The name of Gonarda is cited as that of

1. *Ind. Ant.*, Vol. XV, pp. 81-83.
2. On Pāṇini, III. 1. 26.

a place 'in the land of the Orientals' in the *Candravṛtti*[1] and in the *Kāśikāvṛtti* on Pāṇini, I. 1.75. just for explaining the formation of the derivative *Gonardīya* without *vṛddhi*. Pāṇini says that, by exception, the dipthongs *e* and *o* should be considered as in the degree of *vṛddhi* (which is normally *ai* and *au*) in the names of places of the land of the Orientals (*eṅ prācāṁ deśe*). Candragomin reproduces this *sūtra* adapting it in his system (III. 2. 25: *eṅ-ādy-acaḥ prāg-deśāt*). We are evidently surprised, and even shocked, to see Gonarda, which is in the heart of Malwa, here placed in 'the East' of India. We have not, however, the right to call the assertion of Candragomin and the *Kāśikā* in question. It is not the question of following a blind confidence in the geographical knowledge of these commentaries; but the reasoning, from the grammatical order, is unquestionable. *Gonardīya* is a derivative formed by means of the suffix *cha* (= *īya*). The suffix *cha* joins a stem having *vṛddhi*.[2] *Gonardīya* is treated as a stem of *vṛddhi*, though it has *o*, and not *au*, in the first syllable. The thing then is that it enters into the exception anticipated by Pāṇini.[3] So Gonarda, from which it derived, is the name of a locality of the Orientals.

But what should we mean by 'the Orientals', *prāñcaḥ* ? The grammar, since Pāṇini, knows only two groupings as regards the cardinal points, *viz*., the Northerners (*udañcaḥ*) and the Easterners (*prāñcaḥ*). A traditional verse quoted by the *Kāśikā*[4] and Kṣīrasvāmin[5] establishes that division :

प्रागुदंचौ विभजते हंसः क्षीरोदके यथा ।
विदुषां शब्दसिद्धयर्थं सा नः पातु शरावती ।।

"She separates the East and the North, as the swan separates the milk and the water, for fixing well the usage of the classical language. May the Śarāvatī protect us !"

And the Dictionary of Amara, in describing the earth, stands also on that double division which it completes by the secondary association of the two other directions :

1. On Candragomin, III. 2, 25, etc.
2. *Vṛddhāc=chaḥ* (Pāṇini, IV. 2. 114).
3. I. 1. 75.
4. On Pāṇini, I. 1. 75.
5. On Amara, II. 1. 6-7.

..................शरावत्यास्तु योऽवधेः ।
देशः प्राग्दक्षिणः प्राच्य उदीच्यः पश्चिमोत्तरः ॥[1]

'In starting from the Śarāvatī, the country, which is to the south-east, is the East; that, which is to the north-west, is the North.'

Thus, to Amara, the South is expressly connected with the East, and the North with the West. The glossologist Vandyaghaṭīya [Sarvānanda] writes on the passage that 'the Śarāvatī is a river of India that runs from the north-east towards the Western Ocean' (एतस्मिश्च भारते वर्षे शरावती नाम नदी ऐशान्या दिशः सकाशात् पश्चिमसमुद्रगामिनी वहति). The indication seems plain and clear. Unfortunately real geography does not confirm it. Vandyaghaṭīya, a veritable glossologist, has deduced from the text itself the indication which he appears to have added here. He has borrowed it neither from the modern geography, nor from the ancient, nor from the consecrated nomenclatures of the rivers in the epics and the Purāṇas. The pretended Śarāvatī of Vandyaghaṭīya will be searched in vain. In fact, the tradition has perpetuated, this time also, an appellation that had no more any relation, for a long time, with reality.

There had been a time, when the name Śarāvatī, '[the river] with reeds,' was applied to a course of water, which separated the whole of Aryan India into two parts. Pāṇini expressly teaches the formation of the name.[2] The memory of a frontier indicated by the river Śarāvatī is curiously preserved in a celebrated episode of the Buddhist tradition. When Koṭikarṇa goes to consult the Buddha on the limits of the country of strict observance, the Lord fixes the southern limit at Śarāvatī : "In the South there is a city named Śarāvatī, and beyond that is a river named Śarāvatī. There is the boundary (दक्षिणेन शरावती नाम नगरी, तस्याः परेण शरावती नाम नदी, सोऽन्तः)." Such is at least the tradition of the Mūlasarvāstivādin school in the original text gathered by the compilers of the *Divyāvadāna*.[3] The editors of the text, Cowell and Neil, cite two variants of the name, furnished by some manuscripts of inferior value : *Sarvāvatī* (Ms. A) and *Savārāvatī* (Ms. B). It is this last reading which has been followed by Yi-tsing, the author (responsible, if not

1. II. 1. 6-7.
2. *Sar-ādīnāṁ ca* (VI. 3. 120).
3. Ed. Cowell and Neil, p 21.

actual) of the Mūlasarvāstivādin *Vinaya*. He has rendered the name of the city and the river as *Che-pa* (or *fo*)-*lo-fa-ti* which supposes an original *Śavaravatī*.[1] Yi-tsing reproduces the same form of the name in another treatise of the same *Vinaya*, viz. the *Mūlasarvāstivāda Ekaśatakarma*,[2] wherein the same episode is repeated in an abridged from. The reading *Śavaravatī* is undoubtedly due to an attempt at correction to substitute for the unknown Śarāvatī a name which evoked the idea of the savage tribe of the Śavaras who inhabited the central plateau on the southern border of the basin of the Ganges. The corresponding passage of the *Vinaya* of the Sarvāstivādins is known to us only from the Chinese version due to Puṇyatara, which appears, owing to the fault of the translator or of the original, in a state of inextricable confusion. After having given Mount Uśīra (Yeou-chi-lo) as the northern limit, it adds : 'beyond that mountain, and not far off, there is the tree So-lo by the source with rushes.'[3] The 'source with rushes' seems to be the equivalent of Śarāvatī, '[the water] which has some reeds,' and the tree So-lo=Sārā or Sālā seems to go back to the same original. Through and through, this Vinaya gives 'the river of the Bamboos', as the limit in the North-East which too evokes the Śarāvatī. Such as it is, the passage is not then utilisable. The Pali *Vinaya* of the Sthavira school substitutes *Salalavatī* (with the variants *Sallavatī* and *Salilavatī*, though the evidence of *Jātaka*, I, 49, and the *Sumaṅgalavilāsinī*, I, 173, confirms the reading *Salala*°) for *Śarāvatī*. It makes that river the boundary in the south-east : *puratthima-dakkhiṇāya disāya Salalavatī nāma nadī*.[4] The direction of the south-east in the Pali work partly agrees with the direction of the south in the *Vinaya* of the Mūlasarvāstivādins. It is in absolute contradiction of the Brāhmaṇical conception of the Śarāvatī which, separating the North and the East, must flow to the north-west of the country of the Middle, *i.e.* Madhya-deśa, which is the land of strict observance.[5]

1. Tok., XVII, 4, 108-a, 9.
2. Tok., XVII, 5, 57-b, 2.
3. Tok, XVI, 4, 59-a, 17.
4. *Vinayapiṭaka*, *Mahāvagga*, V, 13, 12.
5. [According to the Brāhmaṇical conception, the Śarāvatī seems to have crossed the land called Madhyadeśa (i. e. the Central region of Northern India) from the north-east to the south-west. This land was originally called Āryāvarta and was later regarded as the heart of it. See Sircar, *Cosm. Geog. Anc. Ind. Lit.*, pp. 17-18.]

The city of Śarāvatī (Śarāvatī-nagarī), which the *Vinaya* of the Mūlasarvāstivādins place just on this side of the river Śarāvatī, is no better known than the river itself. A city of the same name, it is true, appears in the *Raghuvaṁśa*, XV. 97, as the capital of Lava, son of Rāma, while the latter's other son reigned at Kuśāvatī :

स निवेश्य कुशावत्यां रिपुनागांकुशं कुशम् ।
शरावत्यां सतां सूक्तैर्जनिताश्रुलवं लवम् ॥

This is at least the text adopted by Mallinātha and generally accepted on the authority of that commentator. But the commentators Vallabha, Vijayānanda-sūri and Caritravardhana read : *Śrāvastyāṁ ca*, and Hemādri and Summativijaya have : *Śrāvatyāṁ ca*. And in fact, the Uttarakāṇḍa of the *Rāmāyaṇa*, which Kālidāsa follows, calls the capital of Lava *Śrāvastī* in the text of Bombay, CVIII. 5, as also in the text of Calcutta, CXXI. 4 (श्रावस्तीति पुरी रम्या श्राविता च लवस्य च).[1] Actually, according to all the texts of the *Rāmāyaṇa*, Lava reigned over Uttara-Kośala while Kuśa ruled over Kośala proper (कोशलेषु कुशं वीरमुत्तरेषु तथा लवम्). Śrāvastī was the capital of Uttara-Kośala.[2] The city and the river of Śarāvatī on the confines of the North and the East have nothing to do here.

A precise location of the site of Gonarda with reference to Śarāvatī, which is not to be found, must then be given up. But

1. Gorresio's edition, CXIII. 24 has *Śrāvati* :
लवस्य तु पुरीं रम्यां श्रावतिं लोकविश्रुताम्.

2. [The *Raghuvaṁśa* (XVI. 31 ff.) suggests that Kuśāvatī lay in a territory to the south of the Vindhyas, apparently in the present Raipur-Bilaspur-Sambalpur region which was called Kosala at least from before the middle of the fourth century when the Allahabad pillar inscription of Samudragupta was composed. This country was regarded as Kosala proper while the Ayodhyā region was known as Uttara-Kosala. See also *Raghuvaṁśa*, VI. 71, and *Vāyu Purāṇa*, 89. 199-200—

कुशस्य कोशलाराज्यं पुरी वापि कुशस्थली ।
रम्या निवेशिता तेन विन्ध्यपर्वतसानुषु ॥
उत्तराकोशले राज्यं लवस्य च महात्मनः ।
श्रावस्ती लोकविख्यांता . . . . ॥

Cf. Dey, *Geog. Dict.*, s. v. See above, p. 106.]

one fact subsists. In the traditional division of Āryāvarta into two regions, viz. North and East, Gonarda, treated grammatically as a locality of the 'Orientals', is not in the North, however may have come its secondary orientation. It is then not to be surprised if Varāhamihira, the only known author who mentions Gonarda after the texts already cited,[1] places Gonarda among the countries of the south in his astrological chart of India (*Bṛhatsaṁhitā*, XIV. 12 : अथदक्षिणेन लंकाः . . . . भरुकच्छा . . . वनवासि . . . कोंकणाभीराः । आकरवेणाव (sic—न) त्तंकदशपुरगोनर्दके-रलकाः कर्णाट . . . नासिक्य . . . तुम्बवनकार्मणेयकाः). The name of Gonarda appears twice more in the *Bṛhatsaṁhitā*, in the texts of two purely astrological groupings: IX. 13 :

अन्येनाम्राक्रान्ते म्लेच्छाटविकाश्वजीविगोमन्तान् ।
गोनर्दनीचशूद्रान् वैदेहांश्चानयः स्पृशति ॥

and XXXII. 22 : गोनर्दचेदिकुकुरान् किरातवैदेहकान् हन्ति ॥

In his geographical nomenclature, Varāhamihira seems to throw the names at the hazard of the metre, so that nothing about their relative position can be inferred from a stand on the order of classification. Among these names of the southern regions, there is one, however, which deserves detaining attention. This is the name of Tumbavana. But the locality is, as far as I know, completely unknown in literature. Nevertheless, in the *Paramatthajotikā*, his commentary on the *Suttanipāta*, Buddhaghoṣa remarks on some verses of the *Pārāyaṇa*, wherein Gonarda is mentioned : *Ujjeniñ=c=āpi Gonaddhaṁ Vedisaṁ Vana-savhayaṁ*, that 'Vana-savhaya (literally, 'which bears the name of a forest') designates Tumbavana-nagara, others maintaining that this is Vanasāvatthi' (*Vana-savhayan=ti Tumbavana-(na*)garaṁ vuccati, Vanasāvatthin=ti pi eke*). The

1. [The *Mārkaṇḍeya Purāṇa* mentions Gonarda along with the countries of the southern and western parts of India. See Chapter LVIII, verses 20-29, though the stanzas appear to be an adaptation from those of the *Bṛhatsaṁhitā*. But the Purāṇas generally mention Gonarda along with the eastern countries (cf. above, p. 38, note 1). The Gaunardas are mentioned in a list of ancient ruling clans in one of the manuscripts of the *Vāyu Purāṇa* (Pargiter, *The Purāṇa Text*, etc., p. 3). Besides the *Bṛ.* and *Mārk.* lists, Gonarda is found in Parāsara's list *Cosm. Geog.*, p. 94n).]

2. [For *Āvartaka*, the intended reading seems to be *Āvantaka* and not *Ānartaka*, because Ānarta is already mentioned in verse 17 of the same Chapter. The Baṅgabāsī ed. of the work reads *Āvantaka*.]

gloss is quoted in Andersen-Smith's edition of the *Suttanipāta*.[1] Thus, according to the statement of Buddhaghoṣa, Tumbavana would be the step between Vidiśā and Kauśāmbī, on the route from Gonaddha-Gonarda towards the Yamunā.[2] In fact in the *Stūpa* of Sanchi near the site of Vidiśā (lat. 23° 28′ N. ; long. 77° 48′ E.), five inscriptions[3] commemorate the donations made by some inhabitants of Tumbavana. Among the benefactors of the *Stūpa*, are also represented the inhabitants of Māhiṣmatī, Ujjayinī and Vidiśā. It is astonishing not to meet some people of Gonarda.

The authenticity of the geographical nomenclature in the introduction of the *Pārāyaṇa* is confirmed by a decisive trait. Bāvari is here represented as settled on the bank of the Godāvarī, in the region of Aḷaka :

So Assakassa visaye Aḷakassa samāsane ।
vasī Godhāvarī-kule uñchena ca phalena ca ॥ (verse 2)

When his disciples leave him, their first step towards the north is the city of 'Patiṭṭāna of Aḷaka' (*Aḷakassa Patiṭṭhānaṁ*, v. 36). Such is at least the text adopted by the editors, Fausboell and Andersen-Smith. But the material furnished by the latter shows that, in the two passages, the Burmese manuscripts have *Muḷaka* instead of *Aḷaka*. The inscription of Śrī-Puḷumāyi at Nasik, enumerating the provinces annexed by Gautamīputra to his empire, names the country of Muḷaka which it puts together with the country of Asaka (Assaka) exactly as the *Pārāyaṇa* does : *Asika-Asaka-Muḷaka-Suraṭha-Kukur-Āparaṁta-Anūpa-Vidabha-Ākar-Āvati*. The latest editor of the inscription, M. Senart, wrote about the name Muḷaka : "The Muḷakas remain shrouded in obscurity. Bhagwanlal adduced the dynasty of the Muṇḍakas, known from the *Viṣṇu Purāṇa* ; and the way in which they are here mentioned together with the Śakas and Tukhāras is such as to commend the hint. But I am doubtful about the change of *ḷ* into *ṇḍ*."[4] Thanks to the text of the *Pārāyaṇa* the mystery is solved. The site of Muḷaka may also

1. *Op. cit.* p. 194.
2. [Tumbavana is modern Tumain in the Guna District of the former Gwalior State. See the Tumain inscription of the Gupta year 116, which mentions Tumbavana, in *Ep. Ind.*, Vol. XXVI, pp. 115ff.]
3. Lueders' List of Ins., Nos. 201, 202, 449, 450, 520.
4. *Ep. Ind.*, Vol. VIII, p. 62.

be determined with enough precision, since it is between Asaka (Assaka, Aśmaka) where exists the city of Pratiṣṭhāna (Patiṭṭhāna), now-a-days Paiṭhan on the upper Godāvarī, near its source, and Suraṭha (Surāṣṭra) which is the modern peninsula of Kathiawar. Muḷaka should then indicate the portion of the coast with the rear-country lying to the north of Bombay, i.e. Gujarāt.[1]

I shall perhaps be accused of attaching an excessive importance to the order of succession in which the names of provinces are enumerated in the inscription of Puḷumāyi. But the inscription of Rudradāman at Girnar, which goes up to the same epoch, also contains a list of provinces subdued by Rudradāman, the adversary of the Śātakarṇi dynasty, to which Gautamīputra and Puḷumāyi belonged. His domain is partly formed of the territories conquered from the Śātakarṇis. The following are found in this list : Ākar-Āvanti, Anūpa, Surāṣṭra, Kukur-Āparānta, etc. (पूर्वापराकरावन्त्यनूपनिवृदानर्त्तसुराष्ट्रश्वभ्रमरुकच्छसिन्धुसौवीरकुकुरापरान्तनिषघादीनाम्). But here the order of succession is the reverse, since Rudradāman extended his conquests from the north to the south, starting from the region of Ujjayanī, his capital, whereas Gautamīputra, starting from the banks of the Godāvarī, marched conquering from the south to the north.

The memory of the country of Muḷaka is perhaps not entirely effaced from the Purāṇic tradition. In the genealogy of the race of Ikṣvāku, the *Viṣṇu Purāṇa* (IV. 4) assigns a son named Aśmaka to Kalmāṣapāda. Aśmaka's son and successor was Mūlaka, surnamed Nārīkavaca, because some women hid and saved him at the time of the general massacre of the Kṣatriyas. Aśmaka is clearly an eponymous hero, the eponym of the country of Aśmaka, the 'Stony' territory (*aśman* ='stone'), situated to the south of Avanti and closely connected with it. Cf. *Avanty-Aśmakāḥ* in the *Gaṇapāṭha* under the *Kārta-Kaujapādayaḥ* Gaṇa and *Aśmak-Āvanti* in the Sarvāstivādin *Vinaya* episode of Koṭikarṇa. The Mūlasarvāstivādin *Vinaya* has *Aśm-*

1. [Muḷaka was situated close to Aśmaka and the Godāvarī and the city of Pratiṣṭhāna lay in it. Paiṭhan lies in the Aurangabad District of the former Hyderabad State. Muḷaka was therefore the area around that District which was sometimes included in Aśmaka. See above, pp. 189-90, 225ff.; below, p. 274, note 1.]

2. [The expression *pūrv-āpara* seems to mean 'counted from east to west.']

Āparāntaka which Cowell and Neil wrongly read: *asmāt parāntaka.*[1] The relation between Aśmaka and Muḷaka so clearly expressed in the *Pārāyaṇa* (*so Assakassa visaye Muḷakassa samāsane,* v. 2) suggests that the filiation indicated in the Purāṇa between king Aśmaka and king Mūlaka interprets a geographical relation by the genealogy.

It will be less astonishing to meet in a section of the *Suttanipāta* with some details, so precise, of the geography of the regions that encircle the Gulf of Cambay, if it is remembered that one of the pearls of the Collection, viz. the admirable *Dhaniyasutta,*[2] has the bank of the Mahī for its scene. The shepherd Dhaniya is settled here (*anutīre Mahiyā samāna-vāso*). The Buddha passes a night on the bank of the river (*anutīre Mahiy=eka-ratti-vāso*). The audacity of a poet and a local inspiration must have been necessary for bringing the Buddha in these exotic quarters so far from his real activity, in close contact of the sea frequented by the barbarians.[3]

1. *Divyāvadāna,* 1. [Aparānta was the Thana region (Northern Konkan) between the Bombay area and Gujarat and had its capital at Sūrpāraka (modern Sopara). For this country and Lāṭa in the Nausari—Broach region of Southern Gujarat, see above, pp. 225ff.]

2. *Uragavagga,* 2.

3. [See above, p. 273, note 1. For the situation of the Aśmaka country which lay far away from the Mahī valley, see above, pp. 189-90.]

CHAPTER XX

KOKĀMUKHA[1]

Some years ago, no less than five copper-plate charters of the time of three emperors of the Gupta dynasty were discovered at a place called Dāmodarpur near the Phulbari Police Station in the Balurghat Sub-Division of the old Dinajpur District, North Bengal. One of these records refers to the Gupta emperor Budhagupta (477-94 A.D.), the *uparika Mahārāja* Jayadatta, viceroy of the Puṇḍravardhana *bhukti*, and the *āyuktaka* Gaṇḍaka (or Śaṇḍaka) who was in charge of the Koṭivarṣa *viṣaya*. Gaṇḍaka is said to have been helped in the administration of the Koṭivarṣa district by the *nagaraśreṣṭhin* Ṛbhupāla, *sārthavāha* Vasumitra, *prathama-kulika* Varadatta, *prathama-kāyastha* Viprapāla, etc. The *śreṣṭhin* Ṛbhupāla one day approached the *adhiṣṭhān-ādhikaraṇa*, i. e. the office of administration at the headquarters of the district, with the following petition : हिमवच्छिखरे कोकामुखस्वामिनः चत्वारः कुल्यवापाः श्वेतवराहस्वामिनोऽपि सप्त कुल्यवापाः अस्मत्फलाशंसिना पुण्याभिवृद्धये डोङ्गाग्रामे पूर्वं मया अप्रदा अतिसृष्टकाः । तदहं तत्क्षेत्रसामीप्यभूमौ तयोराद्यकोकामुखस्वामिश्वेतवराहस्वामिनोर्नामलिङ्गमेनं देवकुलद्वयमेतत् कोष्ठिकाद्वयञ्च कारयितुमिच्छामि । अहंथ वास्तुना सह कुल्यवापान् यथाक्रयमर्यादया दातुमिति । After having examined Ṛbhupāla's application, the *pustapālas* Viṣṇudatta, Vijayanandin and Sthāṇunandin reported that some *Kulyavāpas* of land at the rate of three *Dīnāras* per *Kulyavāpa* might be sold to the *śreṣṭhin*, because : अनेन हिमवच्छिखरे तयोः कोकामुखस्वामिश्वेतवराहस्वामिनो अप्रदाः क्षेत्रकुल्यवापाः एकादश दत्तकाः । तदर्थञ्च इह देवकुलकोष्ठिकाकरणे युक्तमेतद् विज्ञापितं तत्क्षेत्रसामीप्यभूमौ वास्तु दातुमिति । It should be pointed out that we have quoted above the epigraphic texts with the elimination of minor grammatical errors.[2]

1. For the god Kokāmukhasvāmin, see H. C. Raychaudhuri, *Stud. Ind. Ant.*, 1958, pp. 205 ff.

2. For the actual text of the passages and their interpretation, see *Select Inscriptions*, pp. 328-30; for *Kulyavāpa*, see *Ind. Ep.*, pp. 411 ff. For the other Dāmodarpur inscriptions cited below, see *Sel. Ins.*, pp. 283 ff., 324 ff., 337 ff. The board of administrators headed by the *Nagaraśreṣṭhin*, referred

The expression *Himavac-chikhara* in the passages quoted above literally means 'a peak of the Himalayas'. But the village called Ḍoṅgā-grāma, where the donated lands were situated and where further lands were applied for, does not appear to have been far from Dāmodarpur, the findspot of the charter in question. The village is also mentioned in another Dāmodarpur record that was issued in the Gupta year 124 during the reign of Kumāragupta I. A second Dāmodarpur grant of the time of Budhagupta, dated in the year 163, mentions Vāyi-grāmaka which is no other than the modern Bāigrāṃ in the Bogra District, not far away from the finsdpot of the record. The Dāmodarpur inscription of the Gupta year 224 refers to the purchase of five *Kulyavāpas* of land by the *Kulaputra* Amṛtadeva hailing from Ayodhyā for the following purpose : अत्रारण्ये भगवतः श्वेतवराह-स्वामिनो देवकुले खण्डस्फुटितप्रतिसंस्कारकरणाय बलिचरुसत्रप्रवर्त्तनगन्धधूप-पुष्पप्रापणमधुपर्कदीपाद्युपयोगाय च अप्रदाधर्म्मेण ताम्रपट्टीकृत्य क्षेत्रस्तोकं दातुमिति। In connection with the grant of land to the god Śvetavarāhasvāmin in this case, mention has been made of Svacchanda-pāṭaka, Lavaṅgasikā, Sāṭuvan-āśramaka, Paras-patikā, the Jambū-nadī and Pūraṇavṛndikahari. Of these, Pūraṇavṛndikahari has been identified by some scholars with modern Brindakoori, 14 miles to the north of Dāmodarpur. It is therefore almost certain that the lands granted by Ṛbhupāla to the gods Kokāmukhasvāmin and Śvetavarāhasvāmin lay in the Dāmodarpur region in the southern part of the old Dinajpur District. But the temples of the two gods in question were situated in the Himalayan region far away from Dāmodarapur. It has therefore been suggested by some scholars that the mountainous region about the northern fringe of the modern territory of Bengal formed parts of the Koṭivarṣa *viṣaya* (Dinajpur area) or the Puṇḍravardhana *bhukti* (North Bengal).[1] This hypothesis is apparently due to the fact that nobody has so far been able to determine the exact location of the temples of Kokāmukha and Śvetavarāha in the Himalayas.

to in these records was similar to the Pañcāyat board of the Cauthiās headed by the *Nagar-seth* as prevalent in Rājasthān. See Tod, *An. Ant. Raj.*, ed. Crooke, Vol. I, pp. 171, 231; Vol. II, p. 682. Cf. *Journ. Un. Gauhati*, Vol. VI, pp. 81 ff.

1. *History of Bengal*, Dacca University, Vol. I, p. 24, p. 400 and note 3.

A *tīrtha* called Kokāmukha or Varāha-kṣetra is known from the *Mahābhārata*[1] and the Purāṇas. Chapters 219 and 229 of the *Brahma Purāṇa* locate the holy place of pilgrimage in the Himalayas. The *Brahma Purāṇa* not only refers to the Kokāmukha *tīrtha* in the Himalayas, but also to the Varāha form of Viṣṇu installed there and to the river called Kokā which runs through the place. Cf.

> कोकेति प्रथिता लोके शिशिराद्रिसमाश्रिता ॥ (२२९।१७)
> कोकानदीति विख्याता गिरिराजसमाश्रिता ।
> तीर्थकोटिमहापुण्या मद्रूपपरिपालिता ॥ (२२९।१०६)
> वराहदंष्ट्रासंलग्नाः पितरः कनकोज्ज्वलाः ।
> कोकामुखे गतभयाः कृता देवेन विष्णुना ॥ (२२९।३०९)
> एवं मयोक्तं वरदस्य विष्णोः कोकामुखे दिव्यवराहरूपम् (२२९।११६),etc.

But the *Brahma Purāṇa* does not help us at all in determining the exact location of the temple of Varāha at Kokāmukha in the Himalayas. For this, we have to examine other Purāṇic texts. In this connection, attention of scholars may be drawn to the *Varāha Purāṇa*.

Chapter 140 of the *Varāha Purāṇa* is styled *Kokāmukha-māhātmya-varṇana*. Here Varāha (Viṣṇu in the Boar form) declares to the goddess Earth :

> तव कोकामुखं नाम यन्मया पूर्वभाषितम् ।
> बदरीति च विख्यातं गिरिराजशिलातलम् ॥
> स्थानं लोहार्गलं नाम म्लेच्छराजसमाश्रितम् ।
> क्षणञ्चापि न मुञ्चामि एवमेतन्न संशयः ॥ (१४०।४-५)

According to this passage, only three places on the earth are the abodes of Varāha, and they are: (1) Kokāmukha, (2) Badarī or Badarik-āśrama, and (3) Lohārgala, all in the Himalayas. Chapter 141 of the Purāṇa, styled *Badarikāśrama-māhātmya-varṇana*, refers to a number of holy spots in the region of the *tīrtha* now known as Badarīnārāyaṇa. They are Brahma-kuṇḍa, Agnisatya-pada, Indraloka, Pañcaśikha, Catuḥ-srotaḥ, Vedadhāra, Dvādaśāditya-kuṇḍa, Lokapāla, Sthala-kuṇḍa amongst hills, Meruvara, Mānasodbheda, Pañca-saraḥ, Somābhiṣeka,

1. See III. 48. 158; XIII. 25. 52.

Soma-giri, Urvaśī-kuṇḍa, etc. Chapter 151 of the *Varāha Purāṇa*, styled the *Lohārgala-māhātmya-varṇana*, refers to the holy spots belonging to the Lohārgala *tīrtha* in the land of the Mlecchas in the Himalayas.[1]

The holy spots in Lohārgala are Pañca-saraḥ, Nārada-kuṇḍa, Vasiṣṭha-kuṇḍa (where five streams are falling from the Himakūṭa), Saptarṣi-kuṇḍa (where seven streams fall from the Himavat-parvata), Śarabhaṅga-kuṇḍa (तत्र धारा पतत्येका शरभङ्गाश्रिता नदी), Agnisaraḥ-kuṇḍa, Bṛhaspati-kuṇḍa (where falls a stream which is *Himakūṭa-samāśritā*), Vaiśvānara-kuṇḍa (धारा चैका पतत्यत्र दृश्यते हिमसङ्क्षयात्), Kārttikeya-kuṇḍa (where fifteen streams fall from the Hima-parvata), Umā-kuṇḍa, Maheśvara-kuṇḍa (where three streams fall from the Himavat-parvata), Brahma-kuṇḍa (where four streams fall from the Himālaya), etc.

The holy spots belonging to the Kokāmukha *tīrtha* mentioned in the *Varāha Purāṇa*, Chapter 140, are the following : 1. Jalabindu; 2. Viṣṇu-dhārā; 3. Viṣṇu-pada which is *Kokāmukh-āśrita*; 4. Viṣṇu-saraḥ; 5. Soma-tīrtha—यत्र पञ्चशिला भूमिर्विष्णुनाम्ना तथाङ्किता; 6. Tuṅgakūṭa ; 7. Agni-saraḥ—पञ्च धाराः पतन्त्यत्र गिरिकुञ्जसमाश्रिताः; 8. Brahma-saraḥ; 9. Dhenuvaṭa; 10. Dharmodbhava—गिरिकुञ्जात् पतत्येका धारा भूमितले शुभा; 11. Koṭivaṭa; 12. Pāpapramocana; 13. Yamavyasanaka; 14. Mātaṅga—स्रोतो वहति तत्रैवमाश्रितं कौशिकीं नदीम्; 15. Vajrabhava —स्रोतो वहति तत्रैकमाश्रितं कौशिकीं नदीम्; 16. Śakrarudra which is *Kokā-śilātala-sthita*; 17. Daṁṣṭrāṅkura— यत्र कोका विनिःसृता; 18. Viṣṇu-tīrtha—ततः पर्वतमध्यात्तु कोकायां पतति जलम्; 19. Sarva-kāmikā— अस्ति तत्र वरं स्थानं संगमं कौशि (की*) कोकयोः । सर्वकामिकेति विख्याता शिला तिष्ठति चोत्तरे ।। 20. Matsya-śilā—अस्ति मत्स्यशिला नाम

1. Cf. ततः सिद्धवटं गत्वा त्रिशद्योजनदूरतः ।
म्लेच्छमध्ये वरारोहे हिमवन्तं समाश्रितम् ।।
तत्र लोहार्गले क्षेत्रे निवासो विहितः शुभः ।
गुह्यं पञ्चदशायामं समन्तात् पञ्चयोजनम् ।।
तत्र तिष्ठाम्यहं भद्रे उदीचीं दिशमाश्रितः ।
हिरण्यप्रतिमां कृत्वा जातरूपां न संशयः ।। (१५१।७-१०)

Lohārgala is generally identified with Lohāghāṭ in Kumāon (Kūrmācala). Cf. Dey, *Geog. Dict.*, s. v.

गुह्यं कोकामुखे वरम् । धाराः पतन्ति तिस्रो वै कौशिकीमाश्रिता नदीम् । Besides the above particulars, we have some general observations on the Kokāmukha *tīrtha*, such as— पञ्चयोजनविस्तारं क्षेत्रं कोकामुखं मम, तस्मिन् कोकामुखे रम्ये तिष्ठामि दक्षिणामुखः, वराहरूपमादाय तिष्ठामि पुरुषाकृतिः, वामोन्नतमुखं कृत्वा वामदंष्ट्रासमुन्नतम्, etc. It is quite clear from the description of the Kokāmukha *tīrtha* in the *Varāha Purāṇa* that two rivers named Kokā and Kauśikī as well as their confluence were intimately associated with the holy region.

There are several rivers of the name Kauśikī in different parts of India. But the only Kauśikī that can be associated with a *kṣetra* of Varāha as well as with a river called Kokā is the Kauśikī, modern Kośī, running from Nepal by the western borders of the Purnea District of Bihar. The river is known in Nepal as the Sun-Kośī (*i. e.*, Svarṇa-Kauśikī) and some of its tributaries have names like Arun-Kośī, Dudh-Kośī, etc. The ancient Kokāmukha *tīrtha* or Varāha-kṣetra is situated on the bank of the Sun-Kośī in Nepal and is now popularly known as Barāh-chatra, *chatra* being the common corrupt form derived from Sanskrit *kṣetra*.

It is unfortunate that in most maps of Nepal no locality called Varāha-kṣetra or Barāh-chatra is indicated. Of course Dhankuta a little to the north and Bijapur to the east of the holy region are found even in ordinary maps. E. Thornton's *Gazetteer of India* (London, 1886, s. v.) spells the name wrongly as *Vardhachatra* instead of *Varahachatra* (with *d* wrongly printed for *a*) and says, "Town in Nepal State; situated on the left bank of the San-Kusi river, 124 miles east-south-east of Khatmandu. Lat. 26° 57′, long. 87° 4′." The *Gupta Press Directory Pañjikā* (in Bengali) observes, "Varāhachatra.—The image of Varāhadeva, the third incarnation of Lord Viṣṇu, is installed on the Dhavalāgiri near the kingdom of Bhuṭān forming part of the kingdom of Nepāl. A fair is held at the place every year during the full-moon day of the month of Kārttika. From Calcutta to Jogbani (Kāṭihār-Jogbani branch, B. & A. Ry.), the distance is 331 miles (*via* Rāṇāghāṭ and Lālgolāghāṭ). The foot of the Dhavalāgiri peak is 20 miles from that place by a road along the Kuśī river; the temple of Varāhadeva lies 20 miles above." Although Bhutan and the celebrated Dhavalāgiri in Nepal are far away from the Varāha-kṣetra or Barāh-chatra, the location of the holy place indicated above is fairly correct. The Barāh-chatra and the Kokā river have been placed in the same region in an old book entitled *An Account of the Kingdom of Nepal* (being the substance of observations made during a Mission to that country in the year 1793) by Colonel Kirkpatrick, London, 1811. They are also indicated in the map inserted in the book. While describing the route from Khatmandu to Bijapur, the author says (pp. 324-25): From Ohdhang ghaut 'To the conflux of the Arun and Soan-Kousi—7 [ghurries]; To Ukhurria ghaut (the 2nd)—5 [ghurries]; To the Thumboor, *i. e.* the confluence of the Tumboor and Soan-Kousi at Tambraphede—26 [ghurries]; To Koka-Kola (which falls into the Soan-Kousi according to the map)—28 [ghurries]; To Barah-chatra—28 [ghurries]; To Chattra-ghaut (on the Kousi)—5 [ghurries]; To Bejapour—16 [ghurries]." The distance between the localities has been indicated by time, a ghurry being equal to 22½ minutes. But the estimates are only approximate as the time required in travelling in a hilly region can hardly be uniform. The word *kolā* (probably from Sanskrit *kulyā*) means a stream and Kokā-kolā means the small river called Kokā. Thus we have here not only a Varāha-kṣetra, but also the junction of the rivers Kauśikī (Sun-Kuśī) and Kokā (Kokā-kolā). There is therefore no doubt that the Barāh-chatra in Nepal has to be identified with the Kokāmukha *tīrtha* mentioned in the *Mahābhārata* and the Purāṇas and that the temples of the gods Kokāmukha and Śvetavarāha, associated in the Dāmodarpur inscription with *Himavac-chikhara*, were situated at this place.

The distance of the ancient Kokāmukha *tīrtha* or Barāh-chatra in Nepal from the Dāmodarpur region in the southern part of the old Dinajpur District is about 150 miles by air. The inclusion of the above part of Nepal within the Koṭivarṣa *viṣaya* or the Puṇḍravardhana *bhukti* in the Gupta period seems to be quite out of question at the present state of our knowledge.

People of North Bihar have great regard for the Varāha-kṣetra even to-day. There is again no doubt that the culture of North Bengal was intimately associated with that of North Bihar before the former territory was flooded by Mongolian migrations. It was therefore quite natural for the people of the Dinajpur region to visit the Kokāmukha *tīrtha* even in the Gupta age. The *śreṣṭhin* Ṛbhupāla apparently went to Varāhakṣetra-Kokāmukha on pilgrimage and after having returned home, dedicated a large area of land in his native district in honour of the two gods enshrined at the holy place. But it was not quite easy for him to send the income from it regularly to distant Nepal. The pious Ṛbhupāla therefore made two temples of the same gods near the land originally granted, together with two store-houses.[1] He appears to have installed in these temples two imitation-gods of the same names. That is why the gods in the *Himavac-chikhara* have been styled *ādya*, *i. e.* 'original'. This is further suggested by the distinction of the places indicated by the two expressions *Himavac-chikhare* and *iha* occurring in the second passage quoted above from Ṛbhupāla's record. About half a century after Ṛbhupāla, Amṛtadeva granted lands in favour of the temple of Śvetavarāha. This is no doubt one of the temples founded by Ṛbhupāla in the Dāmodarpur region and not that of the original god in the Himalayas. We have to take note of the fact that the expression *Himavac-chikhara* is conspicuous by its absence in the record of Amṛtadeva. Instead of any reference to the Himalayas, we find here the temple located in *atr = āraṇye*, *i.e.*, 'here in this forest', which no doubt formed a part of the Koṭivarṣa-viṣaya.

1. Cf. तत्क्षेत्रसामीप्यभूमौ तयोराद्यकोकामुखस्वामिश्वेतवराहस्वामिनोर्नाम-लिङ्गमेनं देवकुलद्वयमेतत्कोष्ठिकाद्वयं च, 'these two temples marked by the names of the said original gods, Kokāmukhasvāmin and Śvetavarāhasvāmin, as well as these two store-houses in the vicinity of that [land granted originally at Ḍoṅgā-grāma in the Koṭivarṣa-viṣaya].

Chapter XXI

GAYĀ

The antiquity of the Gayā-tīrtha has been the subject of a controversy for a long time.[1] Recently a paper entitled 'Bihar in the Agni-Purāṇa' has been published in the *Journal of the Bihar Research Society*, Vol. XL, Part i, 1954, pp. 1-7, where the problem has been discussed without reference to the views of earlier writers. In the concluding part of the said paper, the author draws our attention to the well-known stanza :

अयोध्या मथुरा माया काशी कांची अवंतिका ।
पुरी द्वारावती चैव सप्तैता मोक्षदायिकाः ॥

and observes, "This couplet was composed probably in the eighth century A.D., and, from the absence of any mention of Gayā, it appears that any importance it may have had then was only local and that it did not acquire a pan-Indian celebrity before this time. This is strengthened by the fact that while in (of?) the Gupta period there is only one inscription found at Gayā, in (of?) the Pāla period (c. 750-c. 1200) no less than six inscriptions are found here. These things evidently point to the growing importance of Gayā in the period subsequent to 750 A.D., *i.e.*, in the period of the Agni-Purāṇa." We are sorry that we cannot agree with the author's contentions.

In the first place, it is not possible to assign the text of a Purāṇa as a whole, in all cases, to a definite date because it usually contains matter interpolated by various redactors in different ages. Moreover, the incorporation of a tradition in a work at a particular date does not prove that it was non-existent in an earlier age. It is therefore impossible to determine the antiquity of Gayā from its mention in the *Agni Purāṇa* alone.

1. Cf. R. L. Mitra's *Buddha-Gayā*, 1878; Cunningham's *Mahābodhi*, 1892; O'Malley's *Gazetteer of the Gaya District* (also his article in *JASB*, Vol. LXXII, 1903, No. 3, pp. 1-11); B. M. Barua's *Gayā and Buddha-Gayā*, Vols. I-II, 1934; J. C. Ghosh's article in *JBORS*, Vol. XXIV, 1938, pp. 89-111; P. V. Kane's *History of Dharmaśāstra*, Vol. IV, 1953, pp. 642-79.

Secondly, the date of the composition of the stanza, *Ayodhyā*, etc., quoted above, is uncertain. Moreover a theory based on the absence of Gayā in it seems to be merely a case of *argumentum ex silentio*. It is certainly impossible to believe that Prabhāsa, Puṣkara and Prayāga, which are not mentioned in the verse, attained all-India importance after the seventh century. The pilgrimage of a ruler of the North Marāṭhā country to Prabhāsa and Puṣkara is mentioned in an inscription of the second century A.D.,[1] while Prayāga is not only referred to in a *Khila* verse placed in the *Ṛgveda*, X. 75, but is also mentioned in an inscription[2] as a holy place where a sixth century king of East Malwa committed religious suicide according to the injunction of the Dharmaśāstra writers. There is absolutely no doubt that these three *tīrthas*, not represented in the stanza relied on by our author, became widely known many hundred years before the eighth century when the said verse is supposed to have been composed. The verse is found with some modifications once in the *Brahmāṇḍa* (IV. 40. 91) and *Garuḍa* (Preta-khaṇḍa, 38. 5-6) *Purāṇas* and twice in the Kāśī-khaṇḍa (6. 68 and 23. 7) section of the *Skanda Purāṇa*.[3] There is hardly any doubt that the date of its composition is later than that of a traditional stanza regarding the sanctity of Gayā, which is found in the epics as well as in most of the Purāṇas and Dharmaśāstra works. But this is not the only evidence to show that Gayā was already one of the most famous Indian *tīrthas* long before the rise of the Pālas about the middle of the eighth century.

The *Mahābhārata* is believed to have assumed its present form between the fourth century B.C. and the fourth century A.D.,[4] while the *Rāmāyaṇa* is supposed to have had its present extent and contents as early as the second century A.D.[5] The Tīrtha-yātrā section of the Vana-parvan of the *Mahābhārata* contains two subsections on Gayā,[6] which point to the place as a great *tīrtha*. But what is more important is that both these sub-

1. *Select Inscriptions*, pp. 160 ff.
2. *Corp. Ins. Ind.*, Vol. III, p. 200.
3. See N. L. Dey, *Geographical Dictionary*, p. 179; Kane, *op. cit.*, p. 678, note 1935.
4. Cf. *Cambridge History of India*, Vol. I, p. 258; Winternitz, *History of Indian Literature*, Vol. I, p. 465.
5. Winternitz, *op. cit.*, p. 516.
6. Calcutta ed., Ch. 84, verses 82-98; Ch. 87, verses 8-12.

sections have one stanza in common, which is introduced in one case by the passage *kīrtayanti purātanāḥ*[1] showing that it was an old *gāthā* handed down by tradition. The stanza runs as follows :

एष्टव्या बहवः पुत्रा यद्यप्येको गयां व्रजेत् ।
यजेत वाश्वमेधेन नीलं वा वृषमुत्सृजेत् ॥

That it was an old traditional *gāthā* is also clearly demonstrated by the fact that the same verse is quoted, sometimes with slight modifications (not affecting the reference to Gayā), again in the Anuśāsanika-parvan[2] and in the *Rāmāyaṇa*[3] as well as in numerous Purāṇas[4] and Dharmaśāstra works.[5] In many of these works, the verse is introduced as a *gāthā* sung by the Pitṛs or Ṛṣis. There cannot be any doubt that this *gāthā* was composed many centuries before the Pāla age and probably even before the rise of the Guptas in the fourth century A.D.

The *Mahābhārata* (III. 95) describes how, at Brahma-saras (within Gayā), *Rājarṣi* Gaya, son of Amūrtarayas, performed many sacrifices distinguished by 'hundreds of mountains of food and thousands of lakes of clarified butter, many hundred rivers of curds, and streams of curries in thousands' and how people in all lands used to sing *gāthās* on the subject. Elsewhere the same work (VII. 64) speaks of Gaya's sacrificial altar of solid gold that was 30 *yojanas* long, 26 *yojanas* broad and 24 *yojanas* high and how, upon the completion of sacrifice, 25 hills of food remained over together with many lakes and rivulets of savoury drinks. In consequence of this glorious performance, Brahma-saras and the 'eternal-making banyan-tree' (*i. e.* the Akṣaya-vaṭa) at the place are stated to have become famous throughout the three worlds.[6] The same saintly king named Gaya is also known from the *Rāmāyaṇa*,[7] *Bhāgavata*

1. Cf. critical ed., III. 85. 7.
2. Ch. 88. 14.
3. Ayodhyā-kāṇḍa, 107. 13.
4. Cf. *Matsya*, 22. 6; 208. 40; *Vāyu*, 83. 12; 105. 10; *Kūrma*, II. 34. 13; *Brahma*, 220. 32; *Padma*, Sṛṣṭi-khaṇḍa, 11. 65 or 68, and Ādi-khaṇḍa, 38. 17; *Nāradīya*, Uttara-khaṇḍa, 44. 5-6; etc.
5. Cf. Viṣṇu, 85. 63; Atri, verse 55; Bṛhaspati, verse 21; Likhita, verse 10 ; etc. (*Ūnaviṁśatisaṁhitā*, Calcutta, pp. 4, 113, 347, 428).
6. See also S. Soerenson, *An Index to the Names in the Mahābhārata*, pp. 302-03
7. Ayodhyā-kāṇḍa, Ch. 107.

Purāṇa,[1] *Brahmāṇḍa Purāṇa*,[2] *Agni Purāṇa*,[3] *Viṣṇu Purāṇa*,[4] *Vāmana Purāṇa*,[5] etc., although his parentage is sometimes given differently. The antiquity of the tradition is definitely proved by Aśvaghoṣa's *Buddhacarita*[6] composed in the first or second century A.D., during the age of the Kuṣāṇas. This work speaks of the Buddha's visit to the hermitage called 'the city of the royal sage Gaya'. The same person was later conceived as a great giant.[7] The demon's body is said to have been 125 *yojanas* in height and 60 *yojanas* in girth. It will be seen that, while the 25 hills of the *Mahābhārata* story were imagined to explain the existence of sacred hills in the Gayā area, the huge body of the demon was conceived as the hilly region, studded with holy spots, extending from Gayā to the coastal areas of the Bay of Bengal. Gaya's head was believed to be represented by Gaya-śiras or Gaya-śīrṣa (Gayā-śiras or Gayā-śīrṣa) at Gayā while Virajā or Jājpur in the Cuttack District of Orissa and Pithapuram in the E. Godavari District of Andhra Pradesh were taken to represent respectively his navel and feet.[8] The frequent mention of the shaking of the earth in this context in the *Vāyu Purāṇa* (Ch. 106) possibly hints at earthquakes that occasionally depopulated Gayā.[9]

It may be noticed in this connection that the *Yājñavalkya-saṁhitā*, which does not specially deal with *tīrtha-māhātmya*, also recognises the greatness and sanctity of Gayā ((Ācār-ādhyāya, verse 261) :

यद्ददाति गयास्थश्च सर्वमानन्त्यमश्नुते ।
तथा वर्षत्रयोदश्यां मघासु च विशेषतः ॥

The code of Yājñavalkya has to be assigned to the fourth century

1. V, Ch. 15.
2. Ch. 34.
3. Ch. 107.
4. IV, Ch. 11.
5. Ch. 76.
6. XII. 87-88.
7. Cf. *Vāyu Purāṇa*, Ch. 105 ff., etc.
8. N. L. Dey, *Geographical Dictionary*. pp. 64-65.
9. Cf. the deserted appearance of the holy place referred to by the Chinese pilgrims.

A.D.[1] although some scholars are inclined to ascribe it to 'the first two centuries of the Christian era or even a little earlier.'[2]

That Gayā was a famous place of pilgrimage during the Gupta age is also clear from the accounts of the Chinese pilgrims. Fa-hien visited the place about the beginning of the fifth century when the locality offered a rather deserted appearance.[3] According to Hiuen-tsang, who visited Gayā about 637 A.D., "This city was strongly situated, but had few inhabitants; there were about 1000 Brāhmaṇa families, descendants of the original Ṛṣi, and these were not subject to the king, and were treated by all with reverence....From ancient times, sovereigns who have spread their good government to distant peoples and in merit have excelled previous dynasties, all ascend this mountain (the Gayā mountain to the south-west of the city) and solemnly announce what they have done. On the top of the mountain was a stone Tope (*Stūpa*) above 100 feet high built by Aśoka at the place where Buddha uttered the *Pao-yun* and other Sūtras."[4] The *Pao-yun* has been taken to be the *Ratnagarbhasūtra* said to have been communicated to the disciples of the Buddha, assembled on the Gaya-śiras or Gaya-śīrṣa (Gayā-śiras or Gayā-śīrṣa) hill, identified by many scholars with the modern Brahmayoni hill near Gayā. The objections to this identification and to the location of ancient Gayā at the site of the present city of that name are not quite convincing. Moreover, those who object do not say that the ancient Gayā *tīrtha* lay far away from present Gayā in a region outside Bihar. The Brāhmaṇas of Gayā referred to in the Chinese account are no doubt the ancestors of the celebrated Gayāla Brāhmaṇas of today. These Brāhmaṇas are mentioned in the Śaktipur copper-plate inscription[5] of king Lakṣmaṇasena (*circa* 1179-1206 A.D.) of Bengal. They appear to have claimed descent from *Rājarṣi* Gaya. Hiuen-tsang's observations show beyond doubt that Gayā was already a famous holy place long before the seventh century, if not actually in the age of the Maurya emperor Aśoka (*circa* 269-232 B.C.).

1. Cf. *Cambridge History of India*, Vol. I, p. 279.
2. Kane, *op. cit.*, Vol. I, p. 187.
3. Cf. *A Record of the Buddhist Kingdom*, translated by James Legge, pp. 87 ff.
4. Watters, *On Yuan Chwang's Travels in India*, Vol. II, pp. 110-11.
5. *Ep. Ind.*, Vol. XXI, pp. 216 ff.

But there is evidence to show that Gayā is a much older *tīrtha* even earlier than the age of the Buddha.

Gaya as a personal name, with which the holy place called Gaya or Gayā (cf. Gaya-śiras or Gayā° and Gaya-śīrṣa or Gayā°) is intimately associated, is well known from the *Ṛgveda*[1] and *Atharvaveda*.[2] The unanimous testimony of the Purāṇic writers and medieval lexicographers suggests that the Gayā region was known to the Ṛgvedic seers as Kīkaṭa.[3] Although pilgrimage to holy places for the sake of merit seems to be a Non-aryan religious institution which took some time in being fully approved by the Brāhmaṇical society, the importance of the holy hill called Gaya-śiras at Gayā appears to have been recognised in very early times. This is suggested by Yāska's *Nirukta* (12. 19) which, while explaining the Vedic passage *tredhā nidadhe padam*,[4] quotes the view of an earlier authority named Aurṇavābha saying that Viṣṇu plants his foot at Samārohaṇa, Viṣṇupada and Gaya-śiras. These were evidently three holy spots. Since in the whole of India, there is only one Gaya-śiras or Gaya-śīrṣa (Gayā-śiras or Gayā-śīrṣa) near Gayā in Bihar, the tradition, attributed to Aurṇavābha by Yāska who is believed by scholars to have flourished between 700 and 500 B.C.,[5] certainly points to the antiquity of Gayā as a recognised holy place of old. The conclusion is supported by the frequent mention of the holy spots at Gayā and its neighbourhood in early Buddhist literature.

The earliest literary works of the Buddhists very often mention Gayā (also Gayā-tīrtha, Gayā-śīrṣa, Gayā-nadī and Gayā-puṣkariṇī) as well as a festival held there and called Gayā-phalgu or Gayā-phalgunī. See the *Vinayapiṭaka*,[6] *Saṁyuttanikāya*,[7] *Aṅguttaranikāya*,[8] *Suttanipāta*,[9] etc. All these works are to be assigned to dates before the beginning of the Christian era. The Buddha is stated to have stayed at Gayā on several occasions.

1. V. 9. 1; X. 63. 1 and 64. 1.
2. I. 14. 4.
3. Cf. Raychaudhuri *Studies in Indian Antiquities*, pp. 17-19.
4. *Ṛgveda*, I. 22. 17.
5. Winternitz, *op. cit.*, Vol. I, p. 69, note.
6. Ed. Oldenberg, Vol. I, pp. 8, 34.
7. P. T. S., Vol. I, p. 207.
8. P. T. S., Vol. IV, p. 302.
9. P. T. S., p. 47.

It was at Gayā-śīrṣa that the well-known *Gayāsūtra* was preached by him.[1] The evidence of the Buddhist works of the Gupta age supports what has already been said above on the strength of Brāhmaṇical literature. Buddhaghoṣa (5th century) applies the name Gayā both to a locality and a bathing place near it; but Dharmapāla (*circa* 6th century)[2] in his *Udāna* commentary[3] mentions Gayā-nadī and Gayā-puṣkariṇī as two distinct bathing places, both known as Gayā-tīrtha and supposed to possess the power to wash away sins: "People went there, offered sacrifices to the gods, recited the Vedas, and immersed themselves in the water."[4] In Buddhist literature, Gayā is sometimes called Brahma-Gayā (cf. the name of old Brahmasaras and of modern Brahmayoni at Gayā) to avoid its confusion with the Buddha's Gayā.

The facts discussed above do not appear to support the contention that Gayā became an all India *tīrtha* during the age of the Pālas (*circa* 750-1200 A.D.).

1. Cf. *Aṅguttaranikāya*, Vol. IV, pp. 302 ff.; *Mahāvagga*, I. 21. 1 (S.B.E., Vol XIII, p. 134).
2. Cf. Winternitz, *op. cit.*, Vol. II, p. 205.
3. P. T. S., pp. 74-75; cf. *Suttanipāta* commentary, P.T.S., Vol. I, p. 301.
4. Malalasekera, *Dictionary of Pali Proper Names*, Vol. I, p. 752.

Chapter XXII

UDABHĀṆḌA

In the *Vaijayantī*,[1] composed by Yādavaprakāśa in the 11th century A.D., there is a passage which runs : *Gandhārās=tu Dihaṇḍās=syuḥ*, 'the Gandhāras are also known as the Dihaṇḍas'. The name Gandhāra signifying a people and their territory lying in Uttarāpatha, in the north-western region of India, is well known to all students of Indian history. According to some late lexicons,[2] Gandhāra has to be identified with Kāndhāhāra, *i.e.* Kandahār in Southern Afghanistan. But this is a mistake due to the similar sounds of the two names. Although it may be tempting to conjecture that modern Kandahār owes its name to the Gandhāra occupation of Southern Afghanistan in the early period of Indian history, there is little doubt that the name is a corruption of that of the city of Alexandria founded by Alexander the Great (Persian *Iskandar* or *Sikandar*) in the land of the Arachosians near the site of modern Kandahār. There is definite evidence as regards the location of the Gandhāra *janapada* about the present Rawalpindi District of the Punjab and the Peshawar District of the old North-West Frontier in Pakistan.[3] According to epic and Purāṇic traditions, the Gandhāra *viṣaya*, which lay on both sides of the Indus, contained two great cities called Takṣaśilā and Puṣkalāvatī. The remains of Takṣaśilā lie immediately to the east and north-east of the Saraikala Railway Junction, 20 miles north-west of Rawalpindi, in the valley of the river Haro. There are remains of three distinct cities, of which the southernmost and oldest occupied the site of an elevated plateau now known as the Bhir Mound. The ancient city of Puṣkalāvatī or Puṣkarāvatī, 'the city of lotuses', was situated on the Swat in the modern Prang-Charsadda-Mir Ziyarat region, about 17 miles north-east of Peshawar.

1. Paryāya-bhāga, III. 1. 24.
2. Cf. *Śabdakalpadruma*, s. v.
3. Cf. Raychaudhuri, *Pol. Hist. Anc. Ind.*, 1938, pp. 50-52, 124-25.

It is however really curious that Dihaṇḍa, as the name of the famous Gandhāra people, is known from no other source excepting the *Vaijayantī*. This name seems therefore to have been actually based on a wrong reading of the manuscripts of Yādavaprakāśa's work, whose geographical section is full of mistakes.[1] A people is known not only after the land occupied by them, but very often also after their capital city, and we have to see if the form Dihaṇḍa may be associated with the name of the early medieval capital of the Gandhāra country.

Al-Bīrūnī, who was a contemporary of Yādavaprakāśa and wrote his celebrated work on India about 1030 A.D., refers to the 'capital of Al-Kandahār (Gandhāra), *i.e.* Vaihand'.[2] Elsewhere he speaks of 'Waihind, the capital of Kandhāra (Gandhāra), west of the river Sindh',[3] which is said to have been situated 14 *farsakh* (about 52 miles) to the south-east of Purshāwar (Peshawar). Both here and in another context,[4] Al-Bīrūnī locates Vaihand (Waihind) between Peshawar and the river Jhelam. There is no doubt that Vaihand has to be identified with modern Und near Attock on the Indus. As Raychaudhuri points out,[5] a Muslim work called *Hudud-ul 'Alam* (982-83 A.D.) speaks of Waihind as a big city, with some Muslim population, under king Jayapāla who was a feudatory of the Rāy of Kanauj, probably Pratīhāra Vijayapāla. The Sanskrit form of the name is Udabhāṇḍapura (or Udakabhāṇḍapura) which was the capital of the great Śāhi king Lalliya (*c.* 875-90 A.D.) and his successors according to the *Rājataraṅgiṇī* of Kalhaṇa. According to Firishta, the dominions of the Śāhi king Jayapāla (*c.* 965-1001 A.D.), son of Ishtpāl (probably the same as Aṣaṭapāladeva known from his coins), extended in length from Sirhind to Lamghan and in breadth from the kingdom of Kashmir to Mūltān. This ruler is further said to have resided in a fort, the correct reading of whose name is apparently Vaihand, although some writers wrongly take it to

1. Thus *Yavanās=tu Huruṣkarāḥ* (*sic—Turuṣkakāḥ*); *Śākhayo* (sic—*Śāhayo*); *Sūryārak-ādayaḥ* (sic—*Sūrpārak°*); *Traipurās=tu Hahālā°* (sic—*Ḍabhālā°*); etc.
2. Sachau, *Alb. Ind.*, Vol. I, p. 259.
3. *Ibid.*, p. 206.
4. *Ibid.*, p. 317.
5. *Proc. IHC*, 1939, p. 670.

be Bhatinda in the former Patiala State. Firishta frequently refers to Jayapāla as the Rājā of Lahore and, as the king resided in the Vaihand fort 'for the convenience of taking steps for opposing the Muhammadans', it is hardly possible that the historian could have referred to a locality in the Eastern Punjab. It seems that, when the Śāhi possessions lying west of the Indus were threatened by the Turkish Musalmans of Ghaznī, Jayapāla transferred his capital from ancient Udabhāṇḍapura to Lahore. But even then he himself resided at the old capital which now became the advance base of his operations against the Muhammadans. It is thus possible that the passage *Gandhārās = tu Dihaṇḍās = syuḥ* is actually a mistake for *Gandhārās = t = Ūdabhāṇḍās = syuḥ*, 'the Gandhāras are also known as the Udabhāṇḍas'. But Udahāṇḍa was probably another form of the name.

According to traditions, the Kuṣāṇa emperor Kaṇiṣka, who ruled over extensive regions in India and Central Asia, had his capital at the city of Puruṣapura (modern Peshawar) in the Gandhāra country. Al-Bīrūnī[1] says that Kanik (*i.e.* Kaṇiṣka) belonged to a dynasty of Hindu kings called Śāhis who were Turks of Tibetan origin and at first began to reign in Kabul. The last king of this house was Laga Tūrmān (*i.e.* Toramāṇa) who was overthrown by his Brāhmaṇa minister Kallar, a predecessor of Jayapāla. Scholars have suggested the identification of Al-Bīrūnī's Kallar with the great Lalliya Śāhi mentioned in the *Rājataraṅgiṇī*. It is, however, interesting to note that Kalhaṇa represents the Śāhis as Kṣatriyas. It has also to be remembered that the Kashmirian author does not make any distinction between the early Śāhis and the Brāhmaṇa Śāhis who, according to Al-Bīrūnī, succeeded them. We are told[2] that, even before

1. Sachau, *op. cit.*, Vol. II, pp. 10-14.
2. Cf. *Rājatar.*, IV. 140-43 :

प्रीतः पञ्चमहाशब्दभाजनं तं व्यधत्त सः ।
यशोवर्मनृपं तं तु समूलमुदपाटयत् ॥
अष्टादशानामुपरि प्राक्सिद्धानां तदुद्भवैः ।
कर्मस्थानैः स्थितिः प्राप्ता ततः प्रभृति पञ्चभिः ॥
महाप्रतीहारपीडा स महासन्धिविग्रहः ।
महाश्वशालापि महाभाण्डागारश्च पञ्चमः ॥

the reign of king Lalitāditya (*c.* 730-66 A.D.), the Śāhi-mukhyas, or chiefs belonging to the Śāhi family, were employed in the Kashmir administration as *Mahāpratīhāra*, *Mahāsāndhivigrahika*, *Mahāśvaśālika*, *Mahābhāṇḍāgārika* and *Mahāsādhanika*, the *pañca-mahāśabda* later conferred by king Lalitāditya on a single official named Mitraśarman whom he seems to have made his viceroy at Kanauj after having subdued king Yaśovarman. Of the Later Śāhis, described as Brāhmaṇa by Al-Bīrūnī, Kalhaṇa mentions Lalliya's successor and his son Kamaluka Toramāṇa (no doubt the same as Kamlū of Al-Bīrūnī, and Kamlū Rāy of Hindustan mentioned by other Arab writers), a contemporary of 'Amr ibn Layth who ruled in the Khorasan-Kabul region about the last two decades of the ninth century A.D. Kamaluka is said to have been raised to the throne about 902-04 A.D. by a Kashmirian general. Kalhaṇa also speaks of the Śāhi kings Bhīma and Thakkana. The former, who was the maternal grandfather of Diddā, queen of the Kashmirian king Kṣemagupta (950-58 A.D.), built a temple of Viṣṇu in Kashmir during Kṣemagupta's rule. A general of the Kashmirian king Abhimanyu (958-72 A.D.) is said to have defeated the Śāhi king Thakkana who may have been the grandfather of Jayapāla. The latest Śāhi king mentioned by Kalhaṇa is Trilocanapāla (1013-21 A.D.) whose struggle with the Hammīra, *i.e.* Sulṭān Mahmūd of Ghaznī, is also described. Then, after having referred to the final collapse of Śāhi rule in another context, the Kashmirian author says that even in his days, *i. e.* in the middle of the 12th century A.D., 'the appellation *Śāhi* throws its lustre on a numberless host of Kṣatriyas abroad, who trace their origin to that royal family'.[1]

It will be seen that the Kashmirians, who knew the Śāhis from before the middle of the eighth century down at least to the twelfth, regarded them as Kṣatriyas, although Al-Bīrūnī refers to the Hindu Śāhis of Turko-Tibetan origin and their successors of Brāhmaṇa origin. That the early Śāhis were regarded, in spite of their foreign origin, as Kṣatriyas in India is also indicated

महासाधनभागश्चेत्येता यैरभिधाः श्रिताः ।
शाहिमुख्या येष्वभवन्नध्यक्षाः पृथिवीभुजः ॥

1. *Ibid.*, VIII. 3230. Here *Kṣatriya* may mean *Brahma-Kṣatriya*.

by another evidence. In the second quarter of the 7th century A.D., when the Chinese pilgrim Hiuen-tsang was passing through Uttarāpatha, Udakahāṇḍa[1] or Udabhāṇḍapura was the place of residence or a secondary capital of the emperor of Kāpiśa which then dominated over ten neighbouring States and comprised Lampāka (Laghman), Nagara or Nagarahāra (Jalalabad), Gandhāra and Varṇa (Bannu) and probably also Jāguḍa (Southern Afghanistan with Ghaznī as the chief city). About Gandhāra, the pilgrim says that its capital was Puruṣapura ; 'the royal family was extinct and the country was subject to Kāpiśa; the towns and villages were desolate and the inhabitants were very few'.[3] It seems that, under pressure of the Turks from the north and the Arabs from the south and west, the kings of Kāpiśa left their western possessions in the hands of viceroys,[4] and made Udabhāṇḍa their principal seat of residence. The reason why Udabhāṇḍapura was selected in preference to the older capital Peshawar is at present unknown. But it is possible that the new city was built by the Kāpiśa kings for strategic reasons.

The facts that Kalhaṇa speaks of the Śāhis with reference to the period earlier than that of Lalitāditya (*c.* 730-66 A.D.) and of Udabhāṇḍa as the capital of the Śāhis at least from the time of Lalliya (*c.* 875-90 A.D.) and that Chinese evidence refers to the city as the residence of the emperor of Kāpiśa about 645 A.D. would suggest that Hiuen-tsang's king of Kāpiśa was a Śāhi ruler. It is interesting to note that this king has been described by Hiuen-tsang as a Kṣatriya.[5]

1. This seems to be the Indian form that was at the root of the Chinese *Wu-to-kia-han-cha*. Cf. Watters, *On Yuan Chwang's Travels in India*, Vol. I, p. 221.

2. Cf. Ray, *Dynastic History of Northern India*, Vol. I, pp. 60-61.

3. Watters, *op. cit.*, p. 199.

4. These viceroys appear to have been mentioned as the Satraps of Zaranj, capital of Seistan, and as *Rantbil* or *Rutbil* (probably Sanskrit *Prānta-pāla*) or *Zunbil* (probably Sanskrit *Janapāla*) ruling over Southern Afghanistan. Whether the *Kabul-Shāh* was a viceroy or 'the Śāhi of Kabul' indicating the paramount ruler cannot be determined. Cf. Ray, *op. cit.*, pp. 165 ff.

5. Watters, *op. cit.*, p. 123.

Chapter XXIII

TARKĀRI, ŚRĀVASTI AND ḌHĀKĀ

I

The Koḷagallu (967 A.D.) and Kuḍatini (971 A.D.) inscriptions were published respectively in the *Epigraphia Indica*, Vol. XXI, pp. 260 ff., and *South Indian Inscriptions*, Vol. IX, Part i, p. 43, No. 70. Recently a scholar has quoted two stanzas (the first occurring in the Kuḍatini epigraph and the second in both the Koḷagallu and Kuḍatini records) in an interesting note on Tarkāri published in the *Indian Historical Quarterly*, Vol. XXXV, No. 3, September 1959, pp. 271-73.

Both the inscriptions record the installation of the god Skanda-Kārttikeya by Gadādhara described as a good ascetic (*sutapasvin*) apparently belonging to a Brāhmaṇa family of the Sāṇḍilya *gotra*, who was born at Taḍā-grāma and used to sit only on an iron seat (*loh-āsanin*), and as the crest-jewel of the Gauḍa country or people and the illuminator of the Varendrī country.[1] Thus Gadādhara was an inhabitant of Taḍā-grāma (identified with a village near Dinajpur in North Bengal) situated in Varendrī (parts of North Bengal) forming a portion of Gauḍa (western and north-western areas of Bengal). This reminds us of Kullūka's commentary on the *Manusmṛti*, describing the commentator who was originally an inhabitant of a locality in Varendrī within Gauḍa.[2]

The person who composed the Koḷagallu and Kuḍatini records of Gadādhara was another man of a family hailing from

1. *Ep. Ind.*, Vol. XXI, p. 264.
लोहासनी च सांडिल्यस्सुतपस्वी गदाधरः ।
विद्वज्जनाश्रयो विद्वान् गौडचूडामणिर्गुणी ।।
स्वर्गवासनिमित्तार्थं तडाग्रामोद्भवेन तु ।
स्थापितो दिव्यमन्त्रेण वरेंद्युद्द्योतकारिणा ।।

2. गौड़े नंदनवासिनाम्नि सुजनैर्वन्द्ये वरेंद्र्यां कुले ।
श्रीमद्भट्टदिवाकरस्य तनयः कुल्लूकभट्टोऽभवत् ।।

Varendrī. The first of the two stanzas in the description of this person, which occurs only in the Kuḍatini inscription, runs as follows :

गंगापूते सदारम्ये वरेंद्रीपुण्यमंडले ।
दुर्गमोत्तरदिग्भागे ग्रामः पाहुनियोंजने ॥[1]

"There is a village in [the area known as] Pāhuniyojana in the inaccessible northern region in the sacred territory of Varendrī which is purified by the Ganges and is always pleasing." The district of Pāhuniyojana, in which the village in question was situated, thus lay in the northern part of Varendrī. The description of the said part of the territory as *durgama*, i.e. difficult to reach or traverse, reminds us of the fact that the Dāmodarpur (Phulbari P.S., old Dinajpur District) inscription of 542 A. D. describes certain areas of the Koṭivarṣa *viṣaya* (Dinajpur region) in the Puṇḍravardhana *bhukti* (*i. e.* Varendrī) as an *araṇya* or forest territory.[2] As the lands mentioned in the Dāmodarpur plates appear to have been situated in the neighbourhood of Vāyigrāma,[3] *i.e.*, modern Baigram near Hili in the Bogra District, the wild tract seems to have covered the present Hili-Balurghat region of North Bengal, now partly in West Bengal and partly in East Pakistan.

The said verse is followed in the Kuḍatini inscription by another stanza which also occurs in the Koḷagallu record. The text of this verse runs as follows :

श्रीकर्मारिकुलाह्वयद्विजवरग्रामस्तु तर्कारितो[4]
निष्क्रम्य क्रमनिर्मलस्समभवत् तस्मिनृषिपंडितः ।
तत्पुत्रः प्रथितः क्षितावतिबल[5]स्तस्मादभूद्यस्सुतः
तेन श्रीमधुसूदनाख्यकविना शस्ता प्रशस्तिः कृता ॥

The interpretation of the stanza as published in the *Epigraphia Indica* and quoted in the *Indian Historical Quarterly* is defective. As they stand, the first and second sentences in the

1. *SII*, Vol. IX. Part i, p. 43.
2. *Select Inscriptions*, p. 336; above, p. 281.
3. *Sel. Ins.*, p. 325; above, p. 276.
4. This is the reading of the Koḷagallu inscription. The printed text of the Kuḍatini inscription has *grāmasya Tatkārito*.
5. The rest of the stanza s not found in the published text of the Kuḍatini inscription.

first half of the stanza mean to say that the village in Pāhuni-yojana, referred to in the stanza quoted previously, which was the habitation of learned Brāhmaṇas, bore the name Śrī-Karmāra-kula (literally 'the illustrious blacksmiths' family or residence') and, after having separated itself from Tarkāri, became pure in course of time and that the learned Ṛṣi was born in the said village. 'Becoming pure in the course of time' would then mean that the stigma of the association with blacksmiths was gradually removed by the settlement of the learned Brāhmaṇas. In that case, the person in question would appear to have been a Brāhmaṇa. But the mention of this person without the name of his *gotra* renders it doubtful that he was a Brāhmaṇa. Moreover, the above interpretation of the verse involves the unnecessary repetition of the word *grāma* in the second stanza. Considering the number of errors in the engraved text, it is therefore not impossible to think that the first foot of the stanza has to be read as *Śrī-Karmāra-kul-āhvayo dvijavara-grāmāt = tu Tarkārito*. In this case, the subject of the verb *samabhavat* (became) in the passage *Tarkārito niṣkramya krama-nirmalas = samabhavat* (became gradually pure after having come out of Tarkāri) would be the family of the blacksmiths and not the village as suggested by the text as it is. If this is accepted, the word *vaṁśaḥ* has to be understood with *śrī-Karmāra-kul-āhvayaḥ* and the passage would then mean '[a family] known as the community of blacksmiths'.[1]

1. If such is the case, the composition of the *praśasti* by a person of the blacksmith community of Bengal would be an interesting fact. In this connection, it has to be noted that the poet's family is stated to have become pure gradually after having left their original home in a village dominated by Brāhmaṇas. This may suggest that, after settling elsewhere, the members of the family gave up their hereditary profession. Although it was generally the Brāhmaṇas who acquired proficiency in the Sanskrit language and received honour at the royal courts for their learning, a large number of Sanskrit *praśastis* are known to have been composed by members of the non-Brāhmaṇa communities, especially the Kāyasthas. In Bengal, the Vaidya or physician community was always famous for their Sanskrit learning and produced great poets and Sanskritists like the celebrated Umāpatidhara and Bharatamallika, and there are also some *praśastis* composed by poets belonging to this community. During the rule of the Buddhist kings of East India such as those of the Pāla dynasty, Sanskrit learning does not appear to have been confined to the Brāhmaṇas.

See *Ep. Ind.*, Vol. XXXIII, pp. 337, 339, 343; R. D. Banerji, *The Pālas of Bengal* (Mem. A.S.B., Vol. V, No. 3), pp. 78, 82.

The Silimpur inscription[1] describes the village called Bālagrāma, which was situated in the Puṇḍra country (*i.e.* Puṇḍravardhana-bhukti) and was an ornament of Varendrī, as an offshoot (*prasūta*) of Tarkāri, which was attached to Śrāvasti, and as having the Sakaṭī [river] intervening between Tarkāri and Bālagrāma, while the Brāhmaṇa village called Vaigrāma in Sāvathi (Śrāvasti) mentioned in the Guakuchi plate[2] of king Indrapāla of Prāgjyotiṣa, is apparently the modern Baigram near Hili. There is a controversy among scholars some of whom are inclined to locate the places in U. P. while others assign them to North Bengal.[3] It is however now clear that the Hili-Balurghat region is called Śrāvasti in these records and Pāhuniyojana in the Kuḍatini inscription. It appears that a large number of Brāhmaṇas of Śrāvasti in the ancient Kośala country in Madhyadeśa (*i.e.* the Set-Mahet region in the Gonda and Bahraich Districts of U. P.), especially of Tarkāri in that region, were settled in the Hili-Balurghat area in North Bengal and that these Brāhmaṇas named their new settlements after their old habitations in U. P.[4] It may be that Śrāvasti was the new name of what was originally called Pāhuniyojana.

It may be pointed out in this connection that Tarkāri, the home of the Brāhmaṇa donee, is located in some records in Madhyadeśa[5] while, according to the evidence of the Dāmodarpur inscriptions there was enough state land of the uncultivated class for disposal in the Hili-Balurghat region.[6] Another fact to which attention may be drawn is that the Brāhmaṇas of a village named Kroḍāñca, Kroḍañji, Kroḍāñja, Kolañca, Kolāñca, etc., apparently situated in U. P. and very probably in the Śrāvasti area, were highly respected by the Brāhmaṇas of Mithila in the early medieval period.[7] It is not possible to

1. *Ep. Ind.*, Vol. XIII, p. 290.
2. P. N. Bhattacharya, *Kāmarūpaśāsanāvali*, p. 137.
3. See *History of Bengal*, Vol. I, ed. Majumdar, p. 579, note 1.
4. We may refer in this connection to Pāṭaliputtiram in the South Arcot District (*ARE*, p. 1918-19, p. 25), which was apparently so named by some people settling there from Pāṭaliputra (near Patna), called a *Bhaṭṭ-āgrahāra* in early medieval epigraphs (*Ep. Ind.*, Vol. XXXII, pp. 119, 126).
5. Cf. *Ind. Ant.*, Vol. XVII, p. 121.
6. Cf., e. g., *Select Inscriptions*, p. 284, note 10, etc.
7. Cf. *Ep. Ind.*, Vol. XXIX, pp. 52-53.

believe that Maithila Brāhmaṇas could have been so much respectful towards the Brāhmaṇas of North Bengal. Moreover, according to certain traditions, the ancestors of the Kulīna Brāhmaṇas of Bengal hailed from the said locality. The same village called Kroḍāñja is stated to have been situated in Śrāvasti in the Śubhaṅkarapāṭaka grant[1] of king Dharmapāla of Prāgjyotiṣa and it is very probable that this Śrāvasti is identical with the Hili-Balurghat region of North Bengal. Thus several places in the said region appear to have been named after some celebrated Brāhmaṇa villages in the Śrāvasti area of U. P.

II

Dacca is the Anglicised form of the name of the capital of East Pakistan (East Bengal). It is written in Bengali as *Ḍhākā* (cf. the same word literally meaning 'covered'). The principal deity worshipped there is called Ḍhākeśvarī. The name Dacca is also applied to a District and a Division of the State. The real meaning of the name *Ḍhākā* (Dacca) has not been satisfactorily determined, although 'Dacca Muslin' was famous in the medieval world and *Ḍhākā* (apparently derived from the name in question) was the name applied to all muslins, imported through Kābul, throughout Central Asia. Prior to 1801, the annual advances for the Dacca muslin are said to have amounted to £250,000.

According to the *Hobson-Jobson* by Henry Yule and A. C. Burnell, the name is derived from the tree called Ḍhāk or Palāś, and means 'the wood of Ḍhāk trees'.[2] But nobody has taken the explanation seriously. According to some writers, *Dacca* is a corruption of the name of the ancient state of Ḍavāka mentioned in the Allahabad pillar inscription of Samudragupta as a *pratyanta, i. e.* a state bordering on the Gupta empire, about the middle of the fourth century A.D.[3] Unfortunately the theory was always viewed at with a considerable amount of doubt and recent writers on the subject believe that the ancient kingdom of Ḍavāka lay actually about the Ḍabokā region in the valley of the Kapilī river running through the Nowgong District of Assam.[4]

1. P. N. Bhattacharya, *op. cit.*, p. 155 and corrigenda.
2. Ed. W. Crooke, London, 1903, p. 290.
3. *JPASB*, 1910, p. 144.
4. Cf. *Select Inscriptions*, p. 258, note 1.

It is usually believed that the fame and prosperity of Dacca are not older than the days of the Great Mughuls. We know that the city is not traced in any record of the pre-Muslim period of Indian history. For a few centuries before and after the Muslim conquest of West Bengal in the beginning of the thirteenth century, the celebrated city of Vikramapura was the administrative headquarters of East Bengal. There is difference of opinion as regards the location of this city; but some authors appear to be right in holding that it was washed away in the early medieval period by the waters of the Padmā whose erosive activities in the neighbourhood of Dacca have earned for her the name *Kīrtināśā*, literally meaning 'destroyer of fame (*i.e.* the fame-producing works of men)'. Nothing is known about the existence of the city of Dacca even long after the expansion of Muslim rule over East Bengal. During this period the neighbouring city of Suvarṇagrāma or Sonārgāon became the political centre of East Bengal. In the early years of the seventeenth century, during the reign of the Mughul emperor Jahāngīr, Sheikh 'Alāuddīn Islām Khān (1608-13 A.D.) was appointed governor of the Subah of Bengal. Islām Khān transferred the provincial capital from Rājmahal to Dacca where he built a brick fort and a palace. It has been said that Islām Khān's desire to subdue the Portuguese and Arakanese pirates, who were ravaging South-East Bengal about that time, was the main cause of the transference of the provincial headquarters to Dacca. The new capital of the province was styled Jahāngīrnagar after the reigning Mughul emperor. The fame and prosperity of Dacca began from this time.

Although Dacca thus seems to have become a great city only in the early years of the seventeenth century, there is reason to believe that it enjoyed some amount of political importance even in the early Muslim and pre-Muslim periods. This is not only indicated by Islām Khān's choice of the place as the provincial capital,[1] but also by the very name of *Dacca* itself.

Stewart regards Dacca as a modern town since he could not trace its name in Abul Fazl's *Ain-i-Akbarī* which gives an exhaustive account of Akbar's dominions; but H. Blochman point-

1. For some pre-Mughul and even pre-Muhammadan relics in the Dacca area and other interesting observations, see 'An Enquiry into the Origin of Dacca' by N. K. Bhattasali, *JRASB*, Letters, Vol. V, pp. 447-53.

ed out that the Mahall to which it belonged is actually mentioned in the *Ain-i-Akbarī* as Dhakkā Bāzū, although in Gladwin's translation of the work the name is spelt as *Dukha Bazoo*.[1] Blochman further points out that Ḍhākā also occurs in Abul Fazl's *Akbarnāmah* composed in 1584 A.D. and that A. Phayre refers it to 1400 A.D.[2]

The name *Ḍhākā* (Dacca) is apparently a Prakrit corruption of Sanskrit *Ḍhakkā* (or possibly *Ḍhakkaka*). As a matter of fact, this form of the name (together with the variant *Ḍhakka*) is found in a large number of records belonging to the late medieval period. The word *ḍhakka* is no doubt derived from Sanskrit *ḍhakkā*, meaning 'a drum', and it is interesting to note that the celebrated Kashmir chronicle, the *Rājataraṅgiṇī* composed by Kalhaṇa about the middle of the twelfth century, uses *ḍhakka* in the technical sense of a 'drum-station' or 'watch-station'.

The *Rājataraṅgiṇī* (III. 227) has the following verse :

क्रमवर्त्ताभिधाने स प्रदेशे प्राप्तवांस्ततः ।
ढक्कं कांबुवनामानं योऽद्य शूरपुरे स्थितः ।।

"Then he (poet Mātṛgupta) reached, in the province of Kramavarta, the *Ḍhakka* called Kāmbuva, which is at present stationed at Śūrapura." The same *Ḍhakka* is mentioned elsewhere (V. 39) in the following verse :

स्वकृते पत्तनवरे तेन शूरपुराभिधे ।
क्रमवर्त्तप्रदेशस्थो ढक्कोऽभूद्विनिवेशितः ।।

"In the excellent town of Śūrapura, founded by him (*i.e.* Śūra who was a minister of king Avantivarman, 856-83 A.D.), was lodged the *Ḍhakka* of the province of Kramavarta."

In the same section of the *Rājataraṅgiṇī*, we have a third verse (V. 306) that runs as follows :

भ्रष्टश्रीश्चक्रवर्माथ निशि श्रीढक्कवासिनः ।
एकदा डामराग्रयस्य संग्रामस्याविशद् गृहम् ।।

"Then on one occasion Cakravarman (king of Kashmir, 923-37 A.D.), stripped of splendour, entered at night in the house of Saṁgrāma who was the chief of the *ḍāmaras* (landlords) and was living at the beautiful *Ḍhakka*."

1. See *JASB*, 1873, Part I, p. 216, note; Blochman's *Ain-i-Akbari*, text, p. 407.

2. Cf. *JASB*, *op. cit.*, p. 53.

Scholars have suggested that there was in ancient Kashmir a famous watch-station at the village of Śūrapura (modern Hurpor). Drums must have been at the watch-station and were sounded to announce to the people emergencies like the advent of enemies or royal proclamations. It is therefore clear that 'the *Ḍhakka* called Kāmbuva' really means 'the watch-station that was situated at the locality called Kāmbuva'. This *Ḍhakka*, originally stationed at Kāmbuva in the province of Kramavarta, was transferred in the ninth century to the village of Śūrapura in the same province. The expression *Kramavarta-pradeśa-stha* seems to suggest that different *pradeśas* of the ancient kingdom of Kashmir were endowed each with a *Ḍhakka*.

It seems to us that *Ḍhākā* (Dacca) was a similar watch-station during the days of the pre-Muslim rulers of East Bengal and thus had some political importance even in the early period. This suggestion appears to be supported by a passage in the *Prākṛtānuśāsana* by Puruṣottama who is said to have adorned the court of king Lakṣmaṇasena (*circa* 1179-1206 A.D.) of Bengal. The only manuscript of this work so far discovered is known to have been copied in the year 385 of the Newarī era of Nepal, which started from 879 A.D. The date of the copy is therefore 1264 A.D. The *Prākṛtānuśāsana* describes several Prakrit dialects called Vibhāṣā, one of which is given the name *Ḍhakka-bhāṣā*, *i.e.* the dialect spoken in the region known as Ḍhakka. It may, however, be argued that Ḍhakka in this case does not really refer to Dacca in Eastern Bengal, but is a corruption of the name of another land called Ṭakka which was situated in the Punjab. This view can hardly be accepted since the *Ṭakkadeśiya-vibhāṣā*, *i.e.* the dialect spoken in the Ṭakka country, has been separately mentioned in Puruṣottama's *Prākṛtānuśāsana*.[1] There are a few places called *Ḍhakka*, the most famous among them being the one, now called *Ḍhākā* (Dacca) in Eastern Bengal (East Pakistan).[2]

1. See Sircar, *A Grammar of the Prakrit Language*, Calcutta, 1943, pp. 114, 118.

2. Cf. Dhakka on the Kabul near the eastern border of Afghanistan; Dhaka in P. S. Tilhar, Shahjahanpur District, U.P.; Dhaka on the northern border of the Champaran District, Bihar; and Jal-Dhaka in the Nilphamari Sub-Division of the Rangpur District, East Pakistan. See *Journ. As. Soc, Pak.*, Vol. III, pp. 199 ff.

It may also be argued that Dacca could have hardly enjoyed so much cultural influence in the twelfth and thirteenth centuries, when the political centre of the country was not at that place, but at the neighbouring city of Vikramapura (and later at Suvarṇagrāma in the same neighbourhood) to stamp its name on the dialect spoken in the locality. In our opinion, however, such a possibility is not altogether out of question.

Chapter XXIV

KĀLAPRIYA AND BRAHMAŚILĀ

There is an interesting passage in Rājaśekhara's *Kāvyamīmāṁsā*[1] which refers to the Antarvedī tract bounded by the Ganges in the north, the Jamuna in the south, Vinaśana (the same as Kurukṣetra according to the *Trikāṇḍaśeṣa*[2]) in the west and Prayāga (Allahabad) in the east. Thereafter the passage runs—

तदपेक्षया दिशो विभजेत इत्याचार्याः । तत्रापि महोदयं मूलमधिकृत्य इति यायावरीयः । अनियतत्वाद्दिशामनिश्चितो दिग्विभाग इत्येके । तथा हि यो वामनस्वामिनः पूर्वः स ब्रह्मशिलायाः पश्चिमः यो गाधिपुरस्य दक्षिणः स कालप्रियस्योत्तर इति । अवधिनिबन्धनमिदं रूपमिरस्त्वनियतमेव इति यायावरीयः ।
The reference to the four localities—Vāmanasvāmin in the west, Brahmaśilā in the east, Gādhipura in the north and Kālapriya in the south—is very interesting to the student of the historical geography of ancient India, especially when we know that the author of the *Kāvyamīmāṁsā* lived for long at Kanauj, the capital of his Pratīhāra patrons. In the notes appended to the G.O.S. edition of the above work (pp. 243-44), it has been suggested that the said four localities were situated respectively in the western, eastern, northern and southern suburbs of the city of Kanauj.

It has been pointed out that, according to the *Padma Purāṇa*,[3] Rāma built a temple for the god Vāmana at Mahodaya, *i.e.* Kanauj. The author of the notes, referred to above, conjectures that the temple of Vāmana was probably situated at the western end of the city. He admits that Rājaśekhara's own *Bāla-Rāmāyaṇa* (X. 88) applies the name Gādhipura to Kanauj itself, and indeed this identification is supported by a number of authori-

1. G.O.S. edition, 1934, p. 94.

2. III. 14—कुरुक्षेत्रं विनशनम्.

3. Dey, *Geographical Dictionary*, p. 87; *Padma Purāṇa*, Sṛṣṭi-khaṛḍa, Chapter 35 (Baṅgabāsī edition, Ch. 38, 186-87); Uttara-khaṇḍa, Ch. 53.

ties including Hemacandra.[1] Still, however, he concludes that Gādhipura was a locality in the northern suburbs of Kanauj, while Kālapriya was at its southern end. It has been rightly pointed out that the dramas of Bhavabhūti, who lived at the court of king Yaśovarman of Kanauj (*c.* 725-53 A. D.), were staged before the audience assembled in connection with the festivities held in honour of the god Kālapriyanātha, the presiding deity of the locality called Kālapriya. The views of the commentators on Bhavabhūti's works identifying Kālapriyanātha with the god Mahākāleśvara of Ujjain or with the presiding deity of Padmapura, Bhavabhūti's birth place, have, again, been rightly rejected. It has then been suggested that Kālapriyanātha was probably the presiding deity of the city of Kanauj which was the capital of Bhavabhūti's patron and that possibly the god's temple was situated in the southern suburbs of the city. The position of Brahmaśilā at the eastern end of Kanauj has been conjecturally determined on the strength of the identification of the other three localities.

It will be seen that the location of the four places in the suburbs of Kanauj is actually based on mere conjecture. It must be admitted that Gādhipura was either another name of Kanauj or at least the name of a part of the city. We should therefore search for a locality called Kālapriya to the south of Kanauj. According to Bhavabhūti's works, the poet's ancestral home was at Padmapura which lay in the Deccan, *i.e.* to the south with reference to the shrine of the god Kālapriyanātha. It is strange that some scholars have identified this Padmapura with Padam Pawaya (ancient Padmāvatī) near Narwar (ancient Nalapura) in the former Gwalior State.[2] The prelude to Bhavabhūti's *Vīracarita* places Padmapura in Dakṣiṇāpatha or the Deccan, while the *Mālatīmādhava*, with more definiteness, locates it in Vidarbha (modern Berar) in Dakṣiṇāpatha. The identification of Bhavabhūti's birthplace with modern Padampur in the Bhandara District, near the Amgaon Railway Station, seems to be reasonable. The village of Padampur, it may be pointed out, lies directly to the south of Kanauj. It is very

1. *Abhidhānacintāmaṇi*, Bhūmi-kāṇḍa, vv. 39-40—कान्यकुब्जं महोदयम् । कन्यकुब्जं गाधिपुरं कौशं कुशस्थलं च तत् ॥ See Sircar, *Com. Geog.*, pp. 104, 107.

2. Cf. Tripathi, *History of Kanauj*, p. 209.

interesting to note that the only Kālapriya known to history and situated, like both Kanauj and Padampur, to the west of long. 80, appears to be modern Kālpī on the Jamuna in the Jalaon District of U. P., about 75 miles directly to the south of Kanauj. It was an important station between the Ganges-Jamuna Doab and the South not only during the Muslim period, but also in the early medieval age. In the first quarter of the tenth century A.D., when Indra III (915-28 A. D.),[1] the Rāṣṭrakūṭa king of the Deccan, was advancing against Kanauj, the capital of his Pratīhāra enemies, the Rāṣṭrakūṭa army was for a time encamped at Kālpī where it crossed the Jamuna. The Cambay plates of Govinda IV give the story in the following verse :

यन्माद्यद्द्विपदन्तघातविषमं कालप्रियप्राङ्गणं
तीर्णा यत्तुरगैरगाधयमुना सिन्धुप्रतिस्पर्धिनी ।
येनेदं हि महोदयारिनगरं निर्मूलमुन्मूलितं
नाम्नाद्यापि कुशस्थलमिति जनैः ख्यातिं परां नीयते ॥[2]

1. The date of his northern expedition is usually supposed to be 916 A. D. as Indra III is believed to have died in 917 A.D. Both the dates are, however, wrong since the Rāṣṭrakūṭa king is known to have ruled till 928 A.D. Cf. *Ep. Ind.*, Vol. XXXII, pp. 49-50.

2. Cf. Altekar, *The Rāṣṭrakūṭas and their Times*, p. 102. There is a pun on the word *Kuśasthala* which indicates the city of Kanauj as well as a field covered with *kuśa* grass. According to the much exaggerated claim put forward in the verse, Indra III totally destroyed the city of Kanauj which from that time became a field of *kuśa* grass. There is however no reason to believe that the Rāṣṭrakūṭas were successful in razing Kanauj to the ground or to paralyse Pratīhāra power in the Doab even for a short period. That Malwa was occupied by Indra III is concluded from the wrong identification of Kālapriyanātha with Mahākāleśvara of Ujjain. For Pratīhāra possession of Malwa, cf. the Partabgarh inscriptions of Mahendrapāla II dated 946 A. D. The verse—

यस्य परुषेक्षिताखिलदक्षिणदिग्दुर्ग्गविजयमाकर्ण्य ।
गलिता गूर्जरहृदयात् कालंजरचित्रकूटाशा ॥

in the Deoli (940 A. D.) and Karhad (959 A. D.) plates of Kṛṣṇa III does not imply that the Pratīhāra fortresses of Kālañjara and Citrakūṭa (Chitor) were captured by the Rāṣṭrakūṭas who, however, merely threatened them. The claim of Govinda IV having been served by the Ganges and the Jamuna does not, again, prove his mastery over the Doab, but simply refers to the decorative representation of Gaṅgā and Yamunā at his palace gate. The *praśasti* of the Cālukya chief Narasiṁha, probably a feudatory of Indra III, to have defeated Pratīhāra Mahīpāla I and to have bathed his horses

It appears that the Rāṣṭrakūṭa king Kṛṣṇa III (939-67 A.D.) also encamped at Kālapriya in the course of his Bundelkhand expedition. This is suggested by the fact that this ruler developed a fondness for installing gods under the name Kālapriya in different parts of his empire. A god of this name was installed by him in the Kāñcī region near Madras and another at Kandhār in the Nanded District of Maharashtra.[1] The *Varāha Purāṇa* speaks of a temple of the Sun-god at Kālapriya which is located to the south of the Yamunā.[2]

The identity of Kālapriya with Kālpī is proved by epigraphic evidence. In the Khaḍāvadā inscription (1484 A.D.) of king Ghiyās Khaljī of Maṇḍū, Husaṁga Gorī (Hushang Alp Khān Ghūrī) is represented as having defeated Kādira Sāhi (Abdul Qādir), ruler of Kālapriya-pattana, and as having made the latter's son, Salaha, a Khān at Māṇḍū. This Kālapriya-pattana is apparently different from Ujjain and Kanauj and is no doubt the same as Kālpī. The name Kālapriya applied to Kālpī can thus be traced to a date as late as the fifteenth century A.D.[3] Kālpī must have formed a part of the Kanauj kingdom under Yaśovarman. The annual fair and festivities held in honour of Kālapriyanātha were no doubt the most famous in the whole kingdom. It was, therefore, not at all unnatural that Bhavabhūti's dramas were staged on such occasions at Kālpī.

If the *Padma Purāṇa* is to be believed, the temple of Vāmanasvāmin was situated somewhere in the city of Kanauj. We should, therefore, search for a locality called Brahmaśilā to the east of Kanauj. This Brahmaśilā seems to be no other than Barhamshil mentioned by Al-Bīrūnī who says, "A man marching from Kanoj to the south between the two rivers Jaun and

at the junction of the *Gaṅgā-vārdhiayıḷ* as found in the *Pampa-Bhārata* appears to be another exaggeration. The decline of the Gurjara Pratīhāras had nothing especially to do with the northern expedition of Indra III. It was hastened by constant warfare on all fronts and especially by the protracted struggle for the throne (after the death of Mahīpāla I) about the middle of the tenth century A. D. Cf. Ray, *DHNI*, Vol. I, pp. 580-90.

1. *Ep. Ind.*, Vol. XXXV, pp. 109-10. See below, pp. 311-12.

2. Mirashi, *Stud. Indol.*, Vol. I, p. 38.

3. See *JBBRAS*, Vol. XXIII, pp. 12 ff. ; Bhandarkar's List of Inscriptions, No. 859.

Ganges passes the following well-known places :—Jajjamau, 12 *farsakh* from Kanoj, each *farsakh* being equal to four miles or one *Kuroh*; Abhāpurī, 8 *farsakh*; Kuraha, 8 *farsakh*; Barhamshil,8 *farsakh*; the Tree of Prayāga, 12 *farsakh*, the place where the waters of the Jaun join the Ganges, where the Hindus torment themselves with various kinds of tortures which are described in the books about religious sects."[1] As regards the measure of distance, Al-Bīrūnī says that the *krośa* is equal to 'one mile', *yojana* is equal to '8 miles or to 32,000 yards', and '1 *kuroh* = $\frac{1}{8}$ *yojana*.'[2] One *farsakh* was therefore equal to 4 Arabic miles and to $3\frac{771}{1098}$ English miles.[3]

Al-Bīrūnī thus gives the distance from Kanauj to Jaijamau as about 45 miles; from Jajjamau to Abhāpurī as about 30 miles; from Abhāpurī to Kuraha as about 30 miles; from Kuraha to Brahmaśilā as about 30 miles; and from Brahmaśilā to Prayāga or Allahābad as about 45 miles. Brahmaśilā thus lay about 45 miles to the west or north-west of Allahābād and was situated in the Doab between Kanauj and Allahābād. It was therefore to the east or south-east of Kanauj.

1. Sachau, *Alberuni's India*, Vol. I, p. 200; Sircar, *Cosm. Geog. E. Ind. Lit.*, p. 155.
2. Sachau, *op. cit.*, p. 167.
3. Cf. *ibid.*, Vol. II, p. 316.

Chapter XXV

KĀKANDĪNAGARĪ AND KANDHARĀPURA

I

A place called Kākandī or Kākandīnagarī is famous in both Jain and Buddhist traditions. The Jains regarded the locality as the birth place of the Tīrthaṅkara Suvidhinātha,[1] while the Buddhists regarded it as the home of an ancient sage named Kākanda.[2] The locality is also known from early epigraphic records.[3] But it does not appear to have so far been satisfactorily identified.

B. C. Bhattacharya suggested its indentification with the city of Kiṣkindhā celebrated in the story of the *Rāmāyaṇa*.[4] But the equation of *Kākandī* and *Kiṣkindhā* appears to be philologically unsound. Moreover, Kiṣkindhā in the neighbourhood of Pampā (modern Hampi in the Bellary District of Mysore State) is far away from the sphere of activities of the early Buddhists and Jains. B. C. Law, who has ignored Bhattacharya's suggestion, states that the location of Kākandī of the Jain and Buddhist literature is unknown.[5] But there is epigraphical evidence to prove that Kākandī, the traditional birth place of Suvidhinātha, was regarded in the medieval period as identical with a place now called Kākan which lies within the jurisdiction of the Sekandra Police Station in the Jamui Sub-Division of the Monghyr District of Bihar.

About the beginning of the year 1951, I copied three epigraphs in the Jain temple at Kākan.[6] The earliest of the three records is engraved on the pedestal of an image of Pārśvanātha and is dated in Vikrama 1504, Phālguna-sudi 9 (February, 1448 A.D.) while the latest is incised on the back of an *āyāgapaṭa* and is dated in Vikrama 1933 (1876-77 A.D.). The third inscrip-

1. Cf. B. C. Bhattacharya, *The Jaina Iconography*, pp. 64-65.
2. G. P. Malasekera, *Dictionary of Pali Proper Names*, Vol. I, p. 558.
3. See Barua and Sinha, *Barhut Inscriptions*, p. 18.
4. *Loc. cit.*
5. *Historical Geography of Ancient India*, p. 91, s. v. Law locates Kākandī in Northern India, though it was actually situated in Eastern India.
6. See *Annual Report on Indian Epigraphy*, 1950-51, Nos. B 2-4.

tion, dated Vikrama 1822, Vaiśākha-sudi 6 (April, 1765 A.D.), is engraved around two foot-marks fixed in front of the image of Pārśvanātha and records the installation of the said foot-marks. It clearly states that they represent the foot-marks of the Tīrthaṅkara Suvidhinātha and that they were installed at Kākandī or Kākandīnagarī which was a holy place and was the birth place of the said Tīrthaṅkara. Some repairs are also stated to have been carried out apparently in the temple wherein the foot-marks were installed, and one of the inscriptions seems to point to the existence of the temple before the middle of the fifteenth century A.D. Thus the tradition that modern Kākan, where the inscription has been found, is the same as Kākandī or Kākandīnagarī, regarded by the Jains as the birth place of the Tīrthaṅkara Suvidhinātha, can be referred at least to the late medieval period.

The text of the inscription runs as follows :

ॐ नमः ॥ संवत(संवत्) १८२२ वर्षे वैशाष(ख)मासे शुक्लपक्षे षष्ठीतिथौ श्रीसुविधिनाथजिनवरव(च)रणकमले शुभे स्थापिते ॥ श्रीकाकंदीनगरीजन्मकल्याणकस्थाने श्रीसंघेन ॥ जीर्णोद्धारं (च*) करापितं(कारितम्)॥१ चिरं नंदतु तीर्थोयं काकंदीनामको वरः (॥*)[1]

II

The Sanskrit name Kṛṣṇa was often modified in South India as Kandara, Kandāra, Kandhara, Kandhāra, Kanhara, Kanhāra, Kannara and Kannāra.[1a] Sometimes the same name is found in the joint form Kṛṣṇa-Kandhara or Kṛṣṇa-Kandhāra.[2] The Prakrit name Kannara was also sometimes re-Sanskritised as Karṇa.[3] The earliest use of such a Prakrit form of the name is to be found in that of Kandara who was the founder of the Ānanda dynasty of Kandarapura in the Guntur District of Andhra Pradesh and flourished in the fourth century A.D.[4]

1. See also P. C. Nahar, *Jainalekhasaṁgraha*, Vol. I, Calcutta, 1918, p. 41, No. 173.

1a. Cf. *Bomb. Gaz.*, Vol. I, Part ii, pp. 244, 334, 410 (note 1), 468, 526, etc.; *Ep. Carn.*, Vol. III, Ml. 19; Vol. VII, Sk. 197 (Intr., p. 27) and 198 (Intr., p. 36); Hl. 17; Vol. VIII, Sa. 119; Vol. X, Bg. 43; Vol. XI, Dg. 13. A similar form of the name was Kāhnura (cf. Ray, *DHNI*, pp. 340, 565) probably under Bengali influence.

2. *Bomb. Gaz.*, *op. cit.*, pp. 419, 508, 556; *JBBRAS*, Vol. X, p. 241.

3. Altekar, *The Rāṣṭrakūṭas and their Times*, p. 12; *Ep. Ind.*, Vol. XXVIII, p. 314, note 4.

4. Sircar, *The Successors of the Sātavāhanas in the Lower Deccan*, pp. 55ff.

The founder of the Raṭṭa dynasty of Saundatti in the Belgaum District is stated to have been raised to the position of a feudatory chieftain by a king named Kṛṣṇa who has been identified with the Rāṣṭrakūṭa emperor Kṛṣṇa III (939-67 A.D.).[1] An inscription of 1218 A. D. represents the said Raṭṭas as the descendants of the same Kṛṣṇa, called Kṛṣṇa-Kandhāra, while another record of 1209 A.D. (?) from Haṇṇikeri near Sampgaon in the Belgaum District of Mysore mentions the same king as Kṛṣṇa-Kandhara and represents him as *Kandhāra-pura-varādhiśvara*, 'the supreme lord of Kandhārpura, the best of cities.'[2]

The Imperial Rāṣṭrakūṭas had their capital at Mānyakheṭa, modern Mālkheḍ in the Gulbarga District of the former Hyderabad State, now in Mysore. But they did not represent themselves as the lord of 'Mānyakheṭa, the best of cities'. Like the Raṭṭas of Saundatti, the Imperial Rāṣṭrakūṭas, were sometimes represented as the lords of the city of Lattalūra, Lattalūr or Lattanūr, which was claimed to have been the original home of the family and has been identified with modern Lātūr in the Osmanabad District of the present Maharashtra State.[3] The representation of Rāṣṭrakūṭa Kṛṣṇa III, who had his capital at Mānyakheṭa,[4] as the lord of Kandhārapura in the Haṇṇikeri inscription, without reference either to his capital or to the original home of the Rāṣṭrakūṭa family, is interesting to note. It appears that Kṛṣṇa III had a secondary capital at Kandhārapura, built by and named after himself or either of his two predecessors who bore the same name.

Fleet was not inclined to attach any importance to the mention of Kṛṣṇa III in the said epigraph as the lord of Kandhārapura. He draws our attention to the fact that it is an isolated instance and says, "I know of no place that can be identified with an ancient Kandhārapura or Kṛṣṇapura. The name may possibly have been invented from an imaginary Kṛṣṇapura derived from some passage similar to that in which the

1. Cf. *Bomb. Gaz.*, *op. cit.*, pp. 550, 556; *Ind. Ant.*, Vol. XXXII, pp. 216ff. (Nos. 3-4); *JBBRAS*, Vol. X, pp. 240 ff.

2. *Bomb. Gaz.*, *loc. cit.*

3. *Ibid.*, pp. 384, 387; *Ind. Ant.*, Vol. XII p. 22 .

4. *Bomb. Gaz.*, *op. cit.*, p. 419.

5. *Ibid.*, p. 384, note 4.

Eastern Cālukya king Guṇaka-Vijayāditya III is said to have effected the burning of the city of Kṛṣṇa II (*Kṛṣṇa-pura-dahana*; see *Ind. Ant.*, Vol. XX, p. 102, note 26)." But, apart from the fact that the epigraphic passage in question really speaks of Kiraṇapura and not of Kṛṣṇapura, this attitude appears to be rather hypercritical. Since the Raṭṭas of Saundatti, who were used to represent themselves as lords of the city of Lattalūra, could have represented their ancestor as *Mānyakheṭa-pura-var-ādhīśvara* if they wanted to avoid *Lattalūra-pura-var-ādhīśvara*, it is difficult to understand why they should have preferred to bring in the name of an imaginary city. In our opinion, the specific mention of Kṛṣṇa III as the lord of Kandhārapura scarcely raises any reasonable doubt about the existence of a city called Kandhārapura apparently built by and named after a Rāṣṭrakūṭa monarch named Kṛṣṇa (Kandhāra).

Fleet's attitude seems to have been influenced by the fact that he had no knowledge of the existence of a city called Kandhāra which could be ascribed to the Rāṣṭrakūṭa period. It therefore appears that he would have modified his opinion on the subject if he had any knowledge of the town of Kandhār (also spelt Qandhār and Kandahār), the headquarters of the Qandharsharif Taluk of the Nander District of the former Hyderabad State, now in Maharashtra, and of the remains therein of numerous sculptures ascribable to the Rāṣṭrakūṭa age. This place, which appears to be none other than the city of Kandhāra of the Hannikeri inscription, lies about twentyfive miles to the south of Nānded and a hundred miles to the north of Mālkhed.

Sometime ago we published a fragmentary Rāṣṭrakūṭa inscription engraved on a broken pillar lying in the locality called Bahādurpur in the suburbs of the town of Kandhār.[1] The inscription is written in the North Indian alphabet of about the tenth century A.D. The extant parts of the record contain only the beginning of the Rāṣṭrakūṭa genealogy down to Kṛṣṇa I (756-75 A.D.) and mention a number of shrines that existed at the place when the inscription was set up about the tenth century, probably during the reign of Kṛṣṇa III. It is interesting to note that two of the deities mentioned in the inscription are

1. *Ep. Ind.*, Vol. XXXV, pp. 105 ff.

Kālapriya and Kṛṣṇeśvara while the same two gods are also mentioned in the Karhad plates of Kṛṣṇa III.[1]

One of the deities worshipped at Kandhār about the tenth century A.D. bore the name Gojjiga-Somanātha. Since Gojjiga is a well-known name of Rāṣṭrakūṭa Govinda IV (929-33 A.D.), the deity may have been named after that monarch. If such was the case, the city of Kandhāra seems to have existed before the days of Kṛṣṇa III.

1. *Ibid.*, Vol. IV, p. 281.

CHAPTER XXVI

KAPILAVĀSTU AND TĀMRARĀṢṬRA

I

According to the Pali sources, Lord Buddha was born at the Lumbinī-vana near 'Kapilavatthu' of the Śākyas, and Lumbinī is stated to have been situated between Kapilavatthu and Devadaha in the Śākya territory.[1] The site of the Buddha's birth place is determined by the inscribed pillar raised by the Maurya emperor Aśoka (c. 272-232 B.C.), now known as the Rummindei pillar standing in the village of Paḍariyā within the Nepalese Tarai. According to the inscription engraved on the pillar, Aśoka raised it on the occasion of his visit on pilgrimage, when he was anointed twenty years, to the spot where the Lord Buddha was born.[2] It is well known that Aśoka instituted a regular *dharma-yātrā* (pilgrimage) to the Buddhist holy places and visited Lumbinī (the Buddha's birth place) and Sambodhi (Bodhgaya, the place where the Lord obtained *bodhi*). These are two of the four greatest holy places of the Buddhists, associated with the career and activities of the Buddha, the two others being Mṛgadāva (Sarnath where he first preached his doctrine) and Kuśīnagara (Kasia where he breathed his last), both of which also Aśoka may have visited, though no trace of the evidence to prove this is now available. There is no doubt that the Śākya capital was situated in the neighbourhood of the site where Aśoka's Rummindei pillar now stands; but its exact location is disputed.

Kapilavatthu is described as a city near the Himālaya and as having been founded by the sons of Okkāka (Ikṣvāku) on the site of the hermitage of the sage Kapila. It is said that,

1. For references, see Malalasekera, *Dictionary of Pali Proper Names*, s. v. *Lumbinī*, *Kapilavatthu*, *Devadaha*, *Kapila*.

2. Sometimes it is wrongly believed that, according to the Aśokan epigraph on the Rummindei pillar, the people in charge of the Lumbinī-vana raised it to commemorate Aśoka's visit to the Buddha's birth place. See Malalasekera, *op. cit.*, Vol. II, p. 784; cf. Sircar, *Inscriptions of Aśoka*, 1967 ed., p. 69.

when the sons of Okkāka went into voluntary exile and were looking for a spot on which they could found a city, they came upon Kapila in his hermitage in the Himavā (Himavat) by the side of a lake. The sage was versed in the science of *bhūmicāla* and knew the qualities of sites. He advised the sons of Okkāka to build a settlement on the site of his hermitage which, he knew, would become the capital of Jambu-dvīpa and the abode of invincible people.[1] It may be that the story is an echo of the well-known legend of Kapila's association with the sons of king Sagara of the Ikṣvāku clan as noticed in the *Rāmāyaṇa*.[2]

In the Sanskrit Buddhist works, the Pali name Kapilavatthu is generally quoted as Kapilavastu which is also called Kapilapura. The city is mentioned as *Kapilasya vastu* in the *Buddhacarita*[3] and as *Kapil-āhvaya-pura* in the *Lalitavistara*.[4] The *Divyāvadāna*[3a] thrice mentions the city as Kapilavastu. It is therefore intelligible why Childers' *Dictionary* gives the Sanskrit form of Pali 'Kapilavatthu' as 'Kapilavastu,' and N. L. Dey's *Geographical Dictionary of Ancient and Medieval India* and B. C Law's *Historical Geography of Ancient India* adopt the same view.[4a]

It seems, however, that the correct Sanskrit form of Pali *Kapilavatthu* is not *Kapilavastu*, but *Kapilavāstu*, the confusion being due to the fact that the Sanskrit words *vastu* and *vāstu* both become *vatthu* in Pali. The Pali word *vatthu* (neuter gender), standing for Sanskrit *vastu*, is explained as "substance, object, thing, matter; occasion, cause; plot or subject, story, narrative", while the meanings of *vatthu* (masculine gender), the Pali equivalent of Sanskrit *vāstu*, are quoted as "a site, building ground, floor.[5]" It is quite obvious that Pali *Kapilavatthu* can only stand for Sanskrit *Kapilavāstu* because it was a city built, according to tradition, on Kapila's *vāstu* or the site of his hermitage.

1. *Ibid.*, Vol. I, pp. 514, 516. It may be noticed that the Śākya capital never attained the status of the capital of Jambu-dvīpa or India.
2. I. 40. 24-30.
3. I. 2.
3a. Ed. Vaidya, pp. 41, 54 and 249.
4. Ed. S. Lefmann, p. 243; cf. pp. 58, 77, 101-02, 113, 123.
4a. See also 'Kapilavastu' (Chap. VIII) in Law's *Geographical Essays*, pp. 182-93.
5. See Childers, *A Dictionary of the Pali Language*, s. v. *vatthu* (n.) and *vatthu* (m.). The word *vāstu* also means "a house, an abode, a dwelling place" according to Sanskrit lexicons.

II

The Buddhist scholar Anuruddha, author of the *Abhidhammatthasaṅgaha*, *Nāmarūpaparicchedа* and *Paramatthaviniccaya* in Pāli and probably also of the *Anuruddhaśataka* in Sanskrit, is well known to the students of Buddhist literature. He is supposed by scholars to have flourished about the 11th century A.D. In the introduction to the *Nāmarūpaparicchedа*, it is stated that Anuruddha was born at the city of Kāverī within Kāñcīpura in Jambu-dvīpa and that he dwelt at times in Tāmrarāṣṭra in Jambu-dvīpa and in the Mūlasōma-vihāra in Laṅkā-dvīpa or Ceylon.[1] In these cases, the name Jambudvīpa has been applied to the Indian subcontinent and does not signify the Purāṇic island continent of which India was but a small part.

The colophon of the *Paramatthaviniccaya*, recently published in the *Journ. As. Soc.*, also states that Anuruddha was born at Kāverī-nagara in the *rāṣṭra* of Kāñcīpura (also spelt *Kāñcivara*, *Kāñjivara*) and that he was living at the city of Tañja (also written *Rāja* and *Gaja*) in Tāmra-rāṣṭra. Of these, the *rāṣṭra* or district of Kāñcīpura, Kāñcivara or Kāñjivara is no doubt the well-known Conjeeveram near Madras and the city of Kāverī may be one of the following localities, viz. (1) Kāveripur in the Coimbatore District, lat. 11° 55′, long. 77° 47′ ; (2) Kāveripaṭam in the Salem District, lat. 12° 25′, long. 78° 16′ ; and (3) Kāveripak in the North Arcot District, lat. 12° 54′, long. 79° 30′. Out of the above three places, Kāveripak in the North Arcot District is not far away from Conjeeveram and may be the place where Anuruddha was born.

The name of the place where Anuruddha resided in Jambu-dvīpa or India is given as the city of Tañja, some of the manuscripts offering *Rāja* or *Gaja* in place of *Tañja* as we have seen. Since there are cases of the less known being superseded by the well known in literary works[3] and since *Rāja* and *Gaja* are well known Sanskrit words which *Tañja* is not, Tañja seems to be the correct name of the city where Anuruddha lived for

1. *JPTS*, 1913-14, pp. 2-3.
2. *JAS*, 4th Series, Vol. VI, 1964, pp. 49-112.
3. Note how the name of the lesser known river Vaṁkṣu (Oxus) was substituted by that of the Sindhu in the *Raghuvaṁśa*, IV. 67.

sometime in Jambu-dvīpa. It is possible to suggest that the said city of Tañja is modern Thanjavur or Tanjore situated on a branch of the Kāverī, about 180 miles to the south-west of Madras. It is interesting to note that, just as the city of Kāverī has been located in the *rāṣṭra* of Kāñcī, the city of Tañja has been placed in the *rāṣṭra* of Tāmra. Although Tāmra-rāṣṭra as the name of an Indian district is unknown from any other source, names like Tāmraparṇa and Tāmraparṇī are well known.

The island of Ceylon is called Tāmra-dvīpa, Tāmraparṇa or Tāmraparṇī in Indian literature and Taprobane (also Palaesimundu = Pārasamudra) by the Greeks. A story of the *Divyāvadāna* seeks to explain the name Siṁhala applied to Tāmra-dvīpa[1] (Ceylon). A merchant named Siṁhala happened to become the king of Siṁhakalpā in Jambu-dvīpa, which had previously been the capital of king Siṁhakesarin, and soon succeeded in freeing Tāmra-dvīpa from the Rākṣasīs. Thus Tāmra-dvīpa, having now become a settlement of king Siṁhala, came to be known after him as Siṁhala. The Pāli chronicles of Ceylon, however, give a different story on the subject.

According to the *Mahāvaṁsa* and *Dīpavaṁsa*,[2] Tāmraparṇī was the name of a particular area of Ceylon, where Vijaya and his 700 companions landed. It is said that, after disembarking from their ship, they sat down at the place with their hands on the ground and found them coloured with red dust. Vijaya built his capital there and soon the whole island (100 or 300 *yojanas* in extent), originally inhabited by the Yakṣas, became known after it as Tāmraparṇī.[3] According to the chronicles, Vijaya and his companions were called 'the Siṁhalas' because Vijaya's father Siṁhabāhu received the name Siṁhala after killing the lion-king (Siṁha). The name Siṁhala-dvīpa was given to Tāmraparṇī after the Siṁhalas,

1. Ed. P. L. Vaidya, pp. 452-57.
2. See Malalasekera, *Dictionary of Pāli Proper Names*, Vol. I, p. 995.
3. The chronicles and the commentaries also apply the names Laṅkā, Laṅkā-dvīpa and Laṅkātala to Ceylon. Laṅkāpura or Laṅkānagara was one of the cities of the Yakṣas before Vijaya's occupation of Laṅkā-dvīpa. Laṅkāgiri-parvata was the name of the mountainous central province of Ceylon.

i.e. Vijaya and his companions, and the Ceylonese people claim to enjoy the same name as their descendants.

It has to be noticed that Tāmra-rāṣṭra, in which the city of Tañja was situated, was a locality in Jambu-dvīpa or India outside Laṅkā-dvīpa or Ceylon. Unfortunately, in South India, which is near Ceylon, the only geographical name associated with the word *tāmra* seems to be that of the river Tāmraparṇī, modern Tāmbaravari, flowing through the Tirunelveli District into the Gulf of Mannar, a second stream of the same name flowing westwards through the old Travancore State.[1] The valleys of these two rivers are, however, considerably away from the Tanjore region. About the 11th century A.D. when Anuruddha is supposed to have flourished, the Tanjore area formed a part of the metropolitan district of the Cola empire having its capital at the city of Gaṅgaikoṇḍa-coḷapuram.

There was another place of the name of Tañjai, otherwise called Tañjākkūr, in the district of Māra-nāḍu near Madurai.[2] It is difficult to be definite whether this Tañjai in the district of Māra has been represented as Tañja in the District of Tāmra in the Pāli work.

1. For Tāmrapura in Kambuja and Tāmrapattana in Arakan (Burma), see Majumdar, *Hindu Colonies in the Far East*, 1944, pp. 181, 205; and for Tāmraliṅga in the Malay Peninsula, see Coedès, *The Indianised States of South east Asia*, p. 39. See also below, pp. 319-20.

2. See K. A. Nilakanta Sastri, *The Coḷas*, 2nd ed., p. 674.

CHAPTER XXVII

IDENTICAL NAMES OF DIFFERENT LOCALITIES

At the close of Chapter XII above (pp. 210ff.), we have referred to several geographical names which are known to be borne by different localities. It has been pointed out that some of the identical names are accidental cases while some of the places were deliberately named after other well-known localities.[1] Attention is drawn here to a large number of cases of the second type associated with the story of the expansion of Indian culture and comparable with the European geographical names applied to places in the New World by the colonists from Europe.

The geographical names in the territories of South-East Asia, where Indian culture spread, often exhibit Indian influence very considerably. Since Sanskrit or Pali became the language of administration and culture in the countries concerned, it is only natural that a large number of places therein would bear Indian names or would at least have Indian and indigenous names side by side. In some cases, the former are merely translated from the latter. The Indian custom of naming a town after a deity or a ruler or its builder was adopted in the said countries. In some cases, as in respect of the names Campā and Kambuja the Indian names appear to have been selected merely because the sound of the indigenous names suggested them. Such cases and those involving deliberate introduction of the names of well-known Indian places no doubt exhibit Indian cultural influence more clearly.

The eagerness of the colonists to import familiar place names in their land of adoption is quite prominent in Burma wherein we have a very large number of well-known Indian geographical names. The name Maurya applied to Mweyin on the upper course of the Irawadi river is supposed to be the

1. Cf. also *Souvenir* of the Tenth Reunion of AIHC Students, Calcutta University, 1967, p. 1; Sircar, *Stud. Soc. Adm. Anc. Med. Ind.*, Vol. I, p. 20 and note 2.

origin of Mareura of Ptolemy's *Geography* (2nd century A.D.), while Śrīkṣetra (Prome) and Haṁsāvatī (Pegu) are believed to be older than the 5th or 6th century A.D.[1]

Some important old Indian names found in Burma are Aparānta, Avanti, Vārāṇasī, Campānagara, Dvāravatī, Gandhāra, Kāmboja, Kailāsa, Kusumapura, Mithilā, Puṣkara, Puṣkarāvatī, Rājagṛha, etc., and the names Sāṅkāsya (Tagaung on the Upper Irawadi), Utkala (from Rangoon to Pegu) and Vaiśālī (modern Vethali in the Akyab District) also fall in the same category.[2] The name of the well-known river Irawadi reminds us of the Irāvatī (modern Ravi), one of the famous tributaries of the Indus. The list may of course be multiplied to any length. The legends of the Buddha and also scenes of subsequent episodes in the history of Buddhism together with the lives of the previous Buddhas and of holy men have often been located in Burma. This kind of deliberate attempt to create a new India is not noticed in the other Indian colonies.[3]

Among Sanskrit-Pali names in Burma, mention may also be made of Golanagara or Golamṛttikānagara (modern Ayetthema, 20 miles north of Thaton, *Gola* being supposed to stand for *Gauḍa*),[4] Kalasapura (to the south-east of Prome near the mouth of the Sittang),[5] Rāmapura (Moulmein), Rāmaṇyadeśa (Lower Burma)[6] and Śrī (Bhamo or Tagaung).[7] The principalities of Puṣkarāvatī, Trihakumbha, Asitāñjana and Ramyanagara were situated in the region of Rangoon, Rāmāvatī and Dhānyavatī (modern Rakhaingmyu) being situated in Arakan.[8] The city of Śrīkṣetra has been located at Hmawza near Prome. Sudharmapura is modern Thaton and the city of Arimardanapura is Pagan situated in the land of Tattadeśa and the kingdom of Tāmra-dvīpa.[9] King Kyanzittha (1084-1112 A.D.) of Arimardanapura built the Ānanda temple of

1. See R. C. Majumdar, *Hindu Colonies in the Far East*, 1944, pp. 215-16.
2. *Ibid.*, p. 216.
3. *Loc. cit.*
4. *Ibid.*, p. 195.
5. *Ibid.*, p. 197.
6. *Ibid.*, p. 196.
7. *Ibid.*, p. 200.
8. See Coedès, *The Indianised States in Southeast Asia*, pp. 329; cf p. 156; R. C. Majumdar, *op. cit.*, p. 202.
9. Cf. R. C. Majumdar, *op. cit.*, p. 207.

Pagan, which is a masterpiece of Burmese architecture designed on Indian models.[1] Jayasiṁha (Zeyatheinkha), who died in 1210 A.D., built the Mahābodhi temple in imitation of the famous temple at Bodhgaya in the Gaya District, Bihar.[2]

The capital of the Candras of Arakan was Vaiśālī, mentioned above; but king Ānandacandra is mentioned as the ruler of Tāmrapaṭṭana.[3]

The name of the country of Kambuja (Cambodia) reminds us of Kamboja or Kāmboja famous in Indian literature.[4] Of the many cities bearing Sanskrit names, in this land, we may mention Tāmrapura, Āḍhyapura, Dhruvapura, Jyeṣṭhapura, Vikramapura, Ugrapura, etc.[5] Of these places, the name Vikramapura was borne by a city of Eastern India, which was the capital of the Candra (10th and 11th centuries A.D.) and other dynasties and was situated in the present Dacca District of East Pakistan.[6] Some of the cities of Kambuja were named after the kings who founded them, e.g., Śreṣṭhapura, Bhavapura, Īśānapura, etc.[7]

Śreṣṭhapura, capital of Kambuja, was built by Śreṣṭhavarman (probably the second ruler of the same name). It was in the vicinity of the Vat Phu Hill near Bassac in Laos, which bore the Sanskrit name Liṅga-parvata.[8] King Īśānavarman, who flourished in the first half of the 7th century A.D., transferred his capital to Īśānapura (named after him) which is identified with modern Sambor-Prei Kuk.[9] Śambhupura (identified with Sambor on the Mekong), Aninditapura and Vyādhapura are some of the kingdoms that flourished in Kambuja in the 8th century A.D.[10]

King Jayavarman II of Kambuja, who flourished in the 9th century A.D., fixed his capital first at Indrapura and then changed it successively to Kuṭi, Hariharālaya and Amarendrapura. For some time, he also fixed his abode on the

1. *Ibid.*, p. 212.
2. *Ibid.*, p. 213.
3. *Ibid.*, p. 204.
4. See Sircar, *Cosm. Geog. E. Ind., Lit.*, p. 74.
5. R. C. Majumdar, *op. cit.*, p. 181.
6. N. G. Majumdar, *Ins. Beng.*, Vol. III, pp. 2, 15, 59, etc.
7. R. C. Majumdar, *loc. cit.*
8. *Ibid.*, p. 162.
9. *Loc. cit.*
10. *Ibid.*, p. 163.

top of the Mahendra-parvata.[1] Indrapura was in the north-eastern part of Kambuja; Kuṭi is modern Bantay Kdei to the east of Ankor Thom and Hariharālaya is modern Roluos, 13 miles to the south-east, and Amarendrapura in Battamang, about 100 miles to the north-west of Ankor Thom. The Mahendra-parvata is the modern Phnom Kulen hill to the north-west of Ankor Thom.[2] King Yaśovarman (acc. 889 A.D.) built Kambupurī which was later called Yaśodharapura.[3] The city of Īśvarapura (where the temple of Tribhuvanamaheśvara was built in 976 A.D.) is modern Banteay Srei. Jayavarman V built Jayendranagarī (about 978 A.D.) and at its centre stood the Hema-giri or Hemaśṛṅga-giri[4] which is the name of Mt. Meru or Sumeru so famous in Indian mythology.

Bhadravarman and other early kings of Campā, ruled from Campāpura over the districts of Amarāvatī, Vijaya and Pāṇḍuraṅga.[5] Among these places, Campāpura and Amarāvatī are names of well-known Indian localities. Campā or Campāpura was the capital of the ancient Aṅga country and is located in the suburbs of Bhagalpur in Bihar.[6] The name Amarāvatī, which is borne by the mythical capital of the gods, was applied to some Indian cities. Vijayapura, Vijayapurī or Vijayanagara were often found among the names of Indian localities.[7] Pāṇḍuraṅga is found in South Indian history as a personal name.[8] After the Kambuja occupation, Campā

1. R. C. Majumdar, *op. cit.*, p. 166.

2. *Loc. cit.*

3. *Ibid.*, p. 170.

4. Coedès, *op. cit.*, p. 117. Meru or Sumeru was conceived as a mountain of gold, while *Hema-giri* means 'a golden mountain' and *Hemaśṛṅga-giri* 'a mountain with golden peaks'.

5. R. C. Majumdar, *op. cit.*, p. 101. Amarāvatī in Campā has been identified with modern Dong-dūong where sculptures of the Amarāvatī School (Guntur District, Andhra Pradesh) have been discovered. Cf. K. A. Nilakanta Sastri, *South Indian Influences in the Far East*, pp. 52, 119.

6. See Sircar, *Cosm. Geog. E. Ind. Lit.*, p. 35, note 6; cf. pp. 152, 155.

7. An old city named Vijaya is Vijayapurī which was the capital of the Ikṣvākus (3rd and 4th centuries A. D.) in the Nagarjunikonda valley in the Guntur District of Andhra Pradesh (Sircar, *Select Inscriptions*, p. 235, note 3).

8. Cf. Paṇḍaraṅga in N. Venkataramanayya's *The Eastern Cālukyas of Veṅgī*, p. 347; Pāṇḍuraṅgi in K. T. Pandurangi's *Desc. Cat. Sans. Mss.*, Vol. I, p. 240, and Pāṇḍuraṅga, *ibid.*, p. 244.

was divided into two parts, the Northern part having its capital at Vijaya and the Southern at Rājapura (in Panrān).[1]

The streams of Bali are named after celebrated Indian rivers, such as Gaṅgā, Sindhu, Yamunā, Kāverī, Śarayū and Narmadā, though the Balinese admit that the rivers really belong to Kling (the same as 'Kaliṅga', *i.e.* India).[2]

A locality in Borneo bore the name of Vaprakeśvara,[3] and, in Java, there were canals named after the Candrabhāgā (modern Chenab, a well-known tributary of the Indus) and Gomatī (modern Gomal or Goomti).[4]

The name of 'Java' is derived from Sanskrit 'Yavadvīpa' (the Barley Island) which is transliterated in Greek as 'Iabadios' in the *Geography* of Ptolemy.[5] The name of Sumatra is likewise derived from the Sanskrit word *samudra*, 'ocean', and there was, in that country, the city of Śrīvijaya (modern Palembang) which was one of the capitals of the kings of the Śailendra dynasty. The Indra-giri is located in Eastern Sumatra.[6]

The name of the Malay Peninsula is the same as that applied to the *Kulaparvata* identified with the Travancore hills within the southernmost part of the Western Ghāṭs.[7] The land called Tāmraliṅga is located on the eastern coast of the Peninsula.[8] A second capital of the Śailendra kings of Indonesia and Malayasia was Kaṭāha (Tamil *Kaḍāram*, modern Kedah near Penang), the territory around it being known as Kaṭāha-dvīpa in Sanskrit literature.[9]

The early home of the Thai people in Yunnan, which was called Nan-chao by the Chinese, was known in Indochina as Gandhāra, a portion of which was also called Videha-rājya and its capital was Mithilā.[10] The famous Pippala cave, the Bodhi tree, the sacred hill called Gṛdhrakūṭa and many other

1. R. C. Majumdar, *op. cit.*, p. 123.
2. *Ibid.*, p. 87.
3. Sircar, *Select Inscriptions*, 1965, p. 499.
4. *Ibid.*, pp. 502-03; *Cosm. Geog. E. Ind. Lit.*, pp. 83-84.
5. Cf. Si car, *Cosm. Geog. E. Ind. Lit.*, p. 143.
6. See R. C. Majumdar, *op. cit.*, p. 66.
7. Sircar, *Cosm. Geog. E. Ind. Lit.*, pp. 55 (note 88), 70, etc.
8. Coedès, *op. cit.*, p. 39.
9. Sircar, *Cosm. Geog. E. Ind. Lit.*, pp. 55, 67 (note 114).
10. R. C. Majumdar, *op. cit.*, p. 225.

localities associated with Buddhism were located in the area.[1] The Arab author Rashīduddīn (13th century) calls the country Gandhāra and says that its people came from India and China.[2] Among Thai principalities to the west and south of Yunnan, the Chinese speak of a Brāhmaṇa kingdom of Ta-tsin to the east of the mountains bordering Manipur and Assam and another in the east beyond the Chindwin. A group of Thai principalities, united in a loose federation and occupying the region between the Irawadi and Salween, was called Kauśāmbī. Some small States, which extended from the frontier of Yunnan to those of Cambodia and Thailand (Siam), included Suvarṇagrāma, Unmārgaśilā, Yonaka-rāṣṭra and Haripuñjaya.[3] According to the chronicles, the first Thai prince to settle to the south of the Mekong river was Brahman (Prom) who built Jayaprākāra in the Chieng Rai District in the 9th century A.D.[4]

Many of the above-mentioned geographical names, e.g., Gandhāra, Videha, Mithilā and Kauśāmbī, are famous in the history of ancient India. The Gandhāra country had two capitals one at Takṣaśilā in the Rawalpindi District and the other at Puṣkarāvatī or Puṣkalāvatī in the Peshawar District, both now in West Pakistan.[5] Videha was roughly the same as Tīrabhukti (modern Tirhut in North Bihar). It had its capital at Mithilā (modern Janakpur in the Nepalese Tarai).[6] Kauśāmbī was the capital of the Vatsas. It has been identified with Kosam in the Allahabad District of U. P.[7]

The first important Thai kingdom in Siam or Thailand was that of Sukhodaya (Sukhothai) which was founded by a chief named Indrāditya in the 13th century A.D. King Ram Khamheng (c. 1280-95 A.D.), who ascended the throne a few years after Indrāditya's death, annexed Haṁsāvatī (Pegu) in Lower Burma. Besides the city of Sukhodaya, the kingdom had a second capital called Sajjanālaya. The kingdom was

1. *Ibid.*, p. 226.
2. *Loc. cit.*
3. *Loc. cit.* Haripuñjaya comprised modern Lamphun and Chieng Mai in Northern Siam (*ibid.*, p. 197).
4. *Ibid.*, p. 226.
5. Sircar, *Cosm. Geog. E. Ind. Lit.*, p. 73.
6. *Ibid.*, p. 78.
7. *Ibid.*, p. 72.

therefore called Sajjanālaya-Sukhodaya. The State of Lopburi (Lavapurī) acknowledged the supremacy of Sukhodaya.[1] Hiuen-tsang places, to the West of Īśānapura, *i. e.* Cambodia, the kingdom of Dvāravatī, the memory of which is preserved in the official names of the Siamese capitals Ayutthaya (Dvāravatī-śrī-Ayudhyā or Thawarawadi Sri Ayudhya) founded in 1350 A.D. and Bangkok founded in 1782 A.D.[2] The Mons, according to tradition, regarded Sudharmavatī (Thaton at the mouth of the Sittang river in Burma) as the centre of their race. I-tsing places the kingdom of Śrīkṣetra (Prome in Burma)[3] to the west of Dvāravatī.

The names of Lavapurī, Dvāravatī, Ayudhyā and Śrīkṣetra remind us of the Indian cities Lavapura (Lahore),[4] Dvāravatī or Dvārāvatī (*i.e.* Dvārakā, capital of the Yādava-Sātvata-Vṛṣṇis in Kathiawar, Gujarat),[5] Ayodhyā (capital of the ancient Ikṣvākus in the Faizabad District of U.P.)[6] and Śrīkṣetra or Jagannāthakṣetra (*i.e.* Purī in Orissa).[7] It should, however, be noted that Jagannātha-kṣetra or Purī might not have become famous before I-tsing visited South-East Asia about the close of the 7th century A.D.

Almost all important geographical names in Ceylon had their Indian forms. Some of these names are Abhayagiri, Anurādhapura, Guptaśālā (Guptasāla, Buttala), Puṅkhagrāma, Jarjaranadī, Dīrghavāpī Gaṅgāśrīpura (modern Gampāla), Giritaṭa, Jambudroṇi, Jayagaṅga, Jayavardhanapura, Jetavanārāma, Kālavāpī, Kalyāṇī, Kāṇavāpī, Laṅkā, Maṇḍalagiri, Rambhā-vihāra, Nāgadvīpa, Ratnadvīpa, Ratnapura, Rohaṇa, Siṁhagiri, Siṁhapura, Kuśīnārā-vihāra, Tāmraparṇī, Tiṣyavāpī, Tiṣyamahāgrāma, Vātagiri, etc. There are also names like Malaya, Nālandā and Ujjayinī which are borrowed from India.[8]

1. See Majumdar, *op. cit.*, p. 229: cf. p. 171.
2. Cf. Coedès, *The Indianised States of Southeast Asia*, p. 76.
3. *Ibid.*, p. 77.
4. N. L. Dey, *Geographical Dictionary*, 1927, p. 114.
5. Cf. Sircar, *Cosm. Geog. E. Ind. Lit.* pp. 99, 107.
6. *Ibid.*, p. 72.
7. Dey, *op. cit.*, p. 191.
8. See C. W. Nicholas and S. Paranavitana, *A Concise History of Ceylon*, pp. 359 ff.

'Malaya', which is derived from the Dravidian word *malai*, 'a hill', is the name of one of the *Kulaparvatas* of Bhāratavarṣa or Kumārī-dvīpa.[1] It is the source of the Kṛtamālā (modern Vaigai) and Tāmraparṇī (modern Tāmbaravari) and has been identified with the Travancore hills within the Southern fringe of the Western Ghats.[2] The hilly region in the southern part of Ceylon was called Malaya and it was treated in later times as a special province (sometimes including the district of Dakṣiṇadeśa) under an official called Malayarāja (generally the king's younger son). The name Malaya is also applied in the Ceylonese chronicles to the mountainous region of Rāmaṇya or Burma.[3]

Nālandā was one of the most celebrated Buddhist religious establishments located at the site of modern Bargaon near Rajgir (ancient Rājagṛha or Girivraja) in the Patna District of Bihar. The same name was applied to a locality in the Central Province of Ceylon. It is mentioned in the accounts of the wars of king Parākramabāhu I.[4]

Ujjayinī, situated on the Siprā river in West Malwa, is one of the oldest cities of India. The city is famous for its great temple of the god Śiva Mahākāla. It was the capital of the ancient country and people called Avanti. It is now the headquarters of a District of the same name in the Western region of Madhya Pradesh. The name Ujjayinī was applied to a city of Ceylon which is stated to have been founded by Vijaya's minister Acyutagāmin.[5]

1. Sircar, *op. cit.*, p. 70.
2. *Ibid.*, pp. 87-88; above, pp. 243-44.
3. Malalasekera, *DPPN*, s. v. *Malaya.*
4. *Ibid.*, s. v. *Nālandā.*
5. *Ibid.*, s. v. *Ujjeni*. For B. C. Law's confusion between the Indian and Ceylonese Ujjayinīs, see above, pp. 211-12.

Chapter XXVIII

CARTOGRAPHY

There is no special word in Sanskrit for 'a map'. The word *nakśā* (from Arabic *naqshah*) has been adopted in most modern Indian languages in this sense, although it also signifies 'a picture, a plan, a general description, an official report'. In Eastern India, the word *māna-citra* has been coined to indicate the English word 'map'. The absence of any special Sanskrit word raises the question whether map-drawing was at all known to the Indians of old. There is, however, reason to believe that in ancient India a map or chart was regarded as a *citra* or *ālekhya*, *i.e.*, 'a painting, a picture, a delineation'. It will be seen that the Sanskrit word *citra* and its synonyms have practically the same meaning as the Arabic word *naqshah*.

That maps were made in ancient India seems to be quite clear from the evidence of the *New History of the T'ang Dynasty*[1] which gives an account of the Chinese general Wang Hiuen-tse's exploits in India in the year 648 A.D., as to how with the help of 1000 Tibetan soldiers and 7000 Nepalese horsemen he defeated and captured Na-fu-ti O-lo-na-shuen, who had usurped Chi-lo-i-to's (*i.e.* Śīlāditya-Harṣavardhana's) throne, in a battle near the town of Cha-puo-ho-lo, and received the submission of 580 walled towns as well as the offer of friendship of Chi-kieu-mo (*i.e.* Śrī-Kumāra *alias* Bhāskaravarman), king of Eastern India. King Chi-kieu-mo's presents to the Chinese general are stated to have included 30,000 oxen and horses and considerable war material and it is further said in this connection that the said king of Kia-mu-lu (*i.e.* Kāmarūpa) presented to the Chinese emperor some curious articles including 'a map of the country'. This map of Kāmarūpa appears to have been prepared by the artists at king Bhāskaravarman's court.

Act I of the *Uttararāmacarita* by Bhavabhūti, who flourished in the eighth century A.D., is styled 'the inspection of the paint-

1. See *Journ. As. Soc.*, Letters, Vol. XIX, 1953, p. 38.

ing'. It is said that a painter (*citrakara*) painted along a walk (*vīthikā*) the experiences (*carita*) of the Ikṣvāku king Rāma of Ayodhyā in Daṇḍak-āraṇya, Kiṣkindhā, Laṅkā and other places, according to the instructions of the king's brother Lakṣmaṇa who had accompanied Rāma to the forests. These paintings included some which are said to have depicted particular regions and may be regarded as a sort of maps. One of the paintings seems to have been conceived as showing the Prasravaṇa hill as 'extending to the heart of Janasthāna whose darkness is deepened by perpetually pouring clouds and whose caves ring with the flow of the Godāvarī which is embraced by the forests at the skirts that are of a uniformly mild blue colour because of the dense rows of trees."[1] The reference to forests painted in mild blue colour is interesting. Another picture of the nature of a map in the same context is introduced by Lakṣmaṇa to Rāma and Sītā in the following words: "Here is the tract (*bhāga*) of the Daṇḍaka forest, known as Citrakuñjavat, to the west of Janasathāna, haunted by the headless giant Danu; this is the site (*pada*) of the hermitage of Mataṅga on the Ṛṣyamūka hill; this again is the emaciated Śavara woman named Śramaṇā; this is the celebrated lake called Pampā."[2]

The above references to map-like paintings in an eighth century Sanskrit drama remind us of the following remarks of Wilford made about the beginning of the last century: "Besides geographical tracts, the Hindus have also maps of the world both according to the system of the Pauranics and of the astronomers; the latter are very common. They have also maps of India and of particular districts, in which latitudes and longitudes are entirely out of question, and they never make use of a scale of equal parts. The sea shores, rivers and ranges of mountains are represented by straight lines. The best map of this sort I ever saw was one of the kingdom of Nepal presented to Mr. Hastings. It was about four feet long and two and a half broad, of pasteboard, and the mountains raised about an inch above the surface, with trees painted all round. The roads were represented by a red line and the rivers by a blue one. The

1. Ed. S. Ray, Calcutta, 1934, p. 106.
2. *Ibid.*, p. 121.

various ranges were very distinct with the narrow passes through them: in short, it wanted but a scale. The valley of Nepal was accurately delineated; but towards the borders of the map everything was crowded and in confusion."[1] The map of Nepal noticed by Wilford appears to have been of the same type as the painting of the Citrakuñjavat region of Daṇḍak-āraṇya conceived by Bhavabhūti. The Indian maps mentioned by Wilford were probably uninfluenced by foreigners, as otherwise he would have added a note on the point. Unfortunately, Wilford does not say anything about the antiquity of the Indian maps noticed by him. The one of Nepal, of which he has given some details, could not have been very early.

A good deal of information on early Indian cartography under Hindu and Muslim inspiration is given by Francesco I. Pullé in his interesting work in Italian, entitled *La Cartografia Antica dell' India*, Parte I.[2] In section II of the work[3] dealing with Indian sources, there are reproductions of three maps, drawn by ancient Indian cartographers according to the Purāṇic ideas of cosmography and geography. As is well known, the world was regarded as consisting of seven concentric islands, each one of them encircled by a sea. The island at the centre was called Jambu-dvīpa, the southern division of which was called Bhārata-varṣa, bounded by the Himalayas in the north and the waters of the sea in the other directions.[4] Two of the three maps have been reproduced from a manuscript of the *Loka-prakāśa* which seems to have been originally composed by the celebrated Kashmirian polymath Kṣemendra in the eleventh century A.D. but contains a good deal of much later interpolation,[5] while the third[6] from a manuscript of another work entitled *Saṁgrahaṇī*. Figure 2 represents the concentric islands and the encircling seas, while Figures 8 and 9 are representations of Jambu-dvīpa. The map of Jambu-dvīpa published in Gladwyn's translation of the *Ain-i-Akbarī*, Vol. I, was no doubt

1. *Asiatic Researches*, Vol. VIII, 1805, pp. 270-71; cf. pp. 267-334; Vol. X, pp. 127-57.
2. *Studi Italiani Filologia Indo-Iranica*, Vol. IV, Firenze, 1901. See also some reproductions from manuscripts at the end of W. Kirfel's *Die Kosmographie der Inder*, and the sketches in D. C. Sircar's *Cosmography and Geography in Early Indian Literature*.
3. See *op. cit.*, pp. 8-44.
4. Figure 2 at p. 16 and Figure 8 at p. 33. Cf. above, pp. 17ff.
5. Cf. Stein, *Rājatar.*, trans., Vol. II, p. 313.
6. Figure 9 at p. 34.

copied from an original like those found in the manuscripts of the *Lokaprakāśa* and *Saṁgrahaṇī* noticed by Pullé. Similar maps were found by Wilford in the maunscript of a work entitled *Kṣetrasamāsa* and were characterised by him as 'several fanciful delineations of the world'.[1] The cosmographical map from a Tibetan Buddhist source reproduced in Figure 4 at p. 23 of Pullé's work is also interesting in this connection as its origin is certainly Indian. It is, however, difficult to say whether other Tibetan drawings such as the sketch of the Cathedral of Lhāsa published in the *Journal of the Asiatic Society of Bengal*, Vol. LXIV, Part i, 1895, Plate XXVI, were similarly of Indian inspiration.

Figure 5 at p. 25 in Pullé's book is a Hindu map of Jambu-dvīpa and its surroundings. This is of a slightly different kind. In it, the egg-shaped island is surrounded by nine circles in a row, each representing a *graha*. In another row, encircling the *grahas*, are given the twelve *rāśis* each in an oval. Figure 6 (at p. 29) is the representation of the *nava-khaṇḍa* or nine subdivisions of India. In Figures 10 (at p. 36) and 11 (at p. 37), Pullé has copied two maps of Jambu-dvīpa from Wilford. These are drawn after two slightly different cosmographic conceptions of the ancient Indians. On the authority of Rennel and Santarem,[2] Pullé also speaks of an old geographical map incised on a copper plate which was discovered at Monghyr.[3]

In section VII (pp. 139-50) of Pullé's work, dealing with Indian cartography from Persian and Arabic sources, we have several interesting maps. Figure 35 at p. 142 represents India according to an old Persian map of the earth. A map, after that of Ibn Haukal (975 A.D.), is given in Figure 36 at p. 147 and another from Edrisi (1154 A.D.) in Figure 37 at p. 156.

With reference to the knowledge of map-making among the people of India, especially the Dravidians of the South, the following remarks in the *Encyclopaedia Britannica* (14th ed., Vol. XIV, pp. 840-41) are also interesting: "The charts in use by the medieval navigators of the Indian Ocean—Arabs, Persians

1. *Asiatic Researches*, Vol. VIII, p. 269.
2. *Cosmographie et Cartographie*, Paris, 1852, Vol. I, p. 364.
3. *Op. cit.*, p. 12.

or Drāviḍas—were equal in value, if not superior, to the charts of the Mediterranean. Marco Polo (thirteenth century) mentions such charts; Vasco da Gama (1498) found them in the hands of his Indian pilot, and their nature is fully explained in the *Mohit* or 'the Encyclopaedia of the Sea', compiled from ancient sources by the Turkish admiral, Sidi 'Ali Ben Hosein, in 1584. These charts are covered with a close network of lines intersecting each other at right angles. The horizontal lines are parallels, depending upon the altitude of the Pole Star, the Calves of the Little Bear and the Barrow of the Great Bear above the horizon. This altitude was expressed in *isbas* or inches, each equivalent to 1° 42′ 50″. Each *isba* was divided into *zams* or eighths. The intervals between two parallels thus only amounted to 12′ 51″. These intervals were mistaken by the Portuguese occasionally for degrees, which accounts for Malacca, which is in latitude 2′ 13″ N., being placed in Cantino's chart (1502) in latitude 14′ S. It may have been a map of this kind which accounts for Ptolemy's moderate exaggerations (in the second century A.D.) of the size of Taprobane (Ceylon). The first meridian, separating a leeward from a windward region, passed through Ras Kumhari (Comorin) and was thus nearly identical with the first meridian of the Indian astronomer, which passed through the sacred city of Ujjain (Ozéné of Ptolemy) or the meridian of Azīn of the Arabs. Additional meridians were drawn at intervals of *zams*, supposed to be equal to three hours' sail."

The suggestion seems to be that the Indians had the knowledge of making maps as early as the beginning of the Christian era and that the Arabs and Greeks were indebted to them for the charts of the Indian Ocean together with the islands in and the lands bordering on it.

CHAPERR XXIX

THE EIGHT ELEPHANT-FORESTS

The eight cardinal points, with which the *Dik-pālas* and *Dig-gajas* are associated, are enumerated as—(1) *pūrva* (east), (2) *āgneya* (south-east), (3) *dakṣiṇa* (south), (4) *nairṛta* (south-west), (5) *paścima* (west), (6) *vāyava* (north-west), (7) *uttara* (north) and (8) *aiśāna* (north-east).[1] According to the *Amarakoṣa*, the following eight *Dik-pālas* are the lords of the eight quarters beginning with the eastern—(1) Indra, (2) Vahni (Agni), (3) Pitṛpati (Yama), (4) Nairṛta, (5) Varuṇa, (6) Marut (Vāyu), (7) Kubera and (8) Īśa (Īśāna).[2] The list of the eight *Dig-gajas* in the *Amarakoṣa* runs as follows—(1) Airāvata, (2) Puṇḍarīka, (3) Vāmana, (4) Kumuda, (5) Añjana, (6) Puṣpadanta, (7) Sārvabhauma and (8) Supratīka.[3] It is, however, well known that the order of the names has not been maintained in the different lists of the *Dig-gajas*. Thus Kṣīrasvāmin, in his well-known commentary on the *Amarakoṣa* stanza referred to above, points out that Bhāguri enumerated the first five names as Airāvata, Puṇḍarīka, Kumuda, Añjana and Vāmana while the *Mālā* gives the first two names as Airāvata and Supratīka.

The *Viṣṇudharmottara* mentions the names of the *Dig-gajas* in the following order—(1) Airāvaṇa (i.e. Airāvata), (2) Padma (i.e. Puṇḍarīka), (3) Puṣpadanta, (4) Vāmana, (5) Supratīka, (6) Añjana, (7) Nīla (i.e. Sārvabhauma) and (8) Kumuda, which are stated to have been the *vāhanas* of Śakra (Indra) and the other *Dik-pālas* respectively.[4] The same

1. See Apte's *Pract. Sans.-Eng. Dict.*, s. v. *aṣṭan*—
Pūrv=āgneyī dakṣiṇā ca nairṛtī paścimā tathā |
vāyavī c=ottar=aiśānī diśa aṣṭa-vidhāḥ smṛtāḥ ||
2. I.ii.6—
Indro Vahniḥ Pitṛpatir=Nairṛto Varuṇo Marut |
Kubera Īśaḥ patayaḥ pūrv-ādīnāṁ diśāṁ kramāt ||
3. I.ii.8—
Airāvataḥ Puṇḍarīko Vāmanaḥ Kumudo='ñjanaḥ |
Puṣpadantaḥ Sārvabhaumah Supratīkaś=ca dig-gajāḥ ||
4. I. 251.II-12—
Airāvaṇas=tathā Padmaḥ Puṣpadantaś=ca Vāmanaḥ |
Supratīko='ñjano Nīlaḥ Kumudaś=ca mataṅgajāḥ ||
Sakr-ādyānāṁ dig-īśānāṁ yathā-saṅkhyena vāhanāḥ ||

work also speaks of the four classes of elephants born in the clan of each one of the eight, viz. (1) Bhadra the best, (2) Manda the medium, (3) Mṛga the worst and (4) Saṅkīrṇa or the mixed breed.[1] The *Agni Purāṇa* seems to mention the names of the eight *Dig-gajas* as (1) Kumuda, (2) Airāvaṇa, (3) Padma, (4) Puṣpadanta, (5) Vāmana, (6) Supratīka, (7) Añjana and (8) Homa (for Nīla or Sārvabhauma).[2]

The number of the *Dig-gajas* seems to have influenced the ancient Indian writers' classification of the Indian elephants under eight typical groups. Thus the *Kauṭilīya Arthaśāstra* speaks of the elephants of eight countries grouped into three classes—(I) those of Kaliṅga, Aṅga, Karūśa and Prācya constitute the best class; (II) those of Daśārṇa and Aparānta are medium, and (III) those of Surāṣṭra and Pañcajana are the worst.[3] It is said in this connection that the strength, speed and spirit of all the three categories could be enhanced by means of training.

The location of the above eight countries is well-known. Kaliṅga had its ancient capital at Tosalī (in the modern Dhauli region in the Puri District of Orissa); but, from the close of the 5th to the 12th century A.D., its chief city was Kaliṅganagara (modern Mukhalingam near Srikakulam in Andhra Pradesh). The ancient Aṅga country comprised the present Bhagalpur-Monghyr region of East Bihar and had its headquarters at Campā near Bhagalpur. Karūśa (Kārūśa or Kārūṣa) was known to have been identical with the Shahabad region of South-West Bihar as late as the 18th century.[4]

1. I-251. 12-14—
Catvāro jātayas=teṣām=ek-aikasy=ānvaye smṛtāḥ |
Bhadrā Mandā Mṛgā c=aiva Saṅkīrṇā ca janādhipa ||
Bhadrā śreṣṭhā bhavet=tāsāṁ Mandā madhyā kanīyasī |
Mṛgā jñeyā ca bāhulyāt Saṅkīrṇā pārthivottama ||
Bhadra-Manda-Mṛgāṇāñ=ca tathā vakṣyāmi lakṣaṇam ||
2. The *Agni Purāṇa* (291.4) has—
Kumud=Airāvaṇaḥ Padmaḥ Puṣpadanto='tha Vāmanaḥ |
Supratīko='ñjano nāgā aṣṭau Homo='tha dakṣiṇām ||
3. II. 2 (Shama Sastry's ed., p. 50)—
Kaliṅg-Āṅga-gajāḥ śreṣṭhāḥ Prācyāś=c= eti Karūśajāḥ |
Daśārṇāś=c=Āparāntāś=ca dvipānāṁ madhyamā matāḥ ||
Saurāṣṭrikāḥ Pañcajanās=teṣāṁ pratyavarāḥ smṛtāḥ |
sarveṣāṁ karmaṇā vīryaṁ javas=tejaś=ca vardhate ||
4. Cf. *Aṁgareja-rājye varttamāne Kārūṣa-deśe* in the Masār (Shahabad District) inscription of Vikrama 1876 or 1819 A.D. (Bhandarkar's List of Inscriptions, No. 1068).

Prācya or the Eastern Division of ancient India has been described as the territory lying to the east of the Kālaka-vana (probably in the Allahabad region), Prayāga (near Allahabad) or Vārāṇasī by Brāhmaṇical writers, but of Kajaṅgala (near Rajmahal in the Santal Parganas District of Bihar) or Puṇḍravardhana (Mahāsthān in the Bogra District of East Pakistan) by the Buddhist authors.[1] Since Aṅga and Karūśa (which formed part of the territory to the east of Allahabad and Varanasi) are separately mentioned, it may be argued that Prācya of the *Arthaśāstra* in the present case is analogous to the Buddhist Prācya. But the intention of the author of the *Arthaśāstra* may have been to indicate the elephants of Eastern India generally and of the Aṅga and Karūṣa countries in Eastern India in particular. Daśārṇa was East Malwa with its capital at Vidiśā (modern Besnagar in the Vidisha District, M.P.). Aparānta, in a narrow sense, corresponded to the Thana region of Maharashtra and had its capital at Sūrpāraka (modern Sopara). Surāṣṭra comprised Southern Kathiawar with its capital at Girinagara (modern Girnar near Junagarh). *Pañcajana* is the same as *Pañcanada* or the Punjab.[2]

The stanzas quoted above from the *Arthaśāstra* may be compared with the following verses of the *Mānasollāsa* (I.2. 179-81), the celebrated encyclopaedia by king Someśvara III (1126-38 A.D.) of Kalyāṇa—

> *Kāliṅgaṁ Ve(Ce)di-Kārūśaṁ Dāśārṇaṁ ca vanaṁ varam* ।
> *Āṅgireyaṁ tathā Prācyaṁ madhyamaṁ vanam=iṣyate* ॥
> *Aparāntaṁ Pāñcanadaṁ Saurāṣṭraṁ c=ādhamaṁ vanam* ।
> *evam=aṣṭau vanāny=āhur=gajānāṁ janmanaḥ padam* ॥

It will be seen that the names found in the passage quoted from the *Arthaśāstra* are given here with slight modification in a few cases. Thus instead of Karūśa, the *Mānasollāsa* has Vedi-Kārūśa or, correctly, Cedi-Kārūśa,[1] i.e. the Cedi and Kārūśa countries. In the age of Someśvara III, Cedi was the name of the Kalacuri kingdom with its capital at Tripurī,

1. Cf. Sircar, *Cosm. Geog. E. Ind. Lit.*, pp. 16-17.
2. See below. For the location of the eight countries, see Sircar, *op. cit.*, pp. 25, 77, 79-81.
1. The G.O.S. ed. has *Vedi* and the Mysore ed. *Cedi*.

modern Tewar near Jabalpur. Whoever may have been responsible for the above modification of the description, he seems to have been influenced by a desire to distinguish Kārūśa from Prācya. The same desire may have compelled him to change *Aṅga* to *Āṅgireya* which is otherwise unknown. As we have seen above, Prācya was generally supposed to have included both the Kārūśa and Aṅga countries which formed parts of modern Bihar. *Pañcanada* (the Punjab) is of course the same as *Pañcajana*.

What is of considerable importance is that the *Mānasollāsa* classification of elephants is somewhat different from their classification in the *Arthaśāstra*. According to the former, (I) the best elephant forests were Kāliṅga, Cedi-Kārūsa and Dāśārṇa, (II) the medium group comprised the Āṅgireya and Prācya forests, and (III) the Aparānta, Pāñcananda and Saurāṣṭra were the worst. Thus, e.g., of the elephants of Daśārṇa and Aparānta, placed by the *Arthaśāstra* in the medium class, the *Mānasollāsa* places the first group in class I and the second in class III. It is difficult to say whether the reclassification. copied in the *Mānasollāsa*, was a deliberate attempt to improve upon the earlier authorities on the basis of investigation and personal experience. Because, as we shall see below, some medieval writers such as the author of the *Mānasollāsa* had, at least, a vague idea about the location of the Daśārṇa country.

The *Mānasollāsa* (I. 2. 173-74) describes the extent of the Prācya forest in the following stanza—

Gaṅgāsāgara-He(Hi)mādri-Prayāgāṇāṁ ca madhyataḥ ।
vanaṁ Prācyam=iti proktaṁ Lohitā(ty-ā)bdhi(bhde)ś=ca paścime ॥

According to the description, the Prācya forest extended from Gaṅgāsāgara (at the junction of the Gaṅgā and the sea) in the south to the Himādri or Himalayas in the north and from Prayāga (near Allahabad) in the West and the Lohitya Sea (i.e. the lower course of the Brahmaputra which is so mentioned in the epigraphic records of Assam)[1] in the east. It has to be

1. *Ep. Ind.*, Vol. XXIX p. 151. The conception of the Brahmaputra as a sea seems to be associated with the tradition about the existence, in early times, of the Eastern Ocean (Bay of Bengal) near Devīkoṭṭa (modern Bāṇgaḍh in the Dinajpur District, East Pakistan) and with the presence,

noticed that Aṅga in East Bihar and Karūśa or Kārūśa in South-West Bihar lay in this tract and that is why Aṅga has here the doubtful name Āṅgireya and Kārūśa has been bracketed with Cedi and both have been placed outside Prācya as we shall presently see.

The Cedi-Kārūśaka and Āṅgireya forests have been described in the *Mānasollāsa* in the following stanza (I.2.174-75)—

Tripuryāṁ Kosal-ādrau ca Ve(Ce)di-Kārūśakaṁ vanam |
Śrīkṣetraṁ Gauḍa-Vaṅgālam = Āṅgireyaṁ vanaṁ smṛtam ||

Here the Cedi-Kārūśaka forest is stated to have extended from Tripurī (Tewar near Jabalpur) to the mountain range of the Kosala (South Kosala) country in the Raipur-Bilaspur region of M.P. and the Sambalpur-Balangir area of Orissa. Up to the 12th century A.D., South Kosala was under the Somavaṁśīs who had originally their capital at Yayātinagara on the upper Mahānadī. It will be seen that the Kārūśa country comprising the Shahabad District of Bihar seems to be excluded from Cedi-Kārūśaka because Kārūśa really formed a part of Prācya.

The Āṅgireya forest is stated to have covered Śrīkṣetra on one side and Gauḍa-Vaṅgāla (Gaur-wa-Bangāla of the medieval Muslim authors)[1] on the other. Gauḍa and Vaṅgāla may of course be located respectively in the western and south-eastern areas of undivided Bengal. But it is difficult to say whether, by Śrīkṣetra, the author, like the Chinese pilgrim Hiuen-tsang, meant Prome in Burma,[2] or Purī in Orissa, which is now called Śrīkṣetra (i.e. 'the illustrious locality') because it is the *kṣetra* of the celebrated god Puruṣottama-Jagannātha.[3]

in the central regions of Bengal, of large *bils* or lakes like the Chalan in the Rajshahi District (East Pakistan). Wide areas of the Mymensing District of East Pakistan, through which the Brahmaputra now passes, are spoken of as the 'sea' even today. It is a lowlying country which is under water for six or more months of the year.

1. See Gulbadan Begam's *Humāyūnnāma* cited in *Num. Suppl.*, No. XXXIV; also Firishta's work, Nawalkishore Press, Vol. I, p. 21; Vol. II, p. 293. 'Gauḍa-Vaṅgāla' is rare in Sanskrit works.

2. Watters, *On Yuan Chwang's Travels in India*, Vol. II, p. 188.

3. J.M. Das, *Bāṅgālā-bhāṣār Ābhidhān*, s.v. See also Dey, *Geog. Dict.*, s. v.

The Kāliṅga forest is described in the following verse (I.2. 175-76)—

Vindhyādri-Citrakūṭādri-Kaliṅga-Drāviḍ-āśritam ।
vanaṁ Kāliṅgakaṁ nāma samudr-āvadhi kīrtyate ॥

This forest comprised the Vindhya range, the Citrakūṭa hill and the Kaliṅga and Drāviḍa countries and extended upto the sea (i.e. the Indian Ocean). Hiuen-tsang speaks of the Drāviḍa country as having its capital at Kāñcīpura (about 50 miles to the west of Madras).[1] We know two Citrakūṭas; one is modern Citoḍ in the Udaipur Division of Rajasthan and the other is a holy place in the Banda District of U.P. The second of the two is probably meant here.

The description (I.2.176-77) of the Dāśārṇaka forest runs as follows—

Śrīśaile Vedaśaile ca Malay-ādrau tath=aiva ca ।
vanaṁ Dāśārṇakaṁ nāma kariṇāṁ janma-kāraṇam ॥

The Dāśārṇaka forest is thus stated to have covered the Śrīśaila, the Vedaśaila and the Malaya range. Of these Śrīśaila is the well-known hill in the Nallamalur range and is a celebrated *tīrtha* in the Kurnool District of Andhra Pradesh, while the Malaya range comprised the Travancore hills and the southern spurs of the Western Ghāṭs.[2] Vedaśaila is the Vedaparvata near Chingleput in Tamil Nadu. It is famous for the celebrated Pakṣi-tīrtha situated on it.[3] We have to note that this Dāśārṇaka forest is wrongly placed far away from the Daśārṇa country in East Malwa. It may be seen that the original list did not include any elephant forest of South India. The author of the *Mānasollāsa*, being a Karṇāṭīya, was eager to locate at least one name, found in the old list, in the South. Mysore is even now famous for its elephant forests.

The stanza (I.2. 177-78) describing the Aparānta and Saurāṣṭra forests runs as follows :

Sahyādri-Bhṛgukacch-āntam=Aparānta-vanaṁ smṛtam ।
Dvāravatyām=Avantyāṁ ca Saurāṣṭraṁ vanam=ucyate ॥

1. Watters, *op. cit.*, p. 226.
2. See above, pp. 60 (note 6), 243-44.
3. Dey, *op. cit.*, s. v. *Bedaparvata* for *Vedaparvata*.

The Aparānta forest extended from the Sahyādri or Western Ghāṭs upto Bhṛgukaccha (modern Broach at the mouth of the Narmadā) while the Saurāṣṭra forest extended from Dvāravatī (i.e. Dvārakā in Western Kathiawad) upto the Avanti country (West Malwa) which had its capital at Ujjayinī.

The eighth and last forest, called Pāñcanada, is described in a verse (I.2.178-79) which runs—

Kālañjare Kurukṣetre Sindhusāgara-saṅgame I
vanaṁ Pāñcanadaṁ proktaṁ Himālaya-kṛt-āvadhi II

This forest was bounded by Kālañjara (Kaliñjara or Kāliñjara on the borders of Kashmir according to the *Rāja-taraṅgiṇī*)[1] in the west, by Kurukṣetra (Karnal District, Haryana) in the east, by the Himālaya in the north and the junction of the Indus and the Sea (Arabian Sea) in the south.

A similiar description of the elephant-forests is also found in the *Viṣṇudharmottara* (I.251.22ff.), the date of which is regarded as a few centuries earlier than that of the *Mānasollāsa*,[2] although the confusion in the text is even more remarkable. Its account of the *gajānāṁ van-āṣṭakam* begins with the Prācya forest in the following stanza (I.251. 22-23)—

Himavat-Prayāga-Lauhitya[3]*-Gaṅgā-madhye mahad-vanam* I
Prācyam=Airāvaṇasy=oktaṁ vanaṁ yatra mataṅgajāḥ II

It is indicated that the Prācya forest was bounded by the Himavat (Himālaya) in the north, Prayāga (Allahabad) in the west, Lauhitya (Brahmaputra) in the east and the Gaṅgā in the south. These boundaries are the same as those described in the *Mānasollāsa* quoted above. The forest is mentioned

1. Stein, *Kalhaṇa's Rājataraṅgiṇī*, Vol. II, p. 366 (VII. 1256 and note).

2. The *Viṣṇudharmottara* is assigned to a date between 400 and 500 A.D. by some and between 628 and 1000 A.D. by others (cf. Hazra, *Stud. Upap.*, Vol. I, p. 212; Winternitz, *Hist. Ind. Lit.*, Vol. I, p. 580). But mistakes like *Kāleśa* and *Mārgareyaka* (cf. below) seem to suggest that its present text may be later than the present text of the *Mānasollāsa*, though it knew the location of Daśārṇa about which the *Mānasollāsa* had a confused idea.

3. This foot of the stanza has one syllable in excess. The verse is preceded by the line—*ataḥ paraṁ pravakṣyāmi gajānāṁ te van-āṣṭakam.*

here as the abode of [the elephants of the class of] the *Dig-gaja* Airāvaṇa (Airāvata).

The Karūṣa forest is described in another stanza[1] as the abode of [the elephants of the class of] the *Dig-gaja* Padma or Puṇḍarīka and as bounded by the Unmatta-Gaṅgās (probably some tributaries of the Ganges), Tripurī (near Jabalpur), Daśārṇa (East Malwa) and the Mekala (the Amarkantak and Maikal ranges) apparently in the north, south, west and east respectively. These boundaries appear to include the Cedi country, but not Karūṣa in the Shahabad region of Bihar.

The Dāśārṇaka forest is described as the abode of [the elephants of the class of] the *Dig-gaja* Nāga (i.e. Puṣpadanta). It was bounded by the Bilvaśaila (cannot be identified), the Vetravatī (modern Betwa running between Besnagar and Bhilsa), the Daśārṇagiri (apparently in East Malwa) and the Mekala (the Amarkantak and Maikal ranges).[2] *Bilvaśaila* reminds us of the Sun-god Bhāyillasvāmin worshipped at Bhilsa (recently named Vidisha) in the early medieval period.[3]

The name Mārgareyaka-vana is applied to what is called Āṅgireya-vana in the *Mānasollāsa*, the name being modified from *Aṅga* of the *Arthaśāstra* list. It was the abode of [the elephants of the class of] the *Dig-gaja* Vāmana. It is described as bounded by the Pāriyātraka (the Western Vindhyas including the Aravalli), Vaideśya (Vaidiśa or Vidiśā, modern Besnagar), the Narmadā river, and Brahmavardhana which cannot be identified with precision.[4] These appear to have been the boundaries respectively in the west, east, south and north. We are not sure whether Brahmavardhana can be identified with Brahma-tīrtha or the Puṣkara lake.[5]

What the *Mānasollāsa* calls the Kāliṅga forest is men-

1. I.251.24-25—
   Unmattagaṅgās=Tripurī Daśārṇaṁ Mekalās=tathā |
   teṣāṁ madhye Karūṣ-ākhyaṁ vanaṁ Padmasya kīrtitam ||
2. I.251.26-27—
   Bilvaśailaṁ Vetravatī Daśārṇaṁ ca mahāgirim |
   teṣāṁ Daśārṇakaṁ madhye Puṣpadantasya kānanam ||
3. See *Ep. Ind.*, Vol. XXX, pp. 210 ff.
4. I.251.28-29—
   Pāriyātraka-Vaideśya-Narmadā-Brahmavardhanam |
   Vāmanasya vanaṁ madhye teṣāṁ vai Mārgareyakam
5. Dey, *op. cit.*, s. v.

tioned in the *Viṣṇudharmottara*, no doubt wrongly, as Kāleśa. It is stated to have been the abode of [the elephants of the class of] the *Dig-gaja* Supratīka and was bounded by the Vindhya, the Sahya (Western Ghāṭs), Utkala (Coastal Orissa) and the Dakṣiṇa-samudra (Indian Ocean).[1] These boundaries apparently relate to the north, west, east and south.

The Aparāntaka forest is described as the abode of [the elephants of the class of] the *Dig-gaja* Añjana and as bounded by Sevādeśa (possibly a mistake for *Sevaṇadeśa*, i.e. modern Khāndesh), the *samudra* (apparently the Arabian Sea), Premahāra (possibly the same as Preharā (i.e. either the modern Ghaṭaprabhā or the Malaprabhā)[2] and the Narmadā, no doubt respectively in the east, west, south and north.[3]

The Saurāṣṭra forest is stated to have been the abode of [the elephants of the class of] the *Dig-gaja* mentioned as Nīla (i.e. Sārvabhauma) and to have been bounded by Kuśasthalī (i.e. Dvārakā or Dvāravatī in Kathiawar), Avanti (West Malwa), the Arbuda (Mt. Abu in the Sirohi District, Rajasthan) and the Narmadā.[4]

The Pāñcanada or Punjab forest was the abode of [the elephants of the class of] the *Dig-gaja* Kumuda and was bounded by the Himālaya (north), the Kālikā (south ?), the Sindhu or Indus (west) and Kurujāṅgala (east).[5] A.B.L. Awasthi suggests the identification of Kālikā with Kalkā near Simla;[6] but it seems to be a river far to the south. The Kālikā may by the Kālī Sindh, a tributory of the Chambal. Kurujāṅgala was really a part of the Kuru country and lay in the Eastern

1. II. 251.30-31—
Vindhya-Sahy-Otkalānāṁ ca dakṣiṇasy=ārṇavasya ca |
vanaṁ ca madhye Kāleśaṁ Supratīkasya kīrtitam ||
2. Cf. *Select Inscriptions*, 1965 ed., p. 477, note 1.
3. I.251.32-33—
Sevādeśaḥ samudraś=ca Premahāraṁ ca Narmadā |
teṣāṁ madhye='ñjan-ākhyasya vanaṁ khalv=Aparāntakam ||
4. I.251.34-35—
Kuśasthalī mahīpāla Avanty-Arbuda-Narmadāḥ |
teṣāṁ madhye tu Saurāṣṭraṁ vanaṁ Nīlasya kīrtitam ||
5. I.251.36-37—
Himavat-Kālikā-Sindhu-Kurujāṅgalam=eva ca |
teṣāṁ Pañcanadaṁ madhye Kumudasya mahad-vanam ||
6. *Prācīn Bhārat kā Bhaugolik Svarūp*, p. 144.

Punjab though the name was sometimes also applied to the country of the Kurus,[1] which had its chief cities at Hastināpura (Meerut District, U.P.) and Indraprastha (near Delhi).

The *Vāyu* and *Brahmāṇḍa Purāṇas* have a section dealing with four *Dig-gajas* and their respective forests. It is well known that originally only four *Dik-pālas*, viz. (1) Indra (east), (2) Kubera (north), (3) Yama (south) and (4) Varuṇa (west), were recognised and that the number was later raised to eight by the inclusion of the protectors of the south-east, south-west, north-east and north-west.[2] There may have been a similar original conception in respect of the *Dig-gajas* as well. Thus the said Purāṇic section mentions only—(1) Añjana (Saṅkīrṇa, *śveta*), the *vāhana* of Yama (south), (2) Supratīka (Bhadra, *harita*), the *vāhana* of Varuṇa (west), (3) Padma (Puṇḍarīka—Manda, *gaura*), the *vāhana* of Kubera (north), and (4) (Vāmana) (Mṛga, *śyāma*), the *vāhana* of Agni (south-east).[3] These four are called *dig-gajas* and are represented as the sons of Airāvata who is the *vāhana* of Indra, the guardian of the east. It is interesting to note that the above list of four *dig-gajas* includes the elephant of the south-east in place of the one guarding the east.

In this account, the Prācya forest, which was the abode of [elephants of the class of] the *Dig-gaja* Añjana, extended from the Kauśikī (i.e. the Kosī river) to the sea and lay beyond the Gaṅgā (probably to its north and east).[4] The

1. Dey. *op. cit.*, s. v.

2. See Sircar, *The Successors of the Sātavāhanas*, p. 196. Sometimes we have Agni or Soma in place of Kubera (*The Bhakti Cult*, etc., ed. Sircar, p. 43), while the Buddhists mention Kubera along with Dhṛtarāṣṭra of the east, Virūḍhaka of the south and Virūpākṣa of the west (Sircar, *Cosm. Geog.*, etc., p. 49, note).

3. See *Vāyu*, II.8.207ff.—

Tasya putro='ñjanaś=c=aiva Supratīko='tha Vāmanaḥ |
Padmaś=c=aiva caturtho='bhūd=dhastinī c=Ābhramus=tatā
Dig-gajāṁs=tāṁś=ca catvāraḥ śvet=ājanayat=āśugān |
Bhadraṁ Mṛgañ=ca Mandañ=ca Saṅkīrṇaṁ caturaḥ sutān ||
Saṅkīrṇo='py=Añjano yas=tu uparāhyo Yamasya tu |
Bhadro yaḥ Supratīkas=tu haritaḥ sa hy=Apāmpateḥ ||
Padmo Mandas=tu yo gauro dvipo hy=Ailavilasya saḥ |
Mṛgaḥ śyāmas=tu yo hastī (Vāmana) upavāhyaḥ sa Pāvakaiḥ ||

4. (a) *Vāyu*, II.8.232—

Kauśik-ādyāḥ samudrāt=tu Gaṅgāyās=tad-anantaram |
Añjanasy=Aikamūlasya Prācyan=nāga-vanan=tu tat ||

Kosī now joins the Ganges in Bihar, though it is supposed to have flowed through North Bengal in the early period.[1]

The forest of [elephants of the class of] the *Dig-gaja* Supratīka is stated to have been situated between the Gaṅgā in the north and the Vindhya in the south and to have extended from Gaṅgodbheda to the Karūṣa country.[2] Since, however, Gaṅgodbheda, the place where the Gaṅgā rises, lies in the Himalayas, it could hardly have been a locality associated with the territory between the Gaṅgā and the Vindhyas. The text therefore appears to be erroneous.

The language of the description of Vāmana's forest is also defective. It is probably stated to have been situated between Utkala (in Coastal Orissa) and the Cedi country (in the Jabalpeur region).[3]

The forest of [elephants of the class of] the *Dig-gaja* Padma lay to the west of the sea and had the Lauhitya (Brahma-Putra) in its west; that is to say, it was situated between the sea in the east and the Brahmaputra in the west.[4] The area seems to have comprised the south-eastern regions of Bengal and the contiguous areas of Burma.

It appears that the association of a particular *Dig-gaja* with a particular forest was influenced by the different notions

(b) *Brahmāṇḍa*, III. 7.355-56—
Kauśikyā hy=ā samudrāt=tu Gaṅgāyāś=ca yad=uttaram |
Añjanasy=Aikamūlasya vijñeyaṁ gahanaṁ tu tat ||

1. Cf. Sircar, *Cosmography and Geography in Early Indian Literature*, p. 84; also above, p. 50, note 4.

2. (a) *Vāyu*, II.8.283—
Uttarā (raṁ) tasya Vindhyasya Gaṅgāyā dakṣiṇaṁ ca yat |
Gaṅgodbhedāt Karuṣebhyaḥ Supratīkasya tad=vanam ||
(b) *Brahmāṇḍa*, III. 7.356-57—
uttaraṁ c=aiva Vindhyasya Gaṅgāyā dakṣiṇaṁ ca yat |
Gaṅgodbhede sakerubhyaḥ Supratīkasya pattanam ||

3. (a) *Vāyu*, II.8.234—
Apareṇ=Otkalāś=c=aiva hy=ā Ve (Ce)dibhyaś=ca pañca (ści)mam |
Ekabhūt-ātmano='sy=aitad=Vāmanasya vanaṁ smṛtam |
(b) *Brahmāṇḍa*, III.7.357-58—
Apareṇ=Otkalaṁ c=aiva Kāveri (hy=ā Cedi)bhyaś=ca paścimam |
Ekas ūk-ātmajasy=aitad=Vāmanasya vanaṁ smṛtam ||

4. (a) *Vāyu*, II.8.235—
Apareṇa tu Lauhityam=ā sindhoḥ paścimena tu |
Tama (Padma)sy=aitad=vanaṁ proktam=anuparvatam=eva tat ||
(b) *Brahmāṇḍa*, III.7.358-59—
Apareṇa tu Lauhityam=ā sindhoḥ paścimena tu |
Padmasy=aitad=vanaṁ proktam=anuparvatam=eva tat ||

of the early medieval writers regarding the quarter represented by a *Dig-gaja* and that this fact led to the discrepancy among their views.

We have consulted, in this connection, a few other medieval works without getting any addition to our knowledge of the subject. These include the *Mātaṅgalīlā*, Trivandrum Sanskrit Series, 1910 - (cf. *JBORS*, Vol. X, p. 322), the *Gajaśāstra* of Pālakāpya-muni, edited by K. S. Subrahmanya Sastri and S. Gopalan, T.M.S.S.M. Library, Tanjore, 1958, (pp. 32-35) and the *Aparājitapṛcchā*, edited by P. A. Mankad (G.O.S., No. 115, Chap. 79, pp. 199-200). Indeed, they appear to make the confusion worse confounded.

INDEX

Addenda et Corrigenda

Page 12, line 22.—*Read*—Rajputana
„ 13, line 5.—*Read*—beginning
„ 22, line 3.—*Read*—Puṣkara
„ 26, line 11.—Read—*Brahmāṇḍa*
, 28, line 16.—Read—*Śuktimantī, Ujjānaka-maru*, etc.
„ 33, note 3, line 7.—*Read*—Śatadruja
„ 37, note 3, line 5.—*Read*—West
„ „ note 5, line 9.—*Read*—Sec. E,
„ 41, note 7, line 2.—Read—*c=Āntaranarmadāḥ.*
„ 43, line 2.—Read—*Vaidiśās=tathā*[5]
„ 44, note 4, line 7.—*Read*—According
„ 45, note 6.—Add—*Gurgaṇa* reminds us of *Gurjara.*
„ „ note 7.—*Add*—Early Greek writers knew the form *Karṇaprāvaraṇa* (*JAIH*, Vol. III, p. 250).
„ 50, note 4, line 8.—*Read*—Sanskrit
„ 52, note, line 14.—*Read*—Śatladar
„ 57, note 2, line 7.—*Read*—Paisuni
„ 62, note 5, line 2.—*Read*—Śuktimat
„ 64, note 6, line 1.—Read—*Trisandhyā*
„ 70, note 1, line 1.—*Read*—Al-Bīrūnī
„ 72, note, line 3.—*Read*—Suvarṇabhūmi
„ 75.— *Under chapter heading, read*—I
„ 77, line 28.—*Read*—i, *for*—I
„ 78, line 13.—*Read*—ii, *for*—II
„ 107, line 15.—*Read*—Gṛdhrakūṭa, *for*—Gṛdharakūṭa
„ 111, line 20 } —*Read*—Kachar
„ 112, line 20 }
„ 116, line 21.—*Read*—Mānaseśa
„ 128, note 3, line 4.—Read—*Karṇāṭāś=c=aiva*
„ 132, note 6.—Read—'*Ain-i-Akbarī*
„ 161, note 1, line 4.—*Read*—flourished
„ 169, note 1, line 2.—*Read*—Keith
„ 181, line 25.—*Add note*— The name Kaṭaka is, however, older since the Parikud plates (*Ep. Ind.*,

Vol. XI, pp. 281 ff.) of Śailodbhava Madhyamarāja (middle of the 7th century) speaks of the Kaṭaka-bhukti-viṣaya.

Page 186, note 4.—*Add*— which suggests the use of the name Gauḍa in the sense of Eastern India.

„ 187, note 4, line 2.—*Read*— Śivarāja's son Govindarāja is described as the grandson

„ 193, line 10.—*Read*—Kṛṣṇabeṇā

„ 209, line 11.—*Read*—Malwa

„ 214, note 3, line 12.—*Read*—Uśīra-giri

„ 221, line 33.—*Read*—Gaurī's

„ 230, line 12.—*Read*—identification

„ 232, line 29. —*Add note on 'Kola-pattana'*—The name reminds us of Kolaka located by Ptolemy in the Lower Indus Valley (cf. Sircar, *Cosm. Geog.*, p. 130).

„ 246, line 21. —*Read*—Tāmralipta

„ 256, note 1—*Add*—On the basis of epic evidence, Hopkings point out that 84 and 88 are conventional numbers (*EpicMythology*, p. 55).

„ 258, line 14. —*Read*—Jabalpur

„ 260, line 23. —*Add note on 'Tilaṅga-deśa'*—A variant is Atilāṅgala which reminds us of Lang-kie-lo (Lāṅgala) located by Hiuen-tsang in Makran (cf. Sircar, *Cosm. Geog.*, p. 152). The name Lāṅgala is sometimes explained as the 'Date Palm' country (B. D. Mirchandani in *Journ. Ind. Hist.*, Vol. XLVII, 1969, pp. 237 ff.).

„ 271, note 1, line 8.—*Read*—Parāśara's list (*Cosm.*

„ 281, line 10. —*Add note*—For the installation of imitation gods, see Sircar, *The Śākta Pīṭhas*, p. 15, referring to Śivājī's installation of an imitation image of Bhavānī of Tuljāpur (Osmanabad District, Maharashtra) in his fort at Pratāpgaḍh near Javli (Satara District, Maharashtra), and *Ep. Ind.*, Vol. XXVIII, p. 247, to that of Puruṣottama-Jagannātha of Purī at Cuttack by Anaṅgabhīma III (1211-39 A.D.); see also the *Skanda Purāṇa* (Nāgara-khaṇḍa, 67. 71-72) referring to the installation of imitation images of the Sun-god of Muṇḍira, Kālapriya and Mūlasthāna at a place called Hāṭakeśvara in the Ahmedabad District.

„ 300, line 2. —*Read* —Gladwyn's

„ 306, note 2 —*Add*—The *Skanda Purāṇa* (Nāgara-khaṇḍa, Chap. 67) speaks of three great images of the Sun-god, viz. at Muṇḍira in East India, at Kālapriya in Central India and at Mūlasthāna (Mūltān) in West India.

„ 314, note 5. —*Add*—See also I.C. Ghosh, *Jātaka*, Vol. I, p. 280, and numerous other places in all the six volumes of the work.

„ 327, note 2.—*Add*—The *Mahāummaga Jātaka* (VI. 299—not later than the Gupta age) speaks of the construction of a wide and big underground tunnel, the walls of which were beautified with paintings including the representation of the sides of Mount Sumeru, the seas and oceans, the four continents, the Himavat, Lake Anavatapta, Manaḥśilātala and the six heavens beginning with the Cāturmahārājika. This suggests the knowledge of map-like paintings as in the case of the *Uttararāmacarita*.